Nanoparticle for Catalysis

Nanoparticle for Catalysis

Editor

Shaodong Zhou

Basel • Beijing • Wuhan • Barcelona • Belgrade • Novi Sad • Cluj • Manchester

Editor
Shaodong Zhou
College of Chemical and
Biological Engineering
Zhejiang University
Hangzhou
China

Editorial Office
MDPI
St. Alban-Anlage 66
4052 Basel, Switzerland

This is a reprint of articles from the Special Issue published online in the open access journal *International Journal of Molecular Sciences* (ISSN 1422-0067) (available at: www.mdpi.com/journal/ijms/special_issues/Nanoparticle_for_Catalysis).

For citation purposes, cite each article independently as indicated on the article page online and as indicated below:

Lastname, A.A.; Lastname, B.B. Article Title. *Journal Name* **Year**, *Volume Number*, Page Range.

ISBN 978-3-0365-9689-1 (Hbk)
ISBN 978-3-0365-9688-4 (PDF)
doi.org/10.3390/books978-3-0365-9688-4

© 2024 by the authors. Articles in this book are Open Access and distributed under the Creative Commons Attribution (CC BY) license. The book as a whole is distributed by MDPI under the terms and conditions of the Creative Commons Attribution-NonCommercial-NoDerivs (CC BY-NC-ND) license.

Contents

About the Editor . . . . **vii**

Anna A. Strekalova, Anastasiya A. Shesterkina, Alexander L. Kustov and Leonid M. Kustov
Recent Studies on the Application of Microwave-Assisted Method for the Preparation of Heterogeneous Catalysts and Catalytic Hydrogenation Processes
Reprinted from: *Int. J. Mol. Sci.* **2023**, 24, 8272, doi:10.3390/ijms24098272 . . . . **1**

Svetlana Selishcheva, Anastasiya Sumina, Evgeny Gerasimov, Dmitry Selishchev and Vadim Yakovlev
High-Loaded Copper-Containing Sol–Gel Catalysts for Furfural Hydroconversion
Reprinted from: *Int. J. Mol. Sci.* **2023**, 24, 7547, doi:10.3390/ijms24087547 . . . . **14**

Daniel Lach, Błażej Tomiczek, Tomasz Siudyga, Maciej Kapkowski, Rafał Sitko and Joanna Klimontko et al.
Spatially Formed Tenacious Nickel-Supported Bimetallic Catalysts for CO_2 Methanation under Conventional and Induction Heating
Reprinted from: *Int. J. Mol. Sci.* **2023**, 24, 4729, doi:10.3390/ijms24054729 . . . . **34**

Kenneth Fontánez, Diego García, Dayna Ortiz, Paola Sampayo, Luis Hernández and María Cotto et al.
Biomimetic Catalysts Based on Au@TiO_2-MoS_2-CeO_2 Composites for the Production of Hydrogen by Water Splitting
Reprinted from: *Int. J. Mol. Sci.* **2022**, 24, 363, doi:10.3390/ijms24010363 . . . . **47**

Zhenghua Zhao, Mingjie Liu, Kai Zhou, Hantao Gong, Yajing Shen and Zongbi Bao et al.
Zr-Based Metal–Organic Frameworks with Phosphoric Acids for the Photo-Oxidation of Sulfides
Reprinted from: *Int. J. Mol. Sci.* **2022**, 23, 16121, doi:10.3390/ijms232416121 . . . . **63**

Hao Jin, Penghao Liu, Qiaoqiao Teng, Yuxiang Wang, Qi Meng and Chao Qian
Efficient Construction of Symmetrical Diaryl Sulfides via a Supported Pd Nanocatalyst-Catalyzed C-S Coupling Reaction
Reprinted from: *Int. J. Mol. Sci.* **2022**, 23, 15360, doi:10.3390/ijms232315360 . . . . **78**

Viktória Hajdu, Emőke Sikora, Ferenc Kristály, Gábor Muránszky, Béla Fiser and Béla Viskolcz et al.
Palladium Decorated, Amine Functionalized Ni-, Cd- and Co-Ferrite Nanospheres as Novel and Effective Catalysts for 2,4-Dinitrotoluene Hydrogenation
Reprinted from: *Int. J. Mol. Sci.* **2022**, 23, 13197, doi:10.3390/ijms232113197 . . . . **91**

Da Ke and Shaodong Zhou
General Construction of Amine via Reduction of N=X (X = C, O, H) Bonds Mediated by Supported Nickel Boride Nanoclusters
Reprinted from: *Int. J. Mol. Sci.* **2022**, 23, 9337, doi:10.3390/ijms23169337 . . . . **104**

Kewang Zheng, Miao Liu, Zhifei Meng, Zufeng Xiao, Fei Zhong and Wei Wang et al.
Copper Foam as Active Catalysts for the Borylation of α, β-Unsaturated Compounds
Reprinted from: *Int. J. Mol. Sci.* **2022**, 23, 8403, doi:10.3390/ijms23158403 . . . . **117**

Zeng Hong, Xin Ge and Shaodong Zhou
Underlying Mechanisms of Reductive Amination on Pd-Catalysts: The Unique Role of Hydroxyl Group in Generating Sterically Hindered Amine
Reprinted from: *Int. J. Mol. Sci.* **2022**, 23, 7621, doi:10.3390/ijms23147621 . . . . **132**

Min Gao, Yong-Qi Ding and Jia-Bi Ma
Experimental and Theoretical Study of N_2 Adsorption on Hydrogenated $Y_2C_4H^-$ and Dehydrogenated $Y_2C_4^-$ Cluster Anions at Room Temperature
Reprinted from: *Int. J. Mol. Sci.* **2022**, *23*, 6976, doi:10.3390/ijms23136976 . . . . **151**

Baolin Liu, Hao Wu, Shihao Li, Mengjiao Xu, Yali Cao and Yizhao Li
Solid-State Construction of $CuO_x/Cu_{1.5}Mn_{1.5}O_4$ Nanocomposite with Abundant Surface CuO_x Species and Oxygen Vacancies to Promote CO Oxidation Activity
Reprinted from: *Int. J. Mol. Sci.* **2022**, *23*, 6856, doi:10.3390/ijms23126856 . . . . **159**

Viktória Hajdu, Gábor Muránszky, Miklós Nagy, Erika Kopcsik, Ferenc Kristály and Béla Fiser et al.
Development of High-Efficiency, Magnetically Separable Palladium-Decorated Manganese-Ferrite Catalyst for Nitrobenzene Hydrogenation
Reprinted from: *Int. J. Mol. Sci.* **2022**, *23*, 6535, doi:10.3390/ijms23126535 . . . . **172**

About the Editor

Shaodong Zhou

Shaodong Zhou received his Ph.D. in 2013 from Zheijang University. From 2013 to 2017, he worked at the Schwarz Group at TU Berlin as a postdoctoral fellow. Since April 2017, he has worked as a professor in Zhejiang University. His current research interest is focused on the conversion of energy-related small molecules.

Review

Recent Studies on the Application of Microwave-Assisted Method for the Preparation of Heterogeneous Catalysts and Catalytic Hydrogenation Processes

Anna A. Strekalova [1,2], Anastasiya A. Shesterkina [1,3], Alexander L. Kustov [1,3] and Leonid M. Kustov [1,2,3,*]

1. Laboratory of Nanochemistry and Ecology, Institute of Ecotechnologies, National University of Science and Technology MISIS, Leninsky Prospect 4, 119049 Moscow, Russia; anna.strelkova1994@mail.ru (A.A.S.); anastasiia.strelkova@mail.ru (A.A.S.); kyst@list.ru (A.L.K.)
2. Laboratory of Development and Research of Polyfunctional Catalysts, Zelinsky Institute of Organic Chemistry, Russian Academy of Sciences, Leninsky Prospekt 47, 119991 Moscow, Russia
3. Chemistry Department, Lomonosov Moscow State University, Leninskie Gory 1/3, 119234 Moscow, Russia
* Correspondence: lmkustov@mail.ru

Abstract: Currently, microwave radiation is widely used in various chemical processes in order to intensify them and carry out processes within the framework of "green" chemistry approaches. In the last 10 years, there has been a significant increase in the number of scientific publications on the application of microwaves in catalytic reactions and synthesis of nanomaterials. It is known that heterogeneous catalysts obtained under microwave activation conditions have many advantages, such as improved catalytic characteristics and stability, and the synthesis of nanomaterials is accelerated several times compared to traditional methods used to produce catalysts. The present review article is to summarize the results of modern research on the use of microwave radiation for the synthesis of heterogeneous catalytic nanomaterials and discusses the prospects for research in the field of microwave-induced liquid-phase heterogeneous catalysis in hydrogenation.

Keywords: microwave radiation; synthesis of catalysts; selective hydrogenation

1. Introduction

Microwave chemistry is a step toward the future. Today, microwave heating is an integral part of industries such as food processing, pharmaceuticals, biochemistry, materials synthesis, agriculture, and also application in biomedicine or organic electronics [1–5]. The chemistry of materials has recently combined nanotechnology and chemistry, which has led to the search for new approaches to the synthesis of nanomaterials. In the field of synthetic chemistry, the challenge is to improve and optimize methods for obtaining materials. Replacing classical thermal heating with microwave heating is a good alternative in a wide range of chemical syntheses, especially in the cases where long-term synthetic experiments are required [6]. Over the past few years, the use of microwave irradiation has become one of the latest achievements in the field of green chemistry, as microwave heating is considered as a more efficient way to control heating in many processes, as it requires less energy than conventional methods and the microwave technology also allows the use of environmentally friendly solvents, resulting in clean products that do not require additional purification steps [7–11]. The use of microwave heating has expanded with the introduction of a number of environmental techniques through the use of ionic liquids and aqueous media that do not contain organic solvents and catalysts [12,13]. Modern developments in the field of microwave chemistry make it possible to selectively obtain catalytically active materials or nanomaterials as well as organic molecules with almost 100% yield and high reproducibility [14–16].

Currently, most research is focused on the study of microwave radiation, which is due to a number of advantages, such as uniform heating, high selectivity of processes, minimal energy consumption, and the facts that it is more environmentally friendly and more efficient compared to the processes used [17–22]. Choosing the proper catalyst for a heterogeneous catalytic process ideally suited for microwave activation is not an easy task in modern chemistry. The control of the parameters of microwave synthesis (temperature, microwave power, frequency, and pressure) and the choice of a suitable solvent opens new opportunities in the development and production of new materials.

Recently, several review articles on the use of microwave radiation in catalysis have been published, focusing on the mechanism of the effect of microwaves on nanomaterials and a number of catalytic reactions, but the last major review was published in 2020 [23–28]. The review by Gao [23] focuses on the effect of the specific thermal effect of microwaves on liquid-phase catalytic reactions carried out on heterogeneous catalysts. It has been shown that microwave energy is absorbed by materials and converted into thermal energy, which manifests itself as a total loss of microwave energy in materials and causes local heat effects inside the catalyst grain, as shown in Figure 1. The authors report that the higher temperature of the heterogeneous catalyst caused by the microwave field can lead to higher conversion of reagents and higher reaction rates compared to traditional thermal heating.

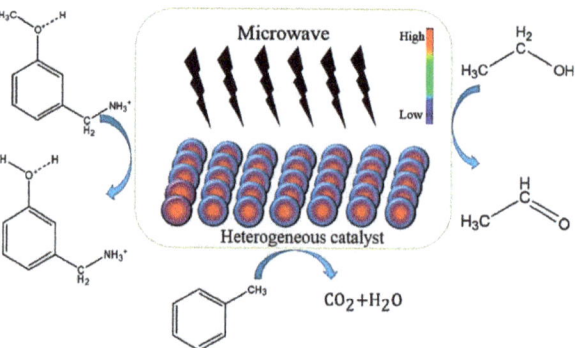

Figure 1. The schematic diagram of the special thermal effect of microwave on the solid catalysts (Reprinted with permission from Ref. [23]).

An extensive review on microwave radiation was published by Kumar [24], which provides a comparison of microwave and traditional thermal heating, types of microwave reaction systems, solvent chemistry, and the role of microwave-assisted strategies for the synthetic chemistry. The published review exhaustively describes the advantages of microwave radiation and the possibilities and prospects of its application for various fields of science, such as solar cells technology, gas sensing, photocatalysis, batteries technology, fuel cells technology, and solvent chemistry. For this reason, in our review, we will not pay attention to these aspects from the point of view of the chemistry and physics of microwave radiation but will instead focus on the practical application of microwave exposure for the synthesis of catalytic nanomaterials, and we will also consider microwave-induced liquid-phase catalytic reactions involving hydrogen.

Our review analyzes and summarizes data on the use and influence of microwave radiation in the production of catalytic nanomaterials as well as on the use of microwave activation in heterogeneous catalytic hydrogenation reactions of various compounds.

2. Synthesis of Catalysts Using Microwave Radiation

Many methods of synthesis of deposited catalysts have been described in the literature, ranging from direct use of material after grinding to other methods of modification [29,30], such as co−precipitation of the carrier, co-sorption of cations, co- and sequential deposition from solutions, deposition of bimetallic colloids, and the method of deposition of components on the carrier [31]. It should be noted that there are difficulties in determining and controlling the structure of obtained bimetallic particles, plus the ratio of one metal to another one in the particle. The advantage of catalyst preparation methods based on supporting the active component onto a carrier is the effective use of the active component due to its high dispersibility. Moreover, it is necessary to carefully select sustainable and renewable sources from which the catalysts will be obtained in order to avoid harmful effects on the environment. However, a number of disadvantages should also be pointed out: restrictions on the concentration of the active component by the pore volume of the carrier, the possibility of uneven distribution of the active component across the cross section of the pellet due to the removal of the solution to the periphery of the grain during drying, etc. Thus, these synthesis methods are replaced by new synthetic approaches and methods, such as microwave synthesis, which allows the synthesis of nanomaterials in one stage, thereby reducing the synthesis time several times and accelerating the crystallization process of materials, allowing one to obtain highly dispersed nanomaterials of given sizes [32]. Research has also focused on the use of resources from waste and environmentally safe methods of catalyst synthesis [33].

The hydrothermal method [34] applied under microwave radiation is used to regulate the morphology of materials to obtain excellent physical and chemical properties. Recently, it has been shown that this method is well suited for the synthesis of molybdates [30]. In this work, six different bismuth molybdate-based catalysts were investigated and synthesized by a fast hydrothermal method using microwave irradiation under different pH conditions. By adjusting the pH value during preparation, the morphology and structure of the synthesized catalysts can be modified as well as the transition of Bi_2MoO_6 crystals to $Bi_{3.2}Mo_{0.8}O_{7.5}$, as shown in Figure 2. Crystals can be obtained selectively. The sample prepared at pH 1 showed excellent activity for the oxidation of sulfur compounds in liquid fuel at 60 °C with a hydrogen peroxide to sulfur molar ratio of four, achieving the removal of dibenzothiophene, 4,6−dimethyldibenzothiophene, and benzothiophene with an efficiency reaching 99.71%, 99.68%, and 77.95%, respectively [35].

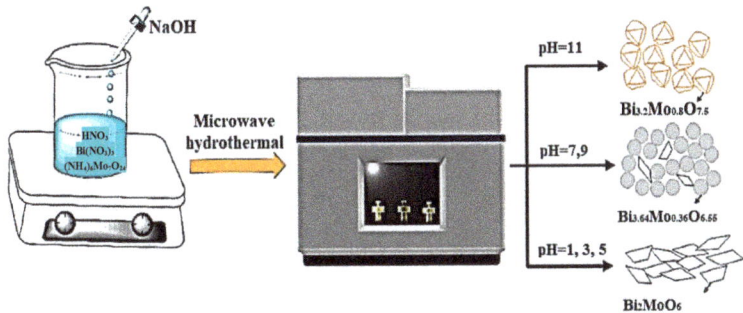

Figure 2. The schematic illustration of the formation of bismuth molybdate catalysts (Reprinted with permission from Ref. [35]).

Composite catalysts $Cu-CeO_2/C$ with a developed porous structure for selective hydrogenolysis of ethylene carbonate were prepared by the carbonization−impregnation method using microwave radiation, and, for comparison, similar catalysts were synthesized by the impregnation method. This reaction drew attention to the environmentally safe synthesis of sustainable chemical raw materials and fuels. The morphology of the Cu-CeO_2/C catalyst series was revealed using physicochemical methods of analysis, and it was shown that $Cu-CeO_2/C$ catalysts with corresponding copper−cerium interactions improve the dispersion of copper particles and provide a higher $Cu^+/(Cu^+ + Cu^0)$ ratio and a higher concentration of oxygen vacancies at the surface. These interactions result in enhanced adsorption of ethylene carbonate and high hydrogenation activity. The reaction was carried out in an autoclave, at 3 MPa H_2, 180 °C, for 5 h, as shown in Figure 3. The developed $Cu-CeO_2/C$ catalyst showed a higher catalytic activity (the conversion reached

92%) in hydrogenation of ethylene carbonate compared to the impregnated Cu−CeO$_2$/C catalyst (the conversion was about 60%), which was explained by the interaction effect created by CeO$_2$ doping. At the same time, the authors proposed a methodology for the synthesis of a porous metal−metal oxide catalyst on a carbon carrier for heterogeneous hydrogenation [36].

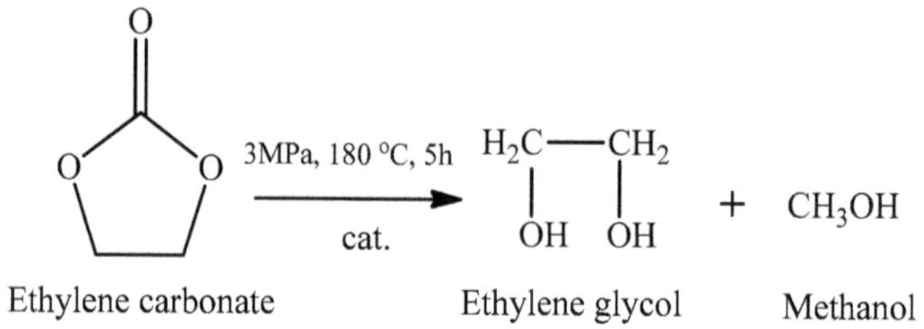

Figure 3. The schematic reaction of hydrogenolysis of ethylene carbonate.

A new microwave synthesis of copper phyllosilicates on a commercial SiO$_2$ carrier was first developed [37]. This technique allows a significant reduction of the sample synthesis time from 9 h to 6 h relative to the method of thermal decomposition of urea. The samples were synthesized in a Multiwave Pro microwave unit under irradiation (2.45 GHz) with urea in four Teflon autoclave-type vessels for 6 h. The morphology of the obtained catalysts was studied by XRD, TEM, and N$_2$ adsorption techniques, and the formation of chrysocolla phases in the samples was confirmed. The catalysts prepared by the microwave method were highly efficient in the selective hydrogenation of the C≡C bond in 1,4−butynediol to 1,4−butenediol and 2−phenylacetylene, with the selectivity of 96.5% and 100% at the complete conversion for 2 and 0.5 h of the reaction, respectively. The resulting method of microwave synthesis showed enough advantages to be considered the most preferable alternative to the traditional methods.

Akay et al. [38–40] considered applied heterogeneous Fe, Co, and Ni catalysts as primary precursors and Ca, Mn, and Cu as precursors for binary SiO$_2$ based catalytic systems, induced by microwave (as well as solar) radiation from the liquid state, and represent new catalysts with such properties as high porosity, surface area, chemical and morphological heterogeneity, oxygen, and cation vacancies. These catalysts have been successfully used to provide extremely high conversions—for example, in ammonia synthesis compared to existing catalysts. The performance of the catalyst was inferior when using thermal methods compared to microwave irradiation. In addition, the use of the classic incipient wetting impregnation method also resulted in poorer catalyst performance.

Jing et al. [41] produced a copper−based catalyst by thermal hydrolysis of urea using microwave radiation to produce hydrogen from methanol decomposition, as shown in Figure 4. The synthesis of the catalysts was performed in such a way that when the temperature of the solution heated uniformly by microwaves reached 80 °C, the hydroxyl ion could be formed during the decomposition of urea, which could form hydroxides with Cu^{2+}, Ni^{2+}, and Zn^{2+} on the surface of the carrier γ-Al$_2$O$_3$. It was found that the content of Cu gradually decreased with increasing microwave heating temperature, while the content of Ni and Zn increased with increasing microwave heating temperature, as shown in Table 1. This result is explained by the difference in solubility of metal hydroxides and solution pH. The characterization results show that there is a well-defined correlation between the catalytic characteristics of the catalysts and the microwave heating temperature.

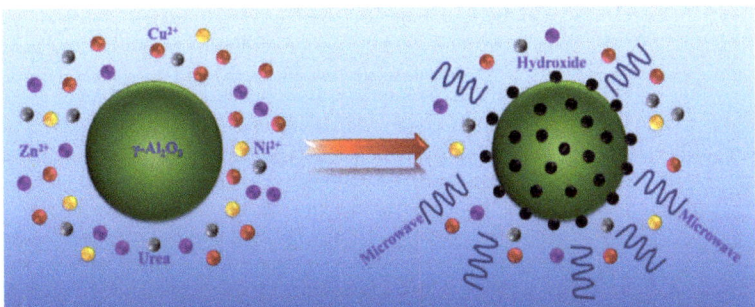

Figure 4. The catalyst generation process (Reprinted with permission from Ref. [41]).

Table 1. Elemental composition and textural properties of catalysts and γ-Al$_2$O$_3$ carrier, crystal size of Cu in reduced catalysts (Adapted with permission from [41]).

Sample	Cu (wt%) [a]	Ni (wt%)	Zn (wt%)	Cu Crystallite Size (nm) [b]	S_{BET} (m^{2}g^{-1})	V (cm^{3}g^{-1})	D_{BJH} (nm)
γ-Al$_2$O$_3$	-	-	-	-	164.1	0.3842	9.37
MW-Cu/Ni-80	47.92	0.21	6.23	22	71.9	0.1670	9.29
MW-Cu/Ni-85	35.33	1.61	31.83	16	56.7	0.1689	11.92
MW-Cu/Ni-90	28.03	3.00	35.30	14	57.9	0.1770	12.22
MW-Cu/Ni-95	27.32	3.40	42.60	10	79.8	0.2522	12.64

[a] Determined by ICP-OES. [b] Calculated by Cu (111) (43,29^0) from the Scherrer formula.

The catalysts synthesized on the basis of Pd on γ-Al$_2$O$_3$, which was developed by a simple and environmentally friendly microwave synthesis, deserve attention. As a model reaction, this catalyst was tested in the hydrogenation of cinnamic aldehyde to hydrocinnamic aldehyde using very "mild" reaction conditions (2 MPa, 100 °C, 3 h) and short irradiation times, without the use of any polymeric stabilizer. This reaction is of significant industrial and pharmacological interest; in fact, hydrocinnamic aldehyde may be an important intermediate product for the production of pharmaceuticals used to treat HIV [42]. The obtained Pd/γ-Al$_2$O$_3$ nanocatalysts showed interesting catalytic characteristics in terms of both activity and selectivity, achieving the complete substrate conversion and selectivity up to 97% for hydrocinnamic aldehyde [43].

Nishida et al. [44] reported a method for a simple and rapid production of size-controlled Rh nanoparticles by microwave chemical reduction using alcohol. The alcohol acts as a reducing agent and solvent, and it is very efficient for changing the size of Rh particles. Using ethanol, which exhibits reducing properties, small Rh particles with high catalytic activity for CO oxidation were obtained. In addition, a Rh catalyst with a Rh particle size of 2.7 nm showed high activity in the hydrogenation of benzonitrile to a secondary imine and showed reusability in the hydrogenation of nitriles (30 °C, 3 bar H$_2$) [45]. Ru, Rh, Pd, Ir, and Pt nanoparticles stabilized by poly(N-vinyl-2-pyrrolidone) (PVP) with a uniform size were prepared by chemical reduction with ethanol in the microwave oven [46]. All metallic nanoparticles had a similar size and the same amount of PVP. The catalytic efficiency of the prepared metal nanoparticles was evaluated for the hydrogenation of benzonitrile under ambient conditions (25 °C, 1 bar H$_2$). Rh nanoparticles showed the highest benzonitrile conversion and the highest selectivity for the secondary imine product.

Lingaya et al. [47] synthesized a series of SiO$_2$-based Pd-Fe catalysts by incipient wetness impregnation of the carrier with Fe and Pd nitrate precursors, the catalysts were dried at 120 °C for 2 h, one part was calcined in an air atmosphere at 450 °C for 5 h, and the remaining part was microwave irradiated (irradiation at 100% power for 5 min). The catalysts prepared by the microwave irradiation method showed higher hydrodechlorination activity compared to the impregnation method. The addition of Fe to Pd reduced the activity, by diluting Pd or forming a Pd-Fe alloy. Microwave irradiation increases the palladium particle size and reduces alloy formation while maintaining the activity. Silica gel provides a clear indication of the change in the morphology of the active particles.

The synthesized bimetallic particles of ruthenium−palladium (Ru−Pd) and ruthenium−nickel (Ru−Ni) nanoalloys with different metal compositions were prepared by solvothermal treatment with microwave irradiation using PVP as a capping agent and ethylene glycol as a solvent and a reducing agent. The synthesized bimetallic nano−alloy particles were then deposited on γ-Al_2O_3 to obtain supported nano-alloy catalysts. The hydrogenation of dibenzo−18−crown−6 ester (DB18C6) was carried out at 9 MPa, 120°C and 3.5 h using the synthesized bimetallic nanoalloy catalysts. It was observed that the bimetallic nanocatalyst synthesized by the microwave method at Ru:Pd 3:1% (wt.) exhibited a higher catalytic activity and resulted in a 98.9% conversion of DB18C6 with a 100% selectivity towards cis−cis dicyclohexano−18−crown−6 ester (CSC DCH18C6), showing better results compared to 4 wt% Ru/γ-Al_2O_3 microwave irradiated (MWI) and 5 wt.% Ru/γ-Al_2O_3 conventionally treated nanocatalysts [48].

The effect of the catalyst carrier, reaction medium, pressure, temperature, and initial concentration of levulinic acid (LA) was investigated [49] to obtain optimal conditions for high yields of γ−valerolactone (GVL). For this purpose, Li et al. proposed the synthesis of a Ru-based catalyst, which was prepared by a one-step microwave thermolytic process using dodecacarbonyltriruthenium [$Ru_3(CO)_{12}$] as a precursor. Generally, the support and $Ru_3(CO)_{12}$ were placed in an agate mortar and grinded for 20 min. The precursor mixture was then placed in a reactor with a quartz tube with an inner diameter of about 10 mm and purged with argon for 2 h at room temperature to remove oxygen in the reactor, and the reaction was performed in an inert atmosphere. The reactor was then placed in a household microwave oven operating at 2.45 GHz and 800 W. Finally, the obtained products were cooled to room temperature under argon. Activated coconut shell carbon, carbon nanotubes, functionalized carbon nanotubes, and γ-Al_2O_3 were used as carriers. For comparison, a sample was synthesized by incipient wet impregnation of γ-Al_2O_3 with ruthenium chloride solution in a sufficient concentration to obtain a solid containing 5% Ru. The sample was vacuum dried at 100 °C for 10 h. GVL was obtained in a high yield (99%) by aqueous−phase hydrogenation of LA in the presence of supported Ru catalysts. The catalyst prepared by the microwave thermolytic method shows the best catalytic performance compared to other systems, at 100 °C, 2.0 MPa. This is due to the high dispersion of Ru particles on the active carbon substrate.

3. Microwave−Assisted Catalytic Hydrogenation

3.1. Hydrogenation of Aldehydes to Alcohols

The vast majority of reactions proceed at elevated temperatures, provided that the process conditions (temperature, MW radiation power, time, solvent, catalyst, ratios, and amounts of reagents) are chosen correctly, thereby proceeding faster and with higher yields. The hydrogenation process is one of the most important and widespread processes in industry, as it makes it possible to obtain a huge range of valuable organic compounds. Selective hydrogenation of carbonyl compounds is of great industrial and scientific interest due to wide application of unsaturated alcohols in pharmaceutical and food industries; in chemical production as intermediates for synthetic polymers, plasticizers, and solvents; and in fine organic synthesis [50]. An additional advantage of the process is the possibility of obtaining aldehydes and ketones from bioavailable materials. Difficulties encountered by researchers in the hydrogenation of carbonyl substrates include the presence of a conjugated C=C bond in aldehydes such as cinnamaldehyde and citral, the presence of an aromatic ring as in benzaldehyde, or steric hindrances in various ketones. One of the most common methods for hydrogenation of carbonyl compounds involves the use of stoichiometric reducing agents such as sodium borohydride and hydrazine hydrate. The obvious disadvantages of this technique are toxicity, explosiveness of the reducing agents, and the large amount of waste produced during the reaction. The use of homogeneous catalysts is complicated by the necessity of their separation from the reaction mixture and the quite severe conditions that are usually required. The use of heterogeneous catalysts is becoming increasingly popular, as this approach is consistent with the principles of "green chemistry". It also makes it easy to separate the catalyst from the reaction mixture and to recycle it. However, until now, heterogeneous hydrogenation proceeds at higher temperature and pressure, which makes the process non-selective, expensive, and requires additional resources.

In recent years, Iqbal et al. [51] have actively studied the catalytic activity of catalysts based on palladium/zirconia in the hydrogenation of cinnamyl aldehyde to cinnamyl alcohol, both in microwave conditions and in an autoclave. The reaction conditions differed slightly: a CEM unit at 1.034 MPa H_2, 120 °C, 40 min was used in microwave hydrogenation, and a Parr autoclave applied under the thermal control also included stirring at the speed of 1000 rpm. Thus, a comparison of thermal and microwave hydrogenation under optimized reaction conditions was made. The conversion percentage with microwave irradiation is much higher compared to the conventional heating system. The conversion for both systems increase linearly with time, and the maximum conversions observed for the microwave and pressurized reactor were 73.5% and 27%, respectively, as shown in Figure 5. Thus, the microwave hydrogenation system demonstrates improved performance compared to the conventional pressurized autoclave system.

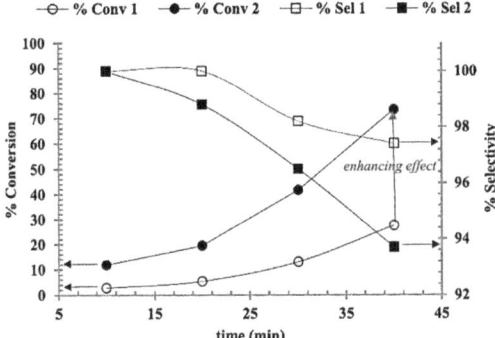

Figure 5. Hydrogenation of CAL ○ % Conv 1: conventional heating, ●% Conv 2: microwave irradiation, □ % Sel 1: conventional heating and ■ Sel 2: microwave irradiation (Reprinted with permission from Ref. [51]).

In the model reaction of hydrogenation of octanal to octanol by molecular hydrogen, microwave irradiation enhances the catalytic activity of tetragonal ZrO_2 [52]. The reaction was carried out under optimal reaction parameters such as microwave power, a temperature of 110 °C, a catalyst (t−ZrO_2), P_{H2} 1 atm, and a reaction time of 1 h 20 min in a solvent-free system. The selectivity to octanol as a target product at a 22% conversion was 99% in the microwave system, which is 1.5 times higher than the value obtained in the traditional heating method. Therefore, the microwave heating is more efficient and more reproducible results are obtained than the conventional heating system due to the safe, environmentally friendly, less labor-intensive, and economical procedure for catalytic conversion of octanal to octanol.

One of the interesting platform chemicals derived from biomass, due to its availability, is furfural. It is used in agrochemistry, perfumery, and plastic production, but it is also applied as a substrate for transformation into value-added products by hydrogenation, oxidation, decarboxylation, and condensation reactions to form furfural derivatives, as shown in Figure 6 [53].

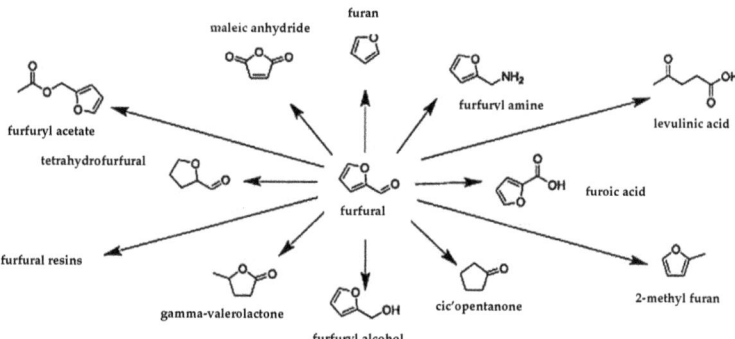

Figure 6. Furfural derivatives (Reprinted with permission from Ref. [53]).

Traditionally, the hydrogenation reaction of furfural to furfuryl alcohol was carried out both in the gas and liquid phase, because the evaporation of furfural requires significant energy and high temperatures (in the gas phase), while in the liquid phase, a high hydrogen pressure is required, which increases the cost of the product [54]. In most cases, good results can be achieved with the use of noble metals, which increases material costs, toxicity, and requires severe reaction conditions.

Ronda-Leal et al. [53] showed that hydroconversion of furfural was first studied using $TiO_2-Fe_2O_3/C$ as a catalyst. The solvent in all experiments was a mixture of isopropanol and formic acid, which was the main hydrogen donor in the chemical reaction. The reaction was carried out in a CEM Discover 2.0 microwave reactor at 200 °C for 15 min and with continuous flow. The selectivity of the formation of the target product at a 70% conversion was 100%, thus representing significant progress in the development of strategies for selective biomass conversion given the uncontrollable reactivity of such molecules, which leads to a number of reaction byproducts.

3.2. Selective Reduction of Nitrobenzene to Aniline

The reduction of aromatic nitro compounds is an important transformation that has been widely studied because anilines are used in the synthesis of pharmaceuticals and agrochemicals. Selective and complete reduction of nitrobenzene in the presence of glycerol used as a hydrogen source has been performed using Raney nickel [55] and in the presence of a recyclable catalyst based on magnetic ferrite−nickel nanoparticles. Despite the disadvantage of high viscosity at room temperature, glycerol is an optimal solvent for catalysis purposes because of its high polarity and ability to remain in the liquid phase over a large temperature range (from 17.8 to 290 °C). It also has a low vapor pressure, which means that it can be used under microwave irradiation conditions. Thus, copper nanoparticles (CuNP) were prepared in glycerol [56], and the efficiency of the glycerol layer interaction with the metal active centers was investigated by HRTEM analysis. Its high polarity, low vapor pressure, long relaxation time, and high acoustic impedance meant that excellent results were also obtained when the reaction medium was subjected to ultrasonic irradiation. Microwave synthesis was shown to play an important role in this process due to its ability to improve CuNP dispersion, promote mechanical depassivation, and increase the catalytically active surface, while MW irradiation reduces the reaction time from hours to minutes. These synergistic combinations contributed to the exhaustive reduction of nitrobenzene to aniline and facilitated the expansion of the protocol for its optimized use in industrial MW reactors.

3.3. Selective Hydrogenation of Levulinic Acid

Levulinic acid and its derivatives are promising platform chemicals that can be obtained from biomass. It can be converted into value-added molecules that can be used as environmentally friendly solvents and additives to biofuels in the pharmaceutical industry and in the synthesis of biopolymers [57]. Currently, most of the research is focused on the hydrogenation of levulinic acid into γ−valerolactone. γ−Valerolactone is a valuable chemical compound, a platform molecule, considered as an intermediate for the synthesis of value-added chemical compounds, components of motor fuels, and biopolymers. This substance is well established as an environmentally friendly solvent, fuel additive, flavoring agent, and food additive [58]. A study was performed on the hydrogenation of levulinic acid using microwave synthesis on gold catalysts (commercial 1 wt% Au/TiO$_2$ using AU-ROlite™ (catalogue number 79-0165, CAS number 7440-57-5, Strem Chemicals INC) and 2.5 wt% Au/ZrO$_2$ prepared by precipitation−sedimentation) to produce 1,4−pentanediol (1,4−PDO) [59]. Hydrogenation of levulinic acid under microwave conditions was tested in a SynthWAVE reactor with a closed microwave cavity. The hydrogenation was carried out both in the absence of a solvent and in the presence of H$_2$O used as a solvent. The mixture was heated under MW and magnetic stirring for 4 h at 150 °C. Interestingly, the selectivity to 1,4−PDO was close to 100% at 200 °C. The extended characterization highlighted the joint role played by the gold nanoparticles and the support on which the activated hydrogen atoms are spilled over to react with LA. This results in remarkable Au/TiO$_2$ activity. Both catalysts showed structural and morphological stability under reaction conditions.

The catalytic activity of the synthesized Al−SBA−15 mesoporous materials was evaluated in the single−reactor conversion of furfuryl alcohol (FAL) using 2−propanol as an H−donor solvent to yield γ−valerolactone (GVL) under both microwave irradiation (microwave synthesizer Discover® 2.0, CEM Corporation's, US) and continuous flow conditions (Phoenix Flow Reactor™, ThalesNano, Budapest), as shown in Figure 7. All materials exhibited conversions up to 99% with GVL selectivities of ca. 20−41% after one hour of the reaction. A study of the recyclability of the materials showed good GVL production over three reaction cycles using both microwave and flow reaction conditions [60].

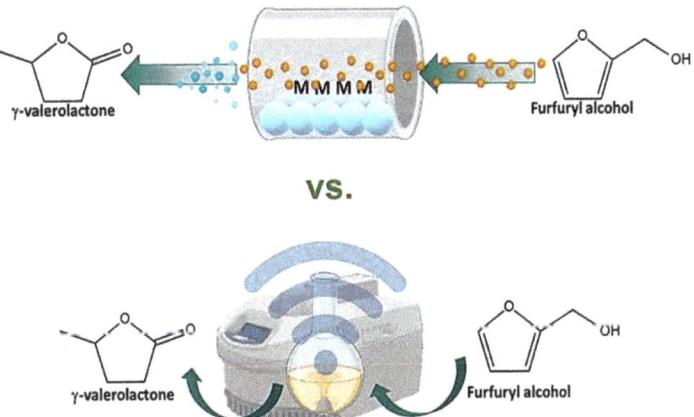

Figure 7. One−pot multi-step synthesis of γ−valerolactone from furfuryl alcohol (Reprinted with permission from Ref. [60]).

3.4. Selective Catalytic Transfer Hydrogenation

Catalytic transfer hydrogenation (CTH) is a new alternative process for selective hydrogenation, and the microwave irradiation is an efficient heating method for the initiation of the organic reaction.

CTH of Jatropha oil biodiesel was performed [61] by microwave heating using Raney nickel as a catalyst and water as a solvent, as well as by conventional heating. The effect of operating parameters on the composition of the upgraded biodiesel was analyzed in detail and optimal conditions for CTH of Jatropha oil biodiesel were found. Under optimal conditions, the mass conversion of methyl linoleate reached 91.98 wt.% with the reaction time of 50 min. For the improved biodiesel obtained after 50 min, the methyl linoleate content, methyl oleate content, methyl stearate content and iodine number were 2.45 wt.%, 76.70 wt.%, 8.45 wt.% and 70.21 wt.%, respectively. Thus, the use of microwave heating in the CTH reaction can shorten the hydrogenation time and speed up the hydrogenation process, which helps to reduce the energy consumption for the reaction.

The CTH of polyunsaturated fatty acid methyl esters was carried out [62] by using a Pd/organobentonite catalyst with water as a solvent and ammonium formate as a hydrogen donor under microwave heating. The effects of CTH reaction conditions, including the amount of ammonium formate, the amount of the solvent, the dosage of the catalyst, reaction temperature, reaction time, and agitation rate, on the hydrogenation process was studied systematically. Meanwhile, the effects of both microwave heating and conventional heating on the CTH reaction have been studied. Under the optimal CTH conditions (40 g as the amount of ammonium formate, 60 g as the water amount, 8 wt.% as the catalyst dosage, 80 °C as the reaction temperature, 140 min as the reaction time, and 350 rpm as the agitation speed), methyl linoleate was successfully hydrogenated into methyl oleate with a high conversion ratio of 78.56%, high methyl oleate yield of 72.22%, and high selectivity for the cis−isomer of methyl oleate of 70.29%. Compared with conventional heating, microwave heating used in the CTH process could enhance the conversion from 60.27% to 78.56% and reduce the hydrogenation time.

4. Conclusions

Considering the great interest of the scientific community over the past 10 years in the application of microwave radiation in various catalytic processes and for the synthesis of nanomaterials and nanotechnologies, we can confidently say that catalysis is at the forefront of microwave research. A huge number of scientific articles, books, and reviews have been devoted to the significant advantages of using microwave radiation, but all known methodologies are used exclusively in lab-scale fundamental research. It is well known that the interaction between microwave irradiation and a heterogeneous catalyst can lead to a local thermal effect inside the catalyst grain, which contributes to the intensification of the catalytic reaction and increases the efficiency of the process. However, the problem of scaling up catalytic processes using microwave radiation to large-scale industrial production has not yet been solved, since this requires an increase in the size of the catalytic equipment and the amount of a catalyst, which will lead to a change in the frequency of microwaves by tens of times and will probably lead to a decrease in the reachable temperature and a drop in efficiency. We expect that the mechanism of the effect of microwave radiation in solutions and on heterogeneous nanomaterials will be actively investigated to improve understanding of the processes that are occurring in heterogeneous catalytic reactions. We believe that the interconnection of rapidly developing modern science and engineering developments will allow one to create more environmentally friendly and efficient catalytic nanomaterials in compliance with the principles of "green" chemistry.

Author Contributions: Conceptualization, L.M.K. and A.A.S. (Anastasiya A. Shesterkina); validation, L.M.K.; formal analysis, A.A.S. (Anastasiya A. Shesterkina); investigation, A.A.S. (Anna A. Strekalova); resources, A.A.S. (Anna A. Strekalova); writing—original draft preparation, A.A.S. (Anna A. Strekalova) and A.A.S. (Anastasiya A. Shesterkina); writing—review and editing, A.A.S. (Anna A. Strekalova) and A.A.S. (Anastasiya A. Shesterkina); visualization, A.L.K.; supervision, A.L.K. and L.M.K.; project administration, L.M.K. and A.L.K. All authors have read and agreed to the published version of the manuscript.

Funding: This research was funded by the Ministry of Science and Higher Education of the Russian Federation, project number 075-15-2021-591. A. L. Kustov thanks the financial support from Russian Science Foundation, grant No. 20-73-10106, L. M. Kustov thanks the «Priority-2030» academic leadership selectivity program, project number K7-2022-062.

Institutional Review Board Statement: Not applicable.

Informed Consent Statement: Not applicable.

Data Availability Statement: Data are contained within the article.

Conflicts of Interest: The authors declare no conflict of interest.

References

1. Tsuji, M. Microwave–Assisted Synthesis of Metallic Nanomaterials in Liquid Phase. *Chem. Select.* **2017**, *2*, 805–819. [CrossRef]
2. Torres-Moya, I.; Harbuzaru, A.; Donoso, B.; Prieto, P.; Ponce Ortiz, R.; Díaz-Ortiz, Á. Microwave Irradiation as a Powerful Tool for the Preparation of n-Type Benzotriazole Semiconductors with Applications in Organic Field-Effect Transistors. *Molecules* **2022**, *27*, 4340. [CrossRef]
3. Chen, Z.; Wu, Q.; Guo, W.; Niu, M.; Tan, L.; Wen, N.; Zhao, L.; Fu, C.; Yu, J.; Ren, X.; et al. Nanoengineered biomimetic Cu-based nanoparticles for multifunational and efficient tumor treatment. *Biomaterials* **2021**, *276*, 121016. [CrossRef]
4. Liang, K.-H.; Som, S.; Gupta, K.K.; Lu, C.-H. Electrochemical characterization of $TiNb_2O_7$ as anode material synthesized using microwave-assisted microemulsion route. *J. Am. Ceram. Soc.* **2022**, *105*, 7446–7454. [CrossRef]
5. Henam, S.D.; Ahmad, F.; Shan, M.A.; Parveen, S.; Wani, A.H. Microwave synthesis of nanoparticles and their antifungal activities. *Spectrochim. Acta A Mol. Biomol. Spectrosc.* **2019**, *213*, 337–341. [CrossRef]
6. Jiang, S.; Daly, H.; Xiang, H.; Yan, Y.; Zhang, H.; Hardacre, C.; Fan, X. Microwave-assisted catalyst-free hydrolysis of fibrous cellulose for deriving sugars and biochemicals. *Front. Chem. Sci. Eng.* **2019**, *13*, 718–726. [CrossRef]
7. Horikoshi, S.; Arai, Y.; Ahmad, I.; DeCamillis, C.; Hicks, K.; Schauer, B.; Serpone, N. Application of Variable Frequency Microwaves in Microwave-Assisted Chemistry: Relevance and Suppression of Arc Discharges on Conductive Catalysts. *Catalysts* **2020**, *10*, 777. [CrossRef]
8. Tompsett, G.A.; Conner, W.C.; Yngvesson, K.S. Microwave Synthesis of Nanoporous Materials. *Chem. Phys. Chem.* **2006**, *7*, 296–319. [CrossRef]
9. Xie, X.; Zhou, Y.; Huang, K. Advances in Microwave-Assisted Production of Reduced Graphene Oxide. *Front. Chem.* **2019**, *7*, 355. [CrossRef]
10. Kostyukhin, E.M.; Kustov, A.L.; Evdokimenko, N.V.; Bazlov, A.I.; Kustov, L.M. Hydrothermal microwave-assisted synthesis of $LaFeO_3$ catalyst for N_2O decomposition. *J. Am. Ceram. Soc.* **2020**, *104*, 492–503. [CrossRef]
11. Jin, J.; Wen, Z.; Long, J.; Wang, Y.; Matsuura, T.; Meng, J. One-Pot Diazo Coupling Reaction Under Microwave Irradiation in the Absence of Solvent. *Synth. Commun.* **2000**, *30*, 829–834. [CrossRef]
12. George, N.; Singh, G.; Singh, R.; Singh, G.; Devi, A.; Singh, H.; Kaur, G.; Singh, J. Microwave accelerated green approach for tailored 1,2,3–triazoles via CuAAC. *Sustain. Chem. Pharm.* **2022**, *30*, 100824. [CrossRef]
13. Zamri, A.A.; Ong, M.Y.; Nomanbhay, S.; Show, P.L. Microwave plasma technology for sustainable energy production and the electromagnetic interaction within the plasma system: A review. *Int. J. Environ. Res.* **2021**, *197*, 111204. [CrossRef]
14. Kustov, L.M.; Kustov, A.L.; Salmi, T. Microwave-Assisted Conversion of Carbohydrates. *Molecules* **2022**, *27*, 1472. [CrossRef] [PubMed]
15. Palanisamy, S.; Wang, Y.-M. Superparamagnetic iron oxide nanoparticulate system: Synthesis, targeting, drug delivery and therapy in cancer. *Dalton Trans.* **2019**, *26*, 9490–9515. [CrossRef] [PubMed]
16. Kustov, L.M.; Kustov, A.L.; Salmi, T. Processing of lignocellulosic polymer wastes using microwave irradiation. *Mendeleev Commun.* **2022**, *32*, 1–8. [CrossRef]
17. Gao, X.; Shu, D.; Li, X.; Li, H. Improved film evaporator for mechanistic understanding of microwave-induced separation process. *Front. Chem. Sci. Eng.* **2019**, *13*, 759–771. [CrossRef]
18. Li, H.; Zhao, Z.; Xiouras, C.; Stefanidis, G.D.; Li, X.; Gao, X. Fundamentals and applications of microwave heating to chemicals separation processes. *Renew. Sust. Energ. Rev.* **2019**, *114*, 109316. [CrossRef]
19. Kostyukhin, E.M.; Kustov, A.L.; Kustov, L.M. One-step hydrothermal microwave-assisted synthesis of $LaFeO_3$ nanoparticles. *Ceram. Int.* **2019**, *45*, 14384–14388. [CrossRef]
20. Vakili, R.; Xu, S.; Al-Janabi, N.; Gorgojo, P.; Holmes, S.M.; Fan, X. Microwave-assisted synthesis of zirconium-based metal organic frameworks (MOFs): Optimization and gas adsorption. *Microporous Mesoporous Mater.* **2018**, *260*, 45–53. [CrossRef]
21. Chia, S.R.; Nomanbhay, S.; Milano, J.; Chew, K.W.; Tan, C.-H.; Khoo, K.S. Microwave-Absorbing Catalysts in Catalytic Reactions of Biofuel Production. *Energies* **2022**, *15*, 7984. [CrossRef]
22. Kostyukhin, E.M. Synthesis of Magnetite Nanoparticles upon Microwave and Convection Heating. *Russ. J. Phys. Chem. A* **2018**, *92*, 2399–2402. [CrossRef]
23. Li, H.; Zhang, C.; Pang, C.; Li, X.; Gao, X. The Advances in the Special Microwave Effects of the Heterogeneous Catalytic Reactions. *Front. Chem.* **2020**, *8*, 355. [CrossRef] [PubMed]
24. Kumar, A.; Kuang, Y.; Liang, Z.; Sun, X. Microwave Chemistry, Recent Advancements and Eco-Friendly Microwave-Assisted Synthesis of Nanoarchitectures and Their Applications: A Review. *Mater. Today Nano* **2020**, *11*, 100076. [CrossRef]
25. Kostyukhin, E.M.; Kustov, L.M. Microwave-assisted synthesis of magnetite nanoparticles possessing superior magnetic properties. *Mendeleev Commun.* **2018**, *28*, 559–561. [CrossRef]
26. Verma, C.; Quraishi, M.A.; Ebenso, E.E. Microwave and ultrasound irradiations for the synthesis of environmentally sustainable corrosion inhibitors: An overview. *Sustain. Chem. Pharm.* **2018**, *10*, 134–147. [CrossRef]

27. Budarin, V.L.; Shuttleworth, P.S.; De Bruyn, M.; Farmer, T.J.; Gronnow, M.J.; Pfaltzgraff, L.; Macquarrie, D.J.; Clark, J.H. The potential of microwave technology for the recovery, synthesis and manufacturing of chemicals from bio-wastes. *Catal. Today* **2015**, *239*, 80–89. [CrossRef]
28. El Khaled, D.; Novas, N.; Gazquez, J.A.; Manzano-Agugliaro, F. Microwave dielectric heating: Applications on metals processing. *Renew. Sust. Energ. Rev.* **2018**, *82*, 2880–2892. [CrossRef]
29. Haruta, M. When Gold Is Not Noble: Catalysis by Nanoparticles. *Chem. Rec.* **2003**, *3*, 75–87. [CrossRef]
30. Kostyukhin, E.M.; Nissenbaum, V.D.; Abkhalimov, E.V.; Kustov, A.L.; Ershov, B.G.; Kustov, L.M. Microwave-Assisted Synthesis of Water-Dispersible Humate-Coated Magnetite Nanoparticles: Relation of Coating Process Parameters to the Properties of Nanoparticles. *Nanomaterials* **2020**, *10*, 1558. [CrossRef]
31. Zhang, Y. Preparation of heterogeneous catalysts based on CWAO technology. *J. Phys. Conf. Ser.* **2020**, *1549*, 032052. [CrossRef]
32. Schutz, M.B.; Xiao, L.; Lehnen, T.; Fischer, T.; Mathur, S. Microwave-assisted synthesis of nanocrystalline binary and ternary metal oxides. *Int. Mater. Rev.* **2017**, *63*, 341–374. [CrossRef]
33. Khan, H.M.; Iqbal, T.; Mujtaba, M.A.; Soudagar, M.E.M.; Veza, I.; Fattah, I.M.R. Microwave Assisted Biodiesel Production Using Heterogeneous Catalysts. *Energies* **2021**, *14*, 8135. [CrossRef]
34. Hare, D.O. Hydrothermal Method. In *Encyclopedia of Materials: Science and Technology*, 2nd ed.; Elsevier: Amsterdam, The Netherlands, 2001; pp. 3989–3992. [CrossRef]
35. Zhang, Z.; Zeng, X.; Wen, L.; Liao, S.; Wu, S.; Zeng, Y.; Zhou, R.; Shan, S. Microwave-assisted rapid synthesis of bismuth molybdate with enhanced oxidative desulfurization activity. *Fuel* **2023**, *331*, 125900. [CrossRef]
36. Song, W.; Cai, W.; Hu, S.; Jiang, X.; Lai, W. Synergistic effect between CeO_2 and Cu for ethylene carbonate hydrogenation. *J. Porous Mater.* **2022**, *29*, 1873–1882. [CrossRef]
37. Shesterkina, A.; Vikanova, K.; Kostyukhin, E.; Strekalova, A.; Shuvalova, E.; Kapustin, G.; Salmi, T. Microwave Synthesis of Copper Phyllosilicates as Effective Catalysts for Hydrogenation of C≡C Bonds. *Molecules* **2022**, *27*, 988. [CrossRef]
38. Akay, G. Co-Assembled Supported Catalysts: Synthesis of Nano-Structured Supported Catalysts with Hierarchic Pores through Combined Flow and Radiation Induced Co-Assembled Nano-Reactors. *Catalysts* **2016**, *6*, 80. [CrossRef]
39. Akay, G. Plasma Generating—Chemical Looping Catalyst Synthesis by Microwave Plasma Shock for Nitrogen Fixation from Air and Hydrogen Production from Water for Agriculture and Energy Technologies in Global Warming Prevention. *Catalysts* **2020**, *10*, 152. [CrossRef]
40. Akay, G. Sustainable Ammonia and Advanced Symbiotic Fertilizer Production Using Catalytic Multi-Reaction-Zone Reactors with Nonthermal Plasma and Simultaneous Reactive Separation. *ACS Sustain. Chem. Eng.* **2017**, *5*, 11588–11606. [CrossRef]
41. Jing, J.; Li, L.; Chu, W.; Wei, Y.; Jiang, C. Microwave-assisted synthesis of high performance copper-based catalysts for hydrogen production from methanol decomposition. *Int. J. Hydrogen Energy* **2018**, *43*, 12059–12068. [CrossRef]
42. Muller, A.; Bowers, J. Processes for Preparing Hydrocinnamic Acid. WO Patent Application WO-9908989-A1, 20 August 1997.
43. Galletti AM, R.; Antonetti, C.; Venezia, A.M.; Giambastiani, G. An easy microwave-assisted process for the synthesis of nanostructured palladium catalysts and their use in the selective hydrogenation of cinnamaldehyde. *Appl. Catal. A Gen.* **2010**, *386*, 124–131. [CrossRef]
44. Nishida, Y.; Sato, K.; Yamamoto, T.; Wu, D.; Kusada, K.; Kobayashi, H.; Matsumura, S.; Kitagawa, H.; Nagaoka, K. Facile Synthesis of Size-controlled Rh Nanoparticles via Microwave-assisted Alcohol Reduction and Their Catalysis of CO Oxidation. *Chem. Lett.* **2017**, *46*, 1254–1257. [CrossRef]
45. Nishida, Y.; Chaudhari, C.; Imatome, H.; Sato, K.; Nagaoka, K. Selective Hydrogenation of Nitriles to Secondary Imines over Rh-PVP Catalyst under Mild Conditions. *Chem. Lett.* **2017**, *47*, 938–940. [CrossRef]
46. Nishida, Y.; Wada, Y.; Chaudhari, C.; Sato, K.; Nagaoka, K. Preparation of Noble-metal Nanoparticles by Microwave-assisted Chemical Reduction and Evaluation as Catalysts for Nitrile Hydrogenation under Ambient Conditions. *J. Jpn. Pet. Inst.* **2019**, *62*, 220–227. [CrossRef]
47. Lingaiah, N.; Sai Prasad, P.; Kanta Rao, P.; Berry, F.; Smart, L. Structure and activity of microwave irradiated silica supported Pd–Fe bimetallic catalysts in the hydrodechlorination of chlorobenzene. *Catal. Commun.* **2002**, *3*, 391–397. [CrossRef]
48. Suryawanshi, Y.R.; Chakraborty, M.; Jauhari, S.; Mukhopadhyay, S.; Shenoy, K.T. Hydrogenation of Dibenzo-18-Crown-6 Ether Using γ-Al_2O_3 Supported Ru-Pd and Ru-Ni Bimetallic Nanoalloy Catalysts. *Int. J. Chem. React. Eng.* **2019**, *17*. [CrossRef]
49. Li, C.; Ni, X.; Di, X.; Liang, C. Aqueous phase hydrogenation of levulinic acid to γ-valerolactone on supported Ru catalysts prepared by microwave-assisted thermolytic method. *J. Fuel Chem. Technol.* **2018**, *46*, 161–170. [CrossRef]
50. Nongwe, I.; Ravat, V.; Meijboom, R.; Coville, N.J. Pt supported nitrogen doped hollow carbon spheres for the catalysed reduction of cinnamaldehyde. *Appl. Catal. A Gen.* **2016**, *517*, 30–38. [CrossRef]
51. Iqbal, Z.; Sadiq, M.; Sadiq, S.; Saeed, K. Selective hydrogenation of cinnamaldehyde to cinnamyl alcohol over palladium/zirconia in microwave protocol. *Catal. Today* **2021**, *397–399*, 389–396. [CrossRef]
52. Iqbal, Z.; Sadiq, S.; Sadiq, M.; Khan, I.; Saeed, K. Effect of Microwave Irradiation on the Catalytic Activity of Tetragonal Zirconia: Selective Hydrogenation of Aldehyde. *Arab. J. Sci. Eng.* **2021**, *47*, 5841–5848. [CrossRef]
53. Ronda-Leal, M.; Osman, S.M.; Jang, H.W.; Shokouhimehr, M.; Romero, A.A.; Luque, R. Selective hydrogenation of furfural using TiO_2-Fe_2O_3/C from Ti-Fe-MOFs as sacrificial template: Microwave vs Continuous flow experiments. *Fuel* **2023**, *333*, 126221. [CrossRef]

54. Wang, X.; Rinaldi, R. Exploiting H-transfer reactions with RANEY® Ni for upgrade of phenolic and aromatic biorefinery feeds under unusual, low-severity conditions. *Energy Environ. Sci.* **2012**, *5*, 8244. [CrossRef]
55. Wolfson, A.; Dlugy, C.; Shotland, Y.; Tavor, D. Glycerol as solvent and hydrogen donor in transfer hydrogenation–dehydrogenation reactions. *Tetrahedron Lett.* **2009**, *50*, 5951–5953. [CrossRef]
56. Moran, M.J.; Martina, K.; Stefanidis, G.D.; Jordens, J.; Gerven, T.V.; Goovaerts, V.; Manzoli, M.; Groffils, C.; Cravotto, G. Glycerol: An Optimal Hydrogen Source for Microwave-Promoted Cu-Catalyzed Transfer Hydrogenation of Nitrobenzene to Aniline. *Front. Chem.* **2020**, *8*, 34. [CrossRef]
57. Rackemann, D.W.; Doherty, W.O. The conversion of lignocellulosics to levulinic acid. *Biofuels Bioprod. Biorefining* **2011**, *5*, 198–214. [CrossRef]
58. Taran, O.P.; Sychev, V.V.; Kuznetsov, B.N. γ-Valerolactone as a promising solvent and basic chemical product. Catalytic synthesis from components of vegetable biomass. *Catal. Prom.* **2021**, *1*, 97–116. [CrossRef]
59. Bucciol, F.; Tabasso, S.; Grillo, G.; Menegazzo, F.; Signoretto, M.; Manzoli, M.; Cravotto, G. Boosting levulinic acid hydrogenation to value-added 1,4-pentanediol using microwave-assisted gold catalysis. *J. Catal.* **2019**, *380*, 267–277. [CrossRef]
60. Lazaro, N.; Ronda-Leal, M.; Pineda, A.; Osman, S.M.; Shokouhimehr, M.; Jang, H.W.; Luque, R. One-pot multi-step synthesis of gamma-valerolactone from furfuryl alcohol: Microwave vs continuous flow reaction studies. *Fuel* **2023**, *334*, 126439. [CrossRef]
61. Wei, G.; Liu, Z.; Zhang, L.; Li, Z. Catalytic upgrading of Jatropha oil biodiesel by partial hydrogenation using Raney-Ni as catalyst under microwave heating. *Energy Convers. Manag.* **2018**, *163*, 208–218. [CrossRef]
62. Lu, C.; Gao, L.; Zhang, L.; Liu, K.; Hou, Y.; He, T.; Zhou, Y.; Wei, G. Selective catalytic transfer hydrogenation of polyunsaturated fatty acid methyl esters using Pd/organobentonite as catalyst under microwave heating. *Chem. Eng. Process.* **2022**, *182*, 109206. [CrossRef]

Disclaimer/Publisher's Note: The statements, opinions and data contained in all publications are solely those of the individual author(s) and contributor(s) and not of MDPI and/or the editor(s). MDPI and/or the editor(s) disclaim responsibility for any injury to people or property resulting from any ideas, methods, instructions or products referred to in the content.

Article

High-Loaded Copper-Containing Sol–Gel Catalysts for Furfural Hydroconversion

Svetlana Selishcheva *, Anastasiya Sumina, Evgeny Gerasimov, Dmitry Selishchev and Vadim Yakovlev

Boreskov Institute of Catalysis, Lavrentiev Ave. 5, Novosibirsk 630090, Russia; sumina@catalysis.ru (A.S.); gerasimov@catalysis.ru (E.G.); selishev@catalysis.ru (D.S.); yakovlev@catalysis.ru (V.Y.)
* Correspondence: svetlana@catalysis.ru; Tel.: +7-(383)326-96-67

Abstract: In this study, the high-loaded copper-containing catalysts modified with Fe and Al were successfully applied for the hydroconversion of furfural to furfuryl alcohol (FA) or 2-methylfuran (2-MF) in a batch reactor. The synthesized catalysts were studied using a set of characterization techniques to find the correlation between their activity and physicochemical properties. Fine Cu-containing particles distributed in an amorphous SiO_2 matrix, which has a high surface area, provide the conversion of furfural to FA or 2-MF under exposure to high pressure of hydrogen. The modification of the mono-copper catalyst with Fe and Al increases its activity and selectivity in the target process. The reaction temperature strongly affects the selectivity of the formed products. At a H_2 pressure of 5.0 MPa, the highest selectivity toward FA (98%) and 2-MF (76%) was achieved in the case of 35Cu13Fe1Al-SiO_2 at the temperature of 100 °C and 250 °C, respectively.

Keywords: hydroconversion; furfural; furfuryl alcohol; 2-methylfuran; Cu-containing catalyst

1. Introduction

Due to a few energy and environmental issues, it is important to find alternative sources of raw materials to produce biofuels, valuable chemicals, and various fuel additives. Such a source can be lignocellulose, which is characterized by a low content of harmful elements (heavy metals, as well as sulfur and nitrogen, resulting in emissions of NO_x and SO_x) and has large volumes of annual production. Hemicellulose is one of the main components of lignocellulose, which can be converted into furfural by acid hydrolysis. Furfural is included in the list of the 30 most biomass-derived valuable chemicals [1,2] and has a wide range of applications. Up to 80 value-added chemicals can be obtained as a result of furfural conversion in various pathways [3–5]. Thus, furfural can be converted into various valuable C_4–C_5 chemicals such as furfuryl alcohol (FA), valerolactone, pentanediols, cyclopentanone, dicarboxylic acids, butanediol, and butyrolactone [6]. Additionally, furfural can be selectively hydrogenated into potential high-octane additives, such as 2-methylfuran (2-MF, sylvan), or can undergo a combination of aldol condensation, esterification, and hydrodeoxygenation to form liquid alkanes [7–9]. The 2-MF can be used in existing internal-combustion engines as a fuel additive [10–13] due to its high octane number (131) and good volatility (boiling point of 65 °C). Corma et al. demonstrated the possibility of obtaining branched alkanes from 2-MF for diesel fraction [14,15]. The 2-MF is used as a solvent in various processes, as a reagent in the production of pesticides, and as chloroquine phosphate for antimalarial drugs [16]. However, ca. 65–85% of furfural is used to produce FA [17–19]. FA is widely used as a solvent in various processes, as a reagent in the production of foundry resins and furan fiber-reinforced plastics, drugs (e.g., ranitidine), flavors, 2-MF, 2-methyltetrahydrofuran, and tetrahydrofurfuryl alcohol [3,20–22]. Additionally, FA is a very important substance for the production of vitamin C, lysine, and lubricants [23,24]. Therefore, special attention is given to the selectivity of catalysts in the hydrogenation of furfural because FA is an intermediate compound in the chain of furfural transformation. Additionally, furfural is used to produce furan resins and selectively purify lubricating oils, to synthesize many commercial products (e.g., tetrahydrofurfuryl alcohol (THFA), furan, tetrahydrofuran (THF)).

Conventional catalysts in the production of FA from furfural are the CuCr catalysts, which provide a high yield of FA (up to 98%) on an industrial scale [25–27]. However, the main disadvantage of the CuCr system is the presence of chromium in its composition, which can contaminate the target products with chromium species. In addition, the coating of Cu active sites with chromium-containing particles formed from copper chromite under process conditions significantly reduces the selectivity of FA formation [28,29]. One of the ways to improve the stability of copper chromite is the atomic layer deposition of alumina, which prevents the formation of coke, sintering, and blocking of copper particles [28]. However, this technology is quite expensive and, for this reason, cannot be applied on an industrial scale. The systems based on noble metals (e.g., Pt, Pd) are also active in the hydroconversion of furfural, but they commonly have high costs and provide insufficient selectivity due to the formation of products from the hydrogenation furan ring [30–32]. Therefore, the development of a highly active, selective, and eco-friendly catalyst for the hydroconversion of furfural is an important task. It is important to note that the efficient catalyst should provide selective hydrogenation of the aldehyde functional group in furfural, while the hydrogenation of the aromatic ring should be suppressed.

Cu-containing systems are the most promising catalysts for selective hydrogenation of furfural to FA or 2-MF. However, a monometallic copper catalyst does not have sufficient activity in the target process because of the accumulation of carbon on the surface of the catalyst and the agglomeration of the active component. High values of hydrogen pressure are commonly used to solve this problem [33]. Another way to increase the activity of copper catalysts is the addition of modifiers (e.g., Fe, Mo, Al, Ni, Co, Zn) [34–40]. However, it is not possible in many cases to achieve the required results due to strong adsorption of reaction products on the catalyst surface (thus providing a decrease in the selectivity toward FA), changes in the oxidation state of the active component, and polymerization of furfural with itself and FA molecules [33,41].

The catalysts based on iron oxide have activity in the oxidative processes, for example, esterification of alcohols and aldehydes [42], decarbonylation of aliphatic aldehydes [43], ammoxidation of aromatic aldehydes [44], and heavy oil cracking in the presence of steam [45]. Ma et al. [46] showed the possibility of using magnetic Fe_3O_4 nanoparticles in the catalytic hydrogenation of furfural to FA. It was concluded that the smaller Fe_3O_4 particles had the larger BET surface area and pore diameter, thus providing better catalytic performance (FA yield of 90% at furfural conversion of 97%, 160 °C, 5 h, 0.1 g of catalyst, 0.67 mmol of furfural, 20 mL of alcohol). Alumina is known to have Lewis acid sites and to promote the activation of C=O bonds in aldehydes or ketones [47]. On the other hand, the Lewis acid sites can contribute to the formation of carbon deposits on the surface, thus providing additional stability of the catalyst to the formation of secondary products under reaction conditions. Thus, the use of copper-containing catalysts modified with iron and aluminum oxides can increase the catalytic performance and provide high selectivity to the formation of FA under reaction conditions.

In this work, homophase and heterophase preparation techniques were used for the synthesis of Cu-containing catalysts to provide selective hydrogenation of furfural [48,49]. Both techniques are based on the sol–gel method using partially hydrolyzed tetraethoxysilane (ethyl silicate, ES), which has a wide range of applications as a binder and hydrophobic agent. Dispersion of the metal precursor in an alcohol solution containing ES and subsequent drying lead to the condensation of ES with the formation of a polysiloxane film directly on the surface of the precursor. The next steps of calcination and reduction result in the formation of a SiO_2 matrix with dispersed metal particles enclosed in its pores. The formed matrix prevents the sintering of metal particles during the reduction and heat treatments.

2. Results and Discussion

2.1. Furfural Hydroconversion in the Presence of Monometallic Copper Catalysts

The monometallic copper catalysts with Cu content of 5–50 wt.% were investigated in the hydroconversion of furfural in a batch reactor. For all samples, only furfuryl alcohol (FA) was detected as a product of this reaction. Thus, the monometallic copper catalysts showed 100% selectivity to the formation of FA. The 5Cu-SiO$_2$ catalyst was the least active, with a furfural conversion of 19% (Figure 1). The low activity of the 5Cu-SiO$_2$ catalyst is probably due to insufficient content of the active component. The samples with a copper content of 10–30 wt.% showed similar activity in the process, while the yield of the target product at the end of the reaction was 82–85%. The most active catalysts in the hydrogenation of furfural were the samples with copper contents of 15 and 40 wt.%. They provided FA yields of 92 and 93%, respectively.

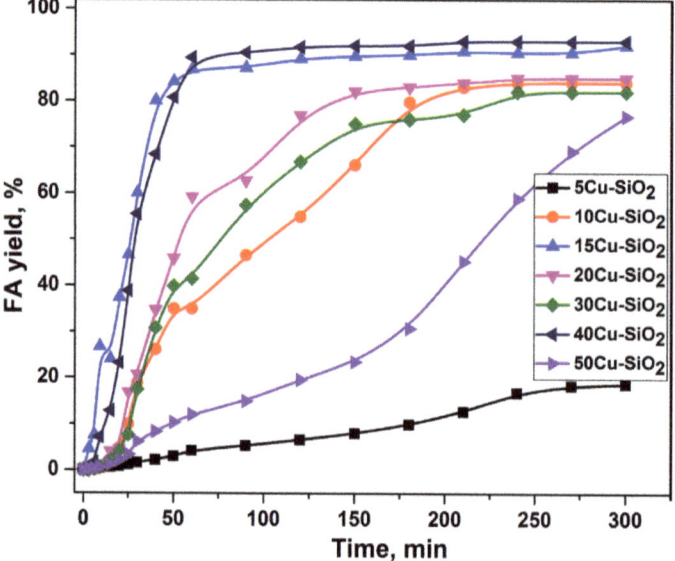

Figure 1. Dependence of FA yield on the reaction time in the presence of monometallic copper catalysts in a batch reactor. Reaction conditions: P(H$_2$) = 5.0 MPa, 100 °C, 7 vol.% of furfural in isopropyl alcohol, 4.5 h, m$_{cat}$ = 1 g.

Table 1 summarizes the FA yield and the amount of CO chemisorbed by the reduced monometallic xCu–SiO$_2$ catalysts at 200 °C. The amount of CO chemisorbed by catalyst samples reflects the number of accessible Cu sites. Although the heat of CO adsorption is extremely low in the case of copper compared, for example, with nickel [50], the measurement of this parameter allows us to draw some conclusions related to the activity of copper-containing systems [51–53].

CO chemisorption data have a good correlation with FA yields for all studied catalysts. A decrease in the activity, as well as a decrease in the amount of chemisorbed CO for 50Cu-SiO$_2$ compared to 40Cu-SiO$_2$, can be explained by an increase in the particle size of the active component on the surface, which is confirmed by XRD and BET analyses (see Table 2).

Table 1. Correlation between the amount of CO chemisorbed by the reduced monometallic xCu–SiO$_2$ catalyst and the FA yield. Reaction conditions: P(H$_2$) = 5.0 MPa, 100 °C, 7 vol.% of furfural in isopropyl alcohol, reaction time is 5 h, m$_{cat}$ = 1 g.

Catalyst	CO Uptake, µmol/g$_{cat.}$	Yield of FA, %
HomSG series [a]		
5Cu-SiO$_2$	26	19
10Cu-SiO$_2$	65	84
15Cu-SiO$_2$	89	92
20Cu-SiO$_2$	80	85
30Cu-SiO$_2$	65	82
HetSG series [b]		
40Cu-SiO$_2$	86	93
50Cu-SiO$_2$	45	77

[a] A series of catalysts obtained via the homophase sol–gel method. [b] A series of catalysts obtained via the heterophase sol–gel method.

Table 2. Textural characteristics and phase composition of Cu-containing catalysts.

Catalyst [a,b]	SSA [c], m^2/g	V$_\Sigma$, cm^3/g	Phase Composition/CSR, Å [d]
HomSG series [e]			
5Cu-SiO$_2$	156	0.71	SiO$_2$ (amorph.[f])/-[g]
10Cu-SiO$_2$	210	0.73	SiO$_2$ (amorph.)/-CuO/40
15Cu-SiO$_2$	290	0.78	SiO$_2$ (amorph.)/-CuO/40
20Cu-SiO$_2$	320	0.78	SiO$_2$ (amorph.)/-CuO/50
30Cu-SiO$_2$	329	0.82	SiO$_2$ (amorph.)/-CuO/90
10Cu10Fe-SiO$_2$ (0.9) [h]	215	1.05	SiO$_2$ (amorph.)/-
15Cu5Fe-SiO$_2$ (2.4)	257	1.03	SiO$_2$ (amorph.)/-
15Cu3Fe-SiO$_2$ (4.4)	262	1.00	SiO$_2$ (amorph.)/-
15Cu2Fe-SiO$_2$ (5.5)	270	0.99	SiO$_2$ (amorph.)/-
HetSG series [i]			
40Cu-SiO$_2$	83	0.45	SiO$_2$ (amorph.)/-CuO/130
50Cu-SiO$_2$	40	0.20	CuO/170
40Cu20Fe-SiO$_2$ (1.8)	360	0.72	SiO$_2$ (amorph.)/-CuO (trace)
35Cu13Fe-SiO$_2$ (2.4)	348	0.73	SiO$_2$ (amorph.)/-CuO (trace)/-
37Cu7Fe-SiO$_2$ (4.4)	330	0.72	SiO$_2$ (amorph.)/-CuO (trace)/-

Table 2. Cont.

Catalyst [a,b]	SSA [c], m²/g	V_Σ, cm³/g	Phase Composition/CSR, Å [d]
40Cu7Fe-SiO₂ (5.5)	323	0.72	SiO₂ (amorph.)/- CuO (trace)/-
40Cu3Fe-SiO₂ (11.6)	310	0.72	SiO₂ (amorph.)/- CuO (trace)/-
35Cu13Fe2.5Al-SiO₂ (1) [j]	235	0.75	CuO/160
35Cu13Fe1Al-SiO₂ (2)	240	0.78	CuO/170
35Cu13Fe0.5Al-SiO₂ (4.6)	260	0.82	CuO/320

[a] Numbers in the catalyst notation correspond to the percentage of Cu, Fe, and Al in the metal form (wt.%). [b] The catalyst composition was determined by atomic emission spectroscopy with inductively coupled plasma (ICP–AES). [c] Specific surface area was determined via the Brunauer–Emmett–Teller method, using nitrogen adsorption isotherms. [d] Determined by the X-ray diffraction method (conditions are described below in Section 3.9). [e] Catalyst series obtained via the homophase sol–gel method. [f] Amorphous. [g] Not determined due to X-ray amorphous. [h] Cu/Fe molar ratio. [i] Catalyst series obtained via the heterophase sol–gel method. [j] Fe/Al molar ratio.

The most active catalysts in this series are 15Cu-SiO₂ (HomSG series) and 40Cu-SiO₂ (HetSG series); however, it is clear from the kinetic plots presented in Figure 1 that after 50 min of reaction, there is no increase in the conversion of furfural until the end of the experiment. One of the approaches to increase the activity of copper catalysts is the introduction of modifying additives (e.g., Fe, Mo, Al), which make it possible to increase the catalytic performance of copper systems by preventing carbon accumulation on the catalyst surface and agglomeration of the active component due to the formation of more dispersed particles. Thus, two samples, namely, 15Cu-SiO₂ and 40Cu-SiO₂, were selected from a series of mono-copper catalysts for further modification and study in the process of furfural hydro conversion.

2.2. Furfural Hydroconversion in the Presence of Copper-Containing Catalysts Modified with Iron and Aluminum

Figure 2 shows the dependence of FA yield on the reaction time in the presence of copper–iron catalysts with different Cu/Fe molar ratios prepared by the homogeneous sol–gel method (15Cu-SiO₂ sample was used as the basis). The modification of the 15Cu-SiO₂ sample with iron does not lead to an increase in the activity of the initial Cu-catalyst but also reduces the FA yield to 70–77%. Apparently, small particles of the active component (CSR of 40 Å, XRD data, Table 2) are shielded by modifier particles (even in the case of a high Cu/Fe molar ratio), which leads to a decrease in the activity of the catalysts. Additionally, according to the XRD analysis (Table 2), the reflections corresponded to CuO disappear in the 15Cu5Fe-SiO₂ catalyst compared to 15Cu-SiO₂, which indicates a decrease in the size of these particles due to the presence of X-ray amorphous iron oxide.

The copper–iron catalysts prepared by the heterophase sol–gel method (40Cu-SiO₂ was used as the basis) were also studied in the hydrogenation of furfural to FA. Figure 3 shows the dependence of FA yield on the reaction time in the presence of copper–iron catalysts with different Cu/Fe molar ratios. The modification of 40Cu-SiO₂ with iron, in some cases, leads to an increase in activity. This is especially noticeable for the 35Cu13Fe-SiO₂, 40Cu7Fe-SiO₂, and 40Cu20Fe-SiO₂ samples. These catalysts make it possible to obtain a higher FA yield (97–99%) under equal process conditions. The 35Cu13Fe-SiO₂ (Cu/Fe = 2.4) turned out to be the catalyst, which completely converted furfural into FA, and the yield of FA was 99%.

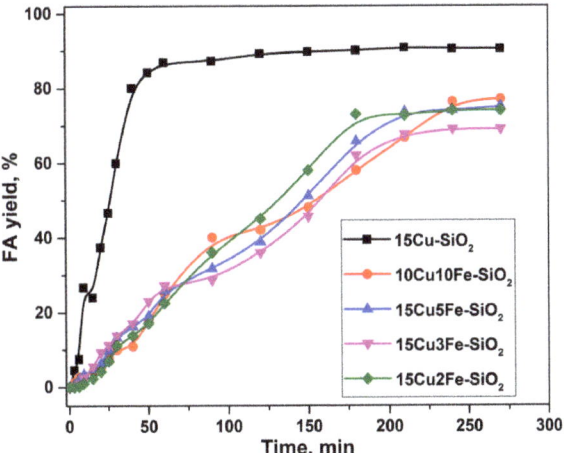

Figure 2. Dependence of FA yield on the reaction time in the presence of CuFe-containing catalysts (HomSG series) in a batch reactor. Reaction conditions: $P(H_2)$ = 5.0 MPa, 100 °C, 7 vol.% of furfural in isopropyl alcohol, 4.5 h, m_{cat} = 1 g.

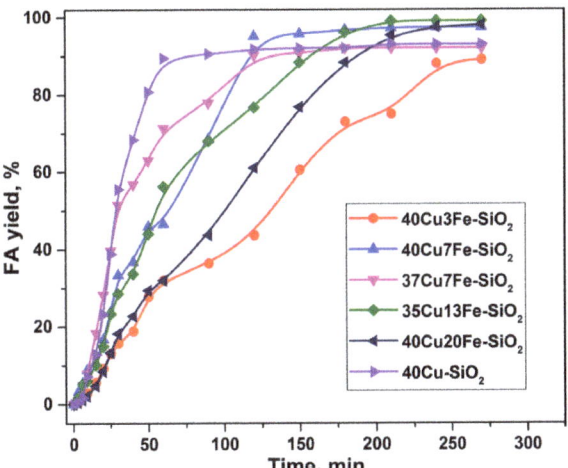

Figure 3. Dependence of FA yield on the reaction time in the presence of CuFe-containing catalysts (HeteroSG series) in a batch reactor. Reaction conditions: $P(H_2)$ = 5.0 MPa, 100 °C, 7 vol.% of furfural in isopropyl alcohol, 4.5 h, m_{cat} = 1 g.

Thus, the incorporation of iron into a high-percentage copper catalyst prepared by the heterogeneous sol-gel method makes it possible to increase the FA yield due to the formation of finer particles of the active component, which are more active in the target process. This assumption is confirmed by the data of XRD analysis (Table 2), which show only traces of crystalline copper oxide, while the major part of the oxide is X-ray amorphous. However, this fact causes an increase in the formation of carbon deposits compared to the pristine catalyst, as will be discussed in Section 2.3.

As stated in Introduction, the Lewis acid sites of alumina can promote the formation of carbon deposits on their own surface, thus providing additional stability of the catalyst to the formation of secondary products under reaction conditions. In this case, an attempt was made to modify the copper–iron catalyst with aluminum with different Fe/Al molar ratios. The dependence of FA yield on the reaction time in the presence of CuFeAl-SiO$_2$ catalysts with different Fe/Al ratios is shown in Figure 4.

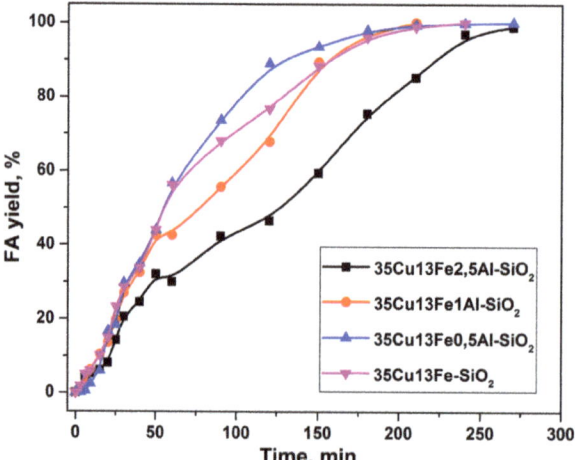

Figure 4. Dependence of FA yield on the reaction time in the presence of CuFeAl-containing catalysts (HeteroSG series) in a batch reactor. Reaction conditions: P(H$_2$) 5.0 MPa, 100 °C, 7 vol.% of furfural in isopropyl alcohol, 4.5 h, m$_{cat}$ = 1 g.

The modification with aluminum (i.e., 35Cu13Fe1Al-SiO$_2$ sample) made it possible to completely convert furfural into FA in a shorter time compared to the copper–iron catalyst. Additionally, the heterophase sol–gel method, used for the synthesis of this catalyst, and modification with iron and aluminum contribute to easier shaping compared pristine sample. This fact is an important advantage for the processes in fixed-bed reactors.

Next, the reaction temperature of furfural hydrogenation was varied in the presence of the 35Cu13Fe1Al-SiO$_2$ catalyst at a hydrogen pressure of 5.0 MPa. The distribution of formed products is shown in Figure 5.

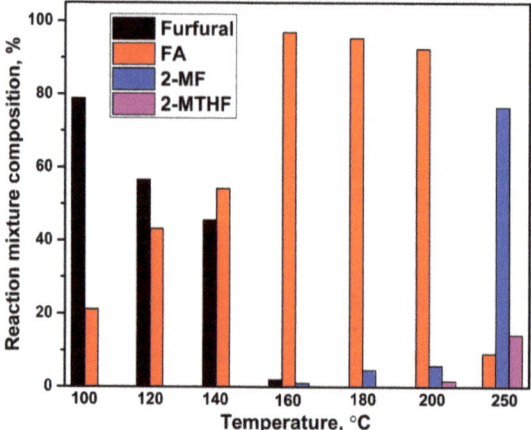

Figure 5. Dependence of reaction mixture composition on temperature for the 35Cu13Fe1Al-SiO$_2$ catalyst in a batch reactor. Reaction conditions: P(H$_2$) = 5.0 MPa, 100 °C, 7 vol.% of furfural in isopropyl alcohol, 2 h, m$_{cat}$ = 0.3 g. FA is furfuryl alcohol, 2-MF is 2-methylfuran, 2-MTHF is 2-methyltetrahydrofuran.

For the 35Cu13Fe1Al-SiO$_2$ catalyst, a temperature of 100–140 °C is optimal to produce FA from furfural because FA selectivity achieves 97–98% at 99% conversion of furfural. At a temperature of 160 °C, FA selectivity is 96% due to the formation of 2-MF. The selectivity toward 2-MF at this temperature is 4%. The hydrogenation of the FA hydroxyl group leads to a more significant formation of 2-MF up to 5–25% at 180–200 °C (at 5 h). The distribution of products changes significantly at 250 °C, when the main product becomes 2-MF with a selectivity up to 76% at 100% conversion of furfural, while 2-MTHF (12% of selectivity) is also formed (Figure 6).

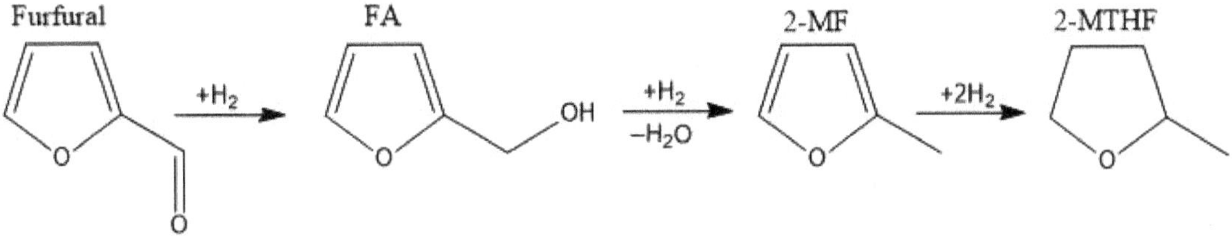

Figure 6. Schema of furfural hydroconversion in the presence of 35Cu13Fe1Al-SiO$_2$ catalyst in a batch reactor (FA is furfuryl alcohol, 2-MF is 2-methylfuran, 2-MTHF is 2-methyltetrahydrofuran).

Thus, the 35Cu13Fe1Al-SiO$_2$ catalyst is highly active and selective in the formation of FA from furfural in a batch reactor at 100–140 °C, a hydrogen pressure of 5.0 MPa, catalyst loading of 0.3–1.0 g, and a reaction time of 2.5–5.0 h. At 200–250 °C, the main product is 2-MF with a selectivity of 76%.

2.3. Catalyst Characterization

The phase composition and textural characteristics were determined for all catalysts in oxide form (see Table 2). All samples are characterized by hysteresis of the H3 type with slit-like pores between particles; the samples are predominantly meso-macroporous. The introduction of iron increases the surface area of the catalysts (HetSG series) and prevents the crystallization of copper oxide, which is confirmed by XRD data (it is not possible to determine the CSR of copper oxide due to its X-ray amorphism) (Table 2). Further addition of aluminum slightly reduces the surface area of the catalysts, but the XPS method determines the Cu state attributed to copper(II) oxide, which allows us to propose that the presence of aluminum promotes the crystallization of CuO.

X-ray diffraction patterns obtained by XRD for the most active 35Cu13Fe-SiO$_2$, 35Cu13Fe1Al-SiO$_2$, and 40Cu-SiO$_2$ catalysts are shown in Supplementary Figure S1. After the reaction, broad peaks appear at 2θ = 36.7°, 42.4°, and 61.8°, attributed to Cu$_2$O [PDF 5–667], and at 2θ = 43.4° and 50.4°, attributed to the metallic Cu [PDF 4–836]. No Fe-containing phases were found, as in the case of the fresh catalysts (Table 2).

Thus, the phase composition of the catalysts after the reaction is represented by the metallic copper and prereduced Cu$_2$O (Table 3). When iron is introduced into the 40Cu-SiO$_2$ catalyst, the size of the metallic copper CSR does not change. However, it increases when aluminum is additionally introduced. The introduction of Al also increases the degree of copper oxide reduction: 50% and 75% of Cu are observed in the 35Cu13Fe-SiO$_2$ and 35Cu13Fe1Al-SiO$_2$ samples after the reaction, respectively.

The study of some catalysts after the reaction was carried out using X-ray photoelectron spectroscopy (XPS). We studied the most active samples: 35Cu13Fe1Al-SiO$_2$ (fresh) and after the reaction, 15Cu-SiO$_2$, 40Cu-SiO$_2$, 50Cu-SiO$_2$, 15Cu5Fe-SiO$_2$, 35Cu13Fe-SiO$_2$, and 35Cu13Fe1Al-SiO$_2$. Relative concentrations of elements (i.e., atomic ratios) in the near-surface layer are shown in Table 4. The binding energies of Si2p, C1s, Cu2p$_{3/2}$, Fe2p$_{3/2}$, and O$_{1s}$ are given in Supplementary Table S1. Due to the low concentration of aluminum and the overlap of the Al2p spectrum with the copper Cu3p spectrum, the surface concentration of aluminum cannot be determined by XPS.

Table 3. Phase composition and corresponding CSR sizes for the catalysts after reaction.

Catalyst after Reaction	Phase Composition, %	CSR, Å
40Cu-SiO$_2$	30% Cu$_2$O 70% Cu	30 140
35Cu13Fe-SiO$_2$	50% Cu$_2$O 50% Cu	50 140
35Cu13Fe1Al-SiO$_2$	25% Cu$_2$O 75% Cu	55 240

Table 4. Atomic ratios of elements in the surface layer of the catalysts.

Catalyst	[Cu]/[Si]	[Fe]/[Si]	[O]/[Si]	[C]/[Si]	%, Cu^{2+}	%, Cu^{1+}	%, Cu^0
15Cu-SiO$_2$	0.082	0.000	2.27	0.51	55	45	0
40Cu-SiO$_2$	0.378	0.000	2.33	1.75	47	53	0
50Cu-SiO$_2$	0.519	0.000	2.40	2.32	36	64	0
15Cu5Fe-SiO$_2$	0.074	0.025	2.32	0.77	47	53	0
35Cu13Fe-SiO$_2$	0.224	0.068	2.44	0.87	43	57	0
35Cu13Fe1Al-SiO$_2$ (fresh)	0.336	0.053	2.27	0	100	0	0
35Cu13Fe1Al-SiO$_2$	0.195	0.073	2.42	0.85	42	58	0

With an increase in Cu content in the monometallic catalysts, the [Cu]/[Si] surface ratio increases; with the introduction of iron, this ratio slightly decreases due to the release of iron to the surface. After the reaction, the [Cu]/[Si] ratio decreases for the 35Cu13Fe1Al-SiO$_2$ catalyst due to an increase in [Fe]/[Si] and carbon formation. At the same time, the metallic copper is not identified by XPS in the 35Cu13Fe1Al-SiO$_2$ catalyst after the reaction, apparently due to its coating with oxides (HRTEM data are given below to confirm this fact). In the fresh 35Cu13Fe1Al-SiO$_2$ catalyst, metallic copper is not detected because this sample is in the oxide form.

In the series of 15Cu-SiO$_2$—40Cu-SiO$_2$—50Cu-SiO$_2$, the content of surface carbon increases (Table 4), which is in good agreement with the results of the CHNS analysis of these samples after the reaction (Table 5). The addition of aluminum to the 35Cu13Fe-SiO$_2$ catalyst slightly reduces the carbon content on the surface, which also agrees with the results of the CHNS analysis (Table 5).

The Si2p spectra of the studied catalysts show a broad symmetrical peak corresponding to silicon in the Si^{4+} state (Supplementary Figure S2). This peak was used as an internal standard (E$_b$ = 103.3 eV) to consider the effect of sample charging. For silicon in the SiO$_2$ structure, the Si2p binding energies are in the range of 103.3–103.8 eV [54–56].

Figure 7 shows the Cu2p spectra of the catalysts. Due to the spin-orbit splitting, the Cu2p spectra exhibit two groups of peaks related to the Cu2p$_{3/2}$ and Cu2p$_{1/2}$ levels, the integrated intensities of which are related as 2:1. The spectrum of the studied catalysts exhibits peaks with Cu2p$_{3/2}$ binding energies in the region of 932.5 and 935.1 eV, as well as peaks in the region of 941.1–943.4 eV—X-ray satellites corresponding to a peak in the region of 935.1 eV.

The shape of the spectra allows us to state that part of the copper is in the Cu^{2+} state in the near-surface layer of the catalysts. Indeed, a characteristic difference between the Cu2p spectra of Cu(II) compounds is the high binding energies of Cu2p$_{3/2}$ in the range of 933.6–935.3 eV and the presence of intense core-level satellites in the region of high binding energies [57]. For example, the integrated intensity of the core-level satellite in the CuO spectrum reaches 55% of the intensity of the main Cu2p$_{3/2}$ line [58]. For copper in the metallic state and Cu(I) compounds, the Cu2p$_{3/2}$ binding energy is in the range of

932.4–932.9 eV, while in the spectrum of metallic copper, there are no core-level satellite peaks. The intensity of the core-level peaks in the spectrum of Cu^{1+} does not exceed 15% of the intensity of the main $Cu2p_{3/2}$ line. The peak at 932.5 eV can be attributed to both metallic copper and copper in the Cu^{1+} state. In the literature, for metallic copper, the $Cu2p_{3/2}$ binding energies are given in the range of 932.5–932.6 eV; for Cu_2O, the binding energy is in the same range as for the metallic state of copper. Since the $Cu2p_{3/2}$ binding energies for Cu^0 and Cu^{1+} are similar, the identification of copper states is a difficult task. The so-called Auger parameter α is usually used to determine the state of copper in such cases. This parameter is equal to the sum of the $Cu2p_{3/2}$ binding energy and the position of the maximum of the CuLMM Auger spectrum on the electron kinetic energy scale [59]. In accordance with the literature data, the Auger parameters for bulk samples of metallic copper, Cu_2O, and CuO are 1851.0–1851.4, 1848.7–1849.3, and 1851.4–1851.7 eV, respectively. For the 15Cu-SiO$_2$ and 15Cu5Fe-SiO$_2$ catalysts, it is not possible to measure the value of the Auger parameter due to the low surface concentration of copper in the analysis zone. The Auger parameter for other catalysts for the peak in the region of 932.5 eV is 1849–1849.3 eV, which corresponds to copper in the Cu^{1+} state (Table 4). For a fresh 35Cu13Fe1Al-SiO$_2$ catalyst, the Auger parameter indicates that the copper is in the Cu^{2+} state (Supplementary Table S1). It is likely that in 15Cu-SiO$_2$ and 15Cu5Fe-SiO$_2$ copper is also in the Cu^{1+} state.

Table 5. Carbon content in the catalysts after reaction determined by CHNS analysis.

Catalyst	Carbon Content, wt.%
5Cu-SiO$_2$	7.6 ± 0.5
10Cu-SiO$_2$	4.5 ± 0.4
15Cu-SiO$_2$	2.80 ± 0.05
20Cu-SiO$_2$	5.68 ± 0.08
30Cu-SiO$_2$	3.16 ± 0.02
40Cu-SiO$_2$	3.0 ± 0.1
50Cu-SiO$_2$	6.3 ± 0.6
10Cu10Fe-SiO$_2$	8.8 ± 0.3
15Cu5Fe-SiO$_2$ (2.4)	5.8 ± 0.1
15Cu3Fe-SiO$_2$ (4.4)	6.82 ± 0.10
15Cu2Fe-SiO$_2$ (5.5)	6.2 ± 0.8
35Cu13Fe-SiO$_2$	4.0 ± 0.2
40Cu20Fe-SiO$_2$	4.33 ± 0.03
37Cu7Fe-SiO$_2$	1.72 ± 0.03
40Cu7Fe-SiO$_2$	3.9 ± 0.1
40Cu3Fe-SiO$_2$	4.8 ± 0.5
35Cu13Fe0,5Al-SiO$_2$	4.2 ± 0.6
35Cu13Fe1Al-SiO$_2$	3.5 ± 0.2
35Cu13Fe2,5Al-SiO$_2$	4.3 ± 0.5

The Fe2p spectra of the studied catalysts are shown in Figure 8. As seen, the Fe2p spectra are a $Fe2p_{3/2}$–$Fe2p_{1/2}$ doublet, the integral intensities of which are in the ratio of 2:1. To determine the state of iron, both the position of the main $Fe2p_{3/2}$ line and the shape of the Fe2p spectrum (intensity and relative position of the lines of core-level satellites due to the manifestation of many-electron processes) are also used. The position and intensity of the line of core-level satellites depend on the chemical state of iron. In the case of the studied catalysts, the spectra of $Fe2p_{3/2}$ represent a peak with binding energy in the region of 711.6–712.3 eV, while core-level satellites are observed. In accordance with the literature data, iron in FeO, Fe$_3$O$_4$, and Fe$_2$O$_3$ is characterized by Fe2p$_{3/2}$ binding energies in the ranges of 709.5–710.2, 710.1–710.6, and 710.7–711.2 eV, respectively [60,61], while the core-level satellites are separated from the main peak Fe2p$_{3/2}$ at 5.7, 8.5, and 8.8 eV. The high value of the binding energy and the presence of core-level satellites allow us to state that iron in the oxidized catalysts is in the Fe^{3+} state.

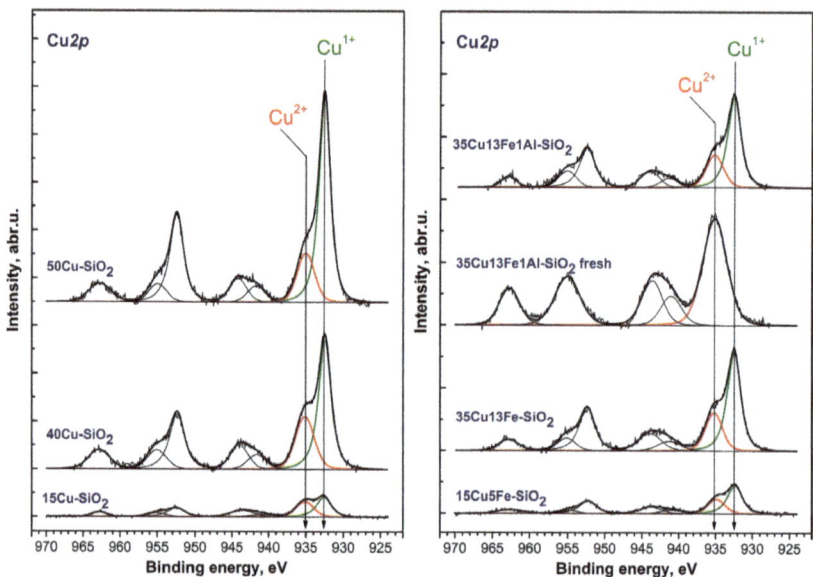

Figure 7. Cu2p spectra of Cu-containing catalysts (the spectra are normalized to the integrated intensity of the corresponding Si2p spectra).

It is known that one of the reasons for catalyst deactivation is the formation of polymerized products on the catalyst surface. These products are formed due to both the self-condensation of furfural molecules and the interaction of furfural with FA [18]. The formation of these products is an undesirable process because they reduce the activity of the catalyst and are coke precursors, carburizing the surface and blocking the access of the substrate to the catalyst's active sites. To determine the carbon amount in the catalysts after the reaction, an elemental CHNS analysis was used. The results of this analysis are shown in Table 5. The carbon content in monometallic catalysts is approximately 3–8 wt.%, while in the case of the most active catalysts (15Cu-SiO$_2$ and 40Cu-SiO$_2$), the carbon content does not exceed 3.5 wt.%. The introduction of iron into the 15Cu-SiO$_2$ catalyst increases the formation of carbon deposits, which block the active centers of the catalysts, thus causing their lower activity compared to the pristine sample. The introduction of iron and aluminum into the 40Cu-SiO$_2$ catalyst has no substantial effect on the carbon content in the samples after the reaction. It can be assumed that iron and aluminum oxides can be covered with carbon deposits while Cu-containing particles remain free, thus providing the high activity of those catalysts.

To confirm this assumption, the structural features and morphology of the most active 35Cu13Fe1Al-SiO$_2$ catalyst (fresh and after reaction) were studied by high-resolution transmission electron microscopy (HRTEM). The fresh 35Cu13Fe1Al-SiO$_2$ catalyst reduced in a tubular reactor at 200 °C and passivated by ethanol was studied by HRTEM mapping. Analysis using this method shows that the catalyst has a matrix of amorphous SiO$_2$ (Supplementary Figure S2), in which Fe$_2$O$_3$ clusters (presumably), less than 1 nm in size, are distributed (Supplementary Figure S3 and Figure 9a). Aluminum oxide is also a finely dispersed phase; it is not possible to determine more accurately due to the small particle size and low concentration. In the "cavities" of matrix space is copper. Copper is represented by several sizes and phases (Figure 9b): small particles (10–50 nm) of metallic copper, covered with an oxide film due to passivation with ethanol after reduction, and large particles up to 100 nm.

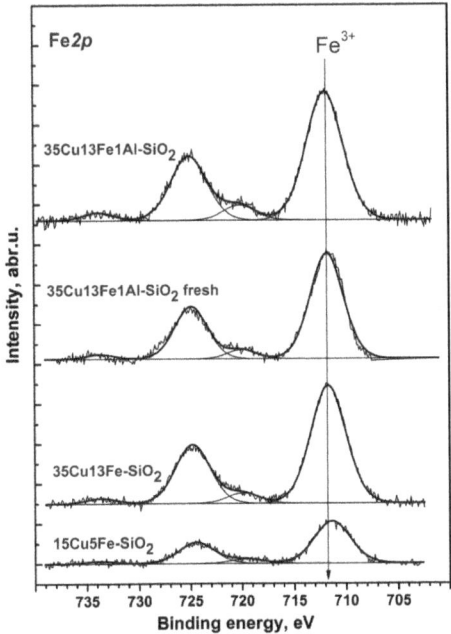

Figure 8. Fe2p spectra of the Cu-containing catalysts (the spectra are normalized to the integrated intensity of the corresponding Si2p spectra).

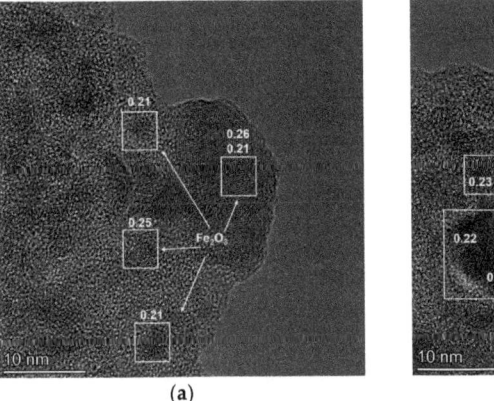 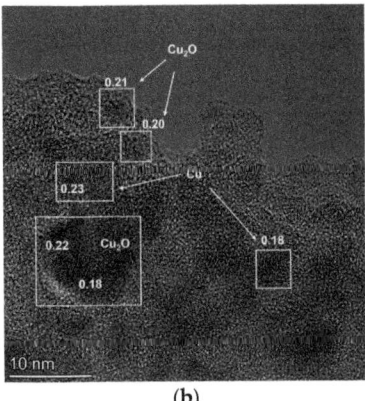

Figure 9. HRTEM images of reduced and passivated 35Cu13Fe1Al-SiO$_2$ catalyst: Fe$_2$O$_3$ particles (**a**), Cu and Cu$_2$O particles (**b**). The numbers indicate the interplanar distances of the indicated phases (nm).

The morphology of the catalyst after the reaction does not significantly change; there is no change in the size of Cu-containing particles, which indicates a stabilization effect due to the modification with iron and aluminum (Figure 10a). However, the size of iron-containing particles increases to 2 nm after the reaction, and it is possible to determine that iron is a phase of hematite Fe_2O_3 (Figure 10b). The amount of formed carbon is 3–4 wt.%, which correlates with the data of the CHNS analysis (Table 5). On the HRTEM images, carbon is identified on the surface of the copper-containing particles, and the coating thickness corresponds to 1–2 monolayers of carbon (Figure 10c).

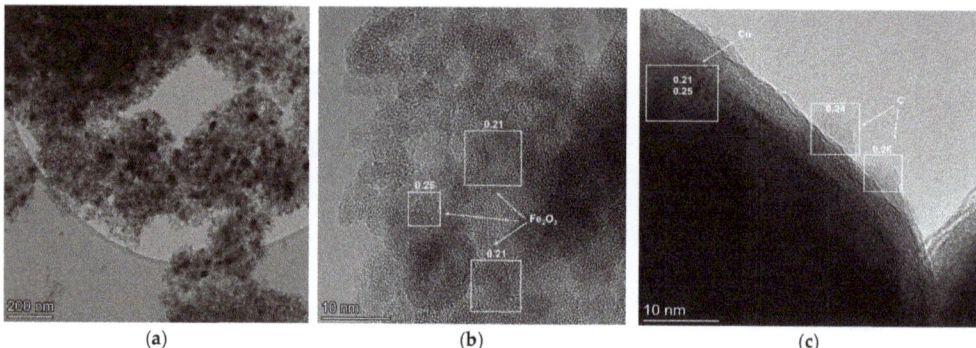

Figure 10. HRTEM images of 35Cu13Fe1Al-SiO$_2$ catalyst after reaction: perspective view (**a**), Fe$_2$O$_3$ particles (**b**), carbon on the surface of copper-containing particles (**c**). The numbers indicate the interplanar distances of the indicated phases (nm).

It is important to note that the distribution of Fe, Al, and C elements in the patterns of 35Cu13Fe1Al-SiO$_2$ are similar (Figure 11). It confirms that carbon deposits are formed on iron- and aluminum-containing particles.

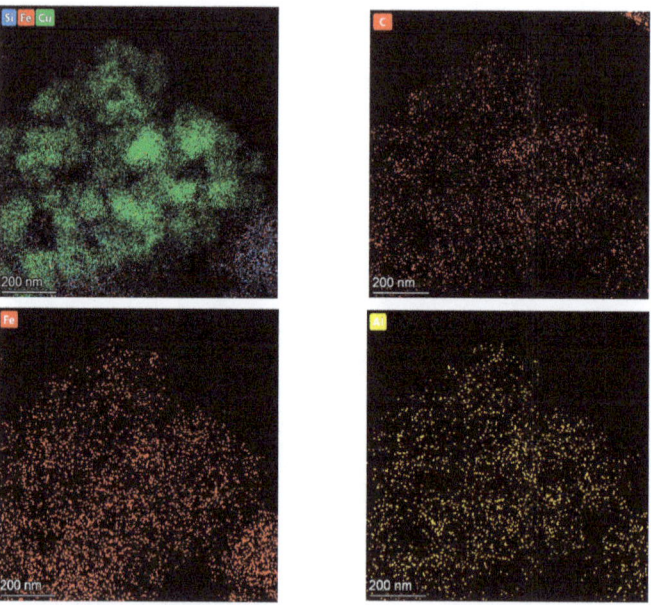

Figure 11. EDS elemental mapping images of 35Cu13Fe1Al-SiO$_2$ catalyst after reaction.

Based on the data of physicochemical methods, the proposed 35Cu13Fe1Al-SiO$_2$ catalyst represents an amorphous SiO$_2$ matrix with a high surface area in which nanosized hematite particles are uniformly distributed. The active component of the catalyst is presented in the form of metallic copper of various sizes (small particles of 10–50 nm and large particles up to 100 nm), which is covered with copper oxides. After the reaction, carbon covers copper particles with a thickness of 1–2 monolayers. It can also be argued that iron and aluminum oxides are strongly covered with carbon deposits, while Cu-containing particles remain relatively free, which can provide the high activity of that catalyst.

3. Materials and Methods

3.1. Chemicals

All chemical reagents used in this study were commercially available, with no additional purification applied. Nickel(II) nitrate hexahydrate Ni(NO$_3$)$_2$·6H$_2$O ($\geq$98%, Reakhim JSC, Moscow, Russia), cupric (II) carbonate basic CuCO$_3$·Cu(OH)$_2$ ($\geq$96%, Reakhim JSC, Moscow, Russia), ammonia solution NH$_4$OH (25%, LenReaktiv JSC, Sankt-Peterburg, Russia), iron nitrate nonahydrate Fe(NO$_3$)$_3$·9H$_2$O ($\geq$98%, Soyuzkhimprom JSC, Novosibirsk, Russia), aluminum nitrate nonahydrate Al(NO$_3$)$_3$·9H$_2$O ($\geq$98%, Reakhim JSC, Moscow, Russia), and ethyl silicate ($\geq$99%, Reakhim JSC, Moscow, Russia) were used for the catalyst preparation. Isopropyl alcohol ($\geq$99.8%, Reakhim JSC, Moscow, Russia), furfural ($\geq$99.5%, Component-Reactiv LLC, Moscow, Russia), argon ($\geq$99.99%, Pure Gases Ltd., Novosibirsk, Russia), and hydrogen ($\geq$99.99%, Pure Gases Ltd., Novosibirsk, Russia) were used in the catalytic tests. Furfuryl alcohol ($\geq$98%, Component Reactiv JSC, Moscow, Russia), 2-methylfuran (99%, Acros Organics, NJ, USA), and 2-methyltetrahydrofuran ($\geq$99%, Acros Organics, NJ, USA) were used to determine the relative response factors.

3.2. Catalyst Preparation

Two approaches based on the sol–gel method (i.e., homophase and heterophase sol–gel methods) were used for the catalyst preparation [48,49]. The homophase sol–gel method was used for the synthesis of samples with copper content up to 30 wt.%. In the case of CuFe-containing catalysts, the required amounts of copper nitrate and iron nitrate were dissolved in 150 mL of distilled water in a glass beaker for the preparation of samples by the homophase sol–gel method (HomSG series). The beaker was mounted on a stand with a tripod, on which a top-drive stirrer with a glass/PTFE propeller stirrer was fixed. Stirring was carried out until the complete dissolution of salts. Then, a pH-meter electrode was immersed in the prepared solution of the precursors, and 75 mL of ethyl silicate (ES) was added under continuous stirring at 700 rpm. Nitric acid was added dropwise until pH $\approx$ 1.3–1.5 and stirring was continued for 40 min to hydrolyze ES (step of sol formation). Then, a solution of ammonia was added dropwise until pH = 7. Upon reaching pH $\approx$ 5–6, the stirring speed was increased up to 14,000–17,000 rpm due to the formation of a viscous pasty mixture (step of gel formation). The resulting precursor was placed in an evaporating bowl and left in a muffle furnace at a temperature of 160 °C overnight. The dried samples were calcined in the following mode: 200 °C for 30 min → 250 °C for 30 min → 300 °C for 30 min → 350 °C for 30 min → 400 °C for 2 h. The prepared catalysts in an oxide form are referred to as xCuyFe-SiO$_2$, where x and y are the mass contents of Cu and Fe in metal form, respectively (see Table 2).

To obtain the catalysts with a high Cu loading (40 and 50 wt.%), the heterophase sol–gel method (HetSG series) was used. The required amounts of copper carbonate, iron nitrate (in the case of CuFe-containing catalysts), and aluminum nitrate (in the case of CuFeAl-containing catalyst) were dissolved in 95 mL of distilled water, placed into a 250 mL low beaker, and intensively mixed at 14,000–17,000 rpm using a top-drive stirrer. Ammonia solution was added dropwise to the resulting suspension until pH = 10. Then, the required amount of ES was added and stirred at 14,000–17,000 rpm for 15 min. The obtained precursor was placed in an evaporating bowl and left in a muffle furnace at a temperature of 160 °C overnight. The calcination was carried out in the following mode: 200 °C for 30 min → 250 °C for 30 min → 300 °C for 30 min → 350 °C for 30 min → 400 °C for 2 h. The prepared catalysts in oxide form are referred to as xCuyFezAl-SiO$_2$, where x, y, and z are the mass contents of Cu, Fe, and Al in metal form, respectively (see Table 2).

3.3. Catalyst Activity Tests

Activity and selectivity of the prepared catalysts were studied in the process of furfural hydro conversion. The experiments were carried out in a 300 mL batch reactor (autoclave) equipped with a mechanical stirrer, an electric furnace, and an operating system for controlling external and internal temperatures, pressure, and stirring rate of the reaction mixture. Before the reaction, 0.3 or 1 g of powdered catalyst was placed in the autoclave and activated in a hydrogen flow (300 mL/min) at a temperature of 200 °C for 1 h. After cooling to room temperature, 60 mL of a solution of 7 vol.% furfural in isopropanol was added, which corresponds to the furfural/catalyst mass ratio of 16.2 or 4.9%, depending on the initial mass of the catalyst. Next, the reactor was sealed, and the mixture was heated to the required temperature (100–250 °C) with a heat rate of 10 °C/min. After reaching the target temperature, hydrogen was supplied to the reactor until a pressure of 5.0 MPa, and the mixture was intensively stirred at 1800 rpm. The start and end times of the reaction corresponded to the moment when the stirring was turned on and off, respectively.

3.4. Product Analysis

Qualitative analysis of liquid reaction products was carried out on an Agilent 7000B GC/MS (Agilent Technologies Inc., Santa Clara, CA, USA) with a triple quadrupole analyzer and an HP-5 MS quartz capillary column ((5%-phenyl)-methylpolysiloxane, length of 30 m, inner diameter of 0.25 mm, phase thickness of 0.25 µm) from Agilent Technologies Inc (Santa Clara, CA, USA). The temperature program was as follows: 50 °C/min for 3 min, then 10 °C/min to 260 °C/min. Mass spectra recording conditions: electron ionization (70 eV), scanning mode in the m/z range of 40–500. Helium was used as a carrier gas. The NIST.11. database was used to identify the components of the analyzed sample.

Quantitative analysis of liquid reaction products was carried out using an Agilent GC-7820A gas chromatograph (Agilent Technologies Inc., Santa Clara, CA, USA) equipped with a flame ionization detector and CM-Wax column (stationary phase polyethylene glycol, 30 m×0,25 mm, phase thickness 0.25 µm) from JSC CXM (Moscow, Russia) and HP-5 capillary column (stationary phase ((5%-phenyl)-methylpolysiloxane, length of 30 m, inner diameter of 0.32 mm, phase thickness of 0.25 µm) from Agilent Technologies Inc. (Santa Clara, CA, USA). Argon was used as a carrier gas with a flow rate of 25 mL/min. The temperature program was as follows: 4 min—constant temperature for 50 °C, then heat at 8 °C/min to 62 °C, 12 °C/min to 146 °C, 20 °C/min to 190 °C and 10 °C/min to 230 °C. The analysis time was 19.2 min. The samples were injected in an amount of 0.1 µL using a chromatographic syringe. The product quantification was determined using the normalization method with the relative response factors of 1.03, 1.00, 1.25, and 1.44 for furfural, furfuryl alcohol, 2-methylfuran, and 2-methyltetrahydrofuran, respectively. The relative yield of the reaction products (%) was estimated as the molar ratio of the amount of formed product to the initial amount of furfural multiplied by 100%. The components of the reaction mixture were identified by retention times. The accuracy of the chromatographic analysis was 5%.

3.5. Elemental Composition of Catalysts

Elemental analysis of the fresh catalysts in oxide form was carried out using atomic emission spectroscopy with inductively coupled plasma (ICP–AES) on an Optima 4300 DV (Perkin Elmer Inc., Waltham, MA, USA).

3.6. Texture Characteristics

The texture properties of catalysts were analyzed by low-temperature nitrogen porosimetry using an automated volumetric adsorption station ASAP-2400 (Micromeritics Instrument Corp., Norcross, GA, USA). The Brunauer–Emmett–Teller method was used for data processing. Before nitrogen adsorption, the samples were degassed for 4 h at 150 °C and a pressure of 0.13 Pa.

3.7. CO Chemosorption

CO pulse chemisorption measurements using a Chemosorb analyzer (JSC SLO, Moscow, Russia) were used to estimate the number of active sites in the synthesized catalysts according to the previously published technique [62]. Before analysis, the catalysts were reduced at 200 °C in a H_2 flow. The samples were cooled to 25 °C in an Ar atmosphere, and CO was pulsed into the reactor until complete sample saturation was observed. The total CO uptake was used to calculate the number of active sites.

3.8. High-Resolution Transmission Electron Microscopy

The structure and microstructure of the catalysts were studied by high-resolution transmission electron microscopy (HRTEM) using a ThemisZ electron microscope (Thermo Fisher Scientific, Waltham, MA, USA) with an accelerating voltage of 200 kV and a limiting resolution of 0.07 nm. Images were recorded using a Ceta 16 CCD array (Thermo Fisher Scientific, Waltham, MA, USA). The instrument is equipped with a SuperX (Thermo Fisher Scientific, Waltham, MA, USA) energy-dispersive characteristic X-ray spectrometer (EDX) with a semiconductor Si detector with an energy resolution of 128 eV. For electron microscopy studies, sample particles were deposited on perforated carbon substrates fixed on copper or molybdenum grids using a UZD-1UCH2 ultrasonic disperser. The sample was suspended in an alcohol solution and placed on an ultrasonic disperser. Ultrasonic treatment resulted in the evaporation of liquid and deposition of sample particles on a copper mesh.

3.9. X-ray Diffraction

The phase composition of the catalysts in the initial oxide state and after the reaction was studied by X-ray diffraction (XRD). An investigation was performed on a Thermo X'tra diffractometer (Bruker, Billerica, MA, USA) in the angle range of 10–72° with a step of $2\theta = 0.02°$ and a speed of $2°$/min with a Mythen2R 1D linear detector (Decstris, Baden, Switzerland) and using monochromatized CuKα radiation (λ = 1.5418 Å). The average sizes of the coherent scattering regions were calculated using the Scherrer formula for the most intense reflections. The refinement of the lattice parameters and phase ratios was carried out by the Rietveld method.

3.10. X-ray Photoelectron Spectroscopy (XPS)

The chemical composition of the catalyst surface was studied by X-ray photoelectron spectrometry (SPECS Surface Nano Analysis GmbH, Berlin, Germany). The spectrometer was equipped with a hemispherical analyzer PHOIBOS-150-MCD-9 and an XR-50 source of X-ray characteristic radiation with a double Al/Mg anode. The spectra were recorded using nonmonochromatized AlKα radiation (1486.61 eV). The binding energy scale (E_b) was calibrated by the internal standard method using the Si2p line of silicon included in the support (E_b = 103.3 eV). The relative concentrations of elements were determined by integral intensities of XPS lines, considering the photoionization cross-section of the corresponding terms [63]. The spectra were decomposed into individual components for a detailed analysis. After subtracting the background by the Shirley method, the experimental curve was decomposed into a series of lines corresponding to the photoemission of electrons from atoms in various chemical surroundings. The XPS data were processed using the CasaXPS 2.3.25 software package [64]. The shape of the peaks was approximated by a symmetric function obtained by convolution of the Gauss and Lorentzian functions.

3.11. Determination of Carbon Content

The carbon content in the catalysts after the reaction was determined using a Vario El Cube elemental (CHNS/O) analyzer (Elementar Analysensysteme GmbH, Langenselbold, Germany) equipped with a high-temperature combustion unit and a thermal conductivity detector. The details on CHNS analysis in solid samples can be found elsewhere [65].

4. Conclusions

A new type of Cu-containing catalysts prepared by homo- and heterophase sol–gel methods using iron and aluminum modifiers has been proposed. The used preparation techniques make it possible to obtain the catalyst with a highly dispersed active component, even at a high Cu content. The synthesized catalysts have a high surface area and exhibit high activity in the hydroconversion of furfural selectively to FA or 2-MF, depending on the reaction temperature.

The active component in these catalysts is copper-containing particles of different sizes, namely, small particles of 10–50 nm and larger particles in size up to 100 nm, which are distributed in the SiO_2 matrix. The incorporation of iron and aluminum in Cu-containing particles leads to a decrease in the particle size of the active component and, consequently, an increase in the activity compared to the activity of the corresponding monometallic catalyst. Modification with iron also increases the surface area of the catalyst by four times and prevents the agglomeration of copper particles during the reaction. In the synthesized catalysts, iron presents in the form of hematite nanoparticles (1–2 nm), while aluminum forms nanoparticles of aluminum oxide.

Considering all studied catalysts, the 35Cu13Fe1Al-SiO_2 sample exhibits the highest activity and selectivity in the formation of FA from furfural in a batch reactor. The 100% conversion of furfural with 97–99% yield of FA is achieved for this catalyst at 100–140 °C, H_2 pressure of 5.0 MPa, catalyst loading of 0.3–1.0 g, and the reaction time of 2.5–5.0 h. At 250 °C, the main product is 2-MF with selectivity up to 76% at 100% conversion of furfural. It is shown that hematite and aluminum oxide activate the C=O bonds of furfural. Additionally, iron and aluminum oxides are covered with carbon deposits during the reaction, while the surface of Cu-containing particles remains relatively free, thus providing the high activity of the 35Cu13Fe1Al-SiO_2 catalyst.

Supplementary Materials: The following supporting information can be downloaded at: https://www.mdpi.com/article/10.3390/ijms24087547/s1.

Author Contributions: Conceptualization, S.S. and V.Y.; methodology, S.S.; formal analysis, S.S. and D.S.; investigation, S.S., A.S., E.G. and D.S.; writing—original draft preparation, A.S.; writing—review and editing, S.S. and D.S.; supervision, S.S. and V.Y. All authors have read and agreed to the published version of the manuscript.

Funding: This research was funded by the Russian Science Foundation, grant number 21-73-00273, https://rscf.ru/project/21-73-00273/ (accessed on 17 April 2023).

Institutional Review Board Statement: Not applicable.

Informed Consent Statement: Not applicable.

Data Availability Statement: Not applicable.

Acknowledgments: The authors are grateful to Dmitry Ermakov (Boreskov Institute of Catalysis, Novosibirsk, Russia) for the TPR and CO chemisorption measurements. The authors also thank Olga Bulavchenko for XRD analysis, Alexandra Leonova for studying textural characteristics of catalysts, Andrey Saraev for XPS experiments, and Maxim Lebedev for technical support.

Conflicts of Interest: The authors declare no conflict of interest.

References

1. Werpy, T.; Petersen, G. *Top Value Added Chemicals from Biomass: Volume I—Results of Screening for Potential Candidates from Sugars and Synthesis Gas*; US Department of Energy: Washington, DC, USA, 2004.
2. Cai, C.M.; Zhang, T.; Kumar, R.; Wyman, C.E. Integrated Furfural Production as a Renewable Fuel and Chemical Platform from Lignocellulosic Biomass. *J. Chem. Technol. Biotechnol.* **2014**, *89*, 2–10. [CrossRef]
3. Mariscal, R.; Maireles-Torres, P.; Ojeda, M.; Sádaba, I.; Granados, M.L. Furfural: A Renewable and Versatile Platform Molecule for the Synthesis of Chemicals and Fuels. *Energy Environ. Sci.* **2016**, *9*, 1144–1189. [CrossRef]

4. Khemthong, P.; Yimsukanan, C.; Narkkun, T.; Srifa, A.; Witoon, T.; Pongchaiphol, S.; Kiatphuengporn, S.; Faungnawakij, K. Advances in Catalytic Production of Value-Added Biochemicals and Biofuels via Furfural Platform Derived Lignocellulosic Biomass. *Biomass Bioenergy* **2021**, *148*, 106033. [CrossRef]
5. Jaswal, A.; Singh, P.P.; Mondal, T. Furfural—A Versatile, Biomass-Derived Platform Chemical for the Production of Renewable Chemicals. *Green Chem.* **2022**, *24*, 510–551. [CrossRef]
6. Xian, M. *Sustainable Production of Bulk Chemicals: Integration of Bio-,Chemo- Resources and Processes (Springerbriefs in Molecular Science)*, 1st ed.; Springer: Berlin/Heidelberg, Germany, 2015; ISBN 94-017-7473-0.
7. Chheda, J.N.; Dumesic, J.A. An Overview of Dehydration, Aldol-Condensation and Hydrogenation Processes for Production of Liquid Alkanes from Biomass-Derived Carbohydrates. *Catal. Today* **2007**, *123*, 59–70. [CrossRef]
8. Huber, G.W.; Chheda, J.N.; Barrett, C.J.; Dumesic, J.A. Production of Liquid Alkanes by Aqueous-Phase Processing of Biomass-Derived Carbohydrates. *Science* **2005**, *308*, 1446–1450. [CrossRef]
9. Chheda, J.N.; Huber, G.W.; Dumesic, J.A. Liquid-Phase Catalytic Processing of Biomass-Derived Oxygenated Hydrocarbons to Fuels and Chemicals. *Angew. Chem. Int. Ed.* **2007**, *46*, 7164–7183. [CrossRef] [PubMed]
10. Yan, K.; Chen, A. Efficient Hydrogenation of Biomass-Derived Furfural and Levulinic Acid on the Facilely Synthesized Noble-Metal-Free Cu–Cr Catalyst. *Energy* **2013**, *58*, 357–363. [CrossRef]
11. Sulmonetti, T.P.; Hu, B.; Ifkovits, Z.; Lee, S.; Agrawal, P.K.; Jones, C.W. Vapor Phase Hydrogenolysis of Furanics Utilizing Reduced Cobalt Mixed Metal Oxide Catalysts. *ChemCatChem* **2017**, *9*, 1815–1823. [CrossRef]
12. Date, N.S.; Hengne, A.M.; Huang, K.-W.; Chikate, R.C.; Rode, C.V. Single Pot Selective Hydrogenation of Furfural to 2-Methylfuran over Carbon Supported Iridium Catalysts. *Green Chem.* **2018**, *20*, 2027–2037. [CrossRef]
13. Varila, T.; Mäkelä, E.; Kupila, R.; Romar, H.; Hu, T.; Karinen, R.; Puurunen, R.L.; Lassi, U. Conversion of Furfural to 2-Methylfuran over CuNi Catalysts Supported on Biobased Carbon Foams. *Catal. Today* **2021**, *367*, 16–27. [CrossRef]
14. Corma, A.; de la Torre, O.; Renz, M.; Villandier, N. Production of High-Quality Diesel from Biomass Waste Products. *Angew. Chem. Int. Ed.* **2011**, *50*, 2375–2378. [CrossRef]
15. Corma, A.; de la Torre, O.; Renz, M. Production of High Quality Diesel from Cellulose and Hemicellulose by the Sylvan Process: Catalysts and Process Variables. *Energy Environ. Sci.* **2012**, *5*, 6328–6344. [CrossRef]
16. Zeitsch, K.J. *The Chemistry and Technology of Furfural and Its Many By-Products*; Elsevier: Amsterdam, The Netherlands, 2000; ISBN 978-0-08-052899-1.
17. Taylor, M.J.; Durndell, L.J.; Isaacs, M.A.; Parlett, C.M.A.; Wilson, K.; Lee, A.F.; Kyriakou, G. Highly Selective Hydrogenation of Furfural over Supported Pt Nanoparticles under Mild Conditions. *Appl. Catal. B Environ.* **2016**, *180*, 580–585. [CrossRef]
18. Yan, K.; Wu, G.; Lafleur, T.; Jarvis, C. Production, Properties and Catalytic Hydrogenation of Furfural to Fuel Additives and Value-Added Chemicals. *Renew. Sustain. Energy Rev.* **2014**, *38*, 663–676. [CrossRef]
19. Wang, Y.; Zhao, D.; Rodríguez-Padrón, D.; Len, C. Recent Advances in Catalytic Hydrogenation of Furfural. *Catalysts* **2019**, *9*, 796. [CrossRef]
20. Schneider, M.H.; Phillips, J.G. Furfuryl Alcohol and Lignin Adhesive Composition. U.S. Patent 6,747,076, 8 June 2004.
21. Barr, J.B.; Wallon, S.B. The Chemistry of Furfuryl Alcohol Resins. *J. Appl. Polym. Sci.* **1971**, *15*, 1079–1090. [CrossRef]
22. An, Z.; Li, J. Recent Advances in the Catalytic Transfer Hydrogenation of Furfural to Furfuryl Alcohol over Heterogeneous Catalysts. *Green Chem.* **2022**, *24*, 1780–1808. [CrossRef]
23. Vaidya, P.D.; Mahajani, V.V. Kinetics of Liquid-Phase Hydrogenation of Furfuraldehyde to Furfuryl Alcohol over a Pt/C Catalyst. *Ind. Eng. Chem. Res.* **2003**, *42*, 3881–3885. [CrossRef]
24. Sitthisa, S.; Sooknoi, T.; Ma, Y.; Balbuena, P.B.; Resasco, D.E. Kinetics and Mechanism of Hydrogenation of Furfural on Cu/SiO$_2$ Catalysts. *J. Catal.* **2011**, *277*, 1. [CrossRef]
25. Vasil'ev, S.N.; Gamova, I.A.; de Vekki, A.V. *Novy Spravochnik Himika i Tehnologa. Syr'e i Produkty Promyshlennosti Organicheskih i Neorganicheskih Veshchestv*, NPO "Professional": Sankt-Peterburg, Russia, 2005; p. 1142.
26. Adkins, H.; Connor, R. The Catalytic Hydrogenation of Organic Compounds over Copper Chromite. *J. Am. Chem. Soc.* **1931**, *53*, 1091–1095. [CrossRef]
27. Wojcik, B.H. Catalytic Hydrogenation of Furan Compounds. *Ind. Eng. Chem.* **1948**, *40*, 210–216. [CrossRef]
28. Zhang, H.; Lei, Y.; Kropf, A.J.; Zhang, G.; Elam, J.W.; Miller, J.T.; Sollberger, F.; Ribeiro, F.; Akatay, M.C.; Stach, E.A.; et al. Enhancing the Stability of Copper Chromite Catalysts for the Selective Hydrogenation of Furfural Using ALD Overcoating. *J. Catal.* **2014**, *317*, 284–292. [CrossRef]
29. Liu, D.; Zemlyanov, D.; Wu, T.; Lobo-Lapidus, R.J.; Dumesic, J.A.; Miller, J.T.; Marshall, C.L. Deactivation Mechanistic Studies of Copper Chromite Catalyst for Selective Hydrogenation of 2-Furfuraldehyde. *J. Catal.* **2013**, *299*, 336–345. [CrossRef]
30. Nguyen-Huy, C.; Kim, J.S.; Yoon, S.; Yang, E.; Kwak, J.H.; Lee, M.S.; An, K. Supported Pd Nanoparticle Catalysts with High Activities and Selectivities in Liquid-Phase Furfural Hydrogenation. *Fuel* **2018**, *226*, 607–617. [CrossRef]
31. Mäkelä, E.; Lahti, R.; Jaatinen, S.; Romar, H.; Hu, T.; Puurunen, R.L.; Lassi, U.; Karinen, R. Study of Ni, Pt, and Ru Catalysts on Wood-Based Activated Carbon Supports and Their Activity in Furfural Conversion to 2-Methylfuran. *ChemCatChem* **2018**, *10*, 3269–3283. [CrossRef]
32. Wang, Z.; Wang, X.; Zhang, C.; Arai, M.; Zhou, L.; Zhao, F. Selective Hydrogenation of Furfural to Furfuryl Alcohol over Pd/TiH$_2$ Catalyst. *Mol. Catal.* **2021**, *508*, 111599. [CrossRef]

33. Sitthisa, S.; Resasco, D.E. Hydrodeoxygenation of Furfural Over Supported Metal Catalysts: A Comparative Study of Cu, Pd and Ni. *Catal. Lett.* **2011**, *141*, 784. [CrossRef]
34. Villaverde, M.M.; Bertero, N.M.; Garetto, T.F.; Marchi, A.J. Selective Liquid-Phase Hydrogenation of Furfural to Furfuryl Alcohol over Cu-Based Catalysts. *Catal. Today* **2013**, *213*, 87–92. [CrossRef]
35. Vetere, V.; Merlo, A.B.; Ruggera, J.F.; Casella, M.L. Transition Metal-Based Bimetallic Catalysts for the Chemoselective Hydrogenation of Furfuraldehyde. *J. Braz. Chem. Soc.* **2010**, *21*, 914–920. [CrossRef]
36. Yan, K.; Chen, A. Selective Hydrogenation of Furfural and Levulinic Acid to Biofuels on the Ecofriendly Cu–Fe Catalyst. *Fuel* **2014**, *115*, 101–108. [CrossRef]
37. Rao, T.U.; Suchada, S.; Choi, C.; Machida, H.; Huo, Z.; Norinaga, K. Selective Hydrogenation of Furfural to Tetrahydrofurfuryl Alcohol in 2-Butanol over an Equimolar Ni-Cu-Al Catalyst Prepared by the Co-Precipitation Method. *Energy Convers. Manag.* **2022**, *265*, 115736. [CrossRef]
38. Zhang, J.; Wu, D. Aqueous Phase Catalytic Hydrogenation of Furfural to Furfuryl Alcohol over In-Situ Synthesized Cu–Zn/SiO$_2$ Catalysts. *Mater. Chem. Phys.* **2021**, *260*, 124152. [CrossRef]
39. Yan, X.; Zhang, G.; Zhu, Q.; Kong, X. CuZn@N-doped Graphene Layer for Upgrading of Furfural to Furfuryl Alcohol. *Mol. Catal.* **2022**, *517*, 112066. [CrossRef]
40. Smirnov, A.A.; Shilov, I.N.; Alekseeva, M.V.; Selishcheva, S.A.; Yakovlev, V.A. Study of the Composition Effect of Molybdenum-Modified Nickel–Copper Catalysts on Their Activity and Selectivity in the Hydrogenation of Furfural to Different Valuable Chemicals. *Catal. Ind.* **2018**, *10*, 228–236. [CrossRef]
41. Vargas-Hernández, D.; Rubio-Caballero, J.M.; Santamaría-González, J.; Moreno-Tost, R.; Mérida-Robles, J.M.; Pérez-Cruz, M.A.; Jiménez-López, A.; Hernández-Huesca, R.; Maireles-Torres, P. Furfuryl Alcohol from Furfural Hydrogenation over Copper Supported on SBA-15 Silica Catalysts. *J. Mol. Catal. A Chem.* **2014**, *383–384*, 106–113. [CrossRef]
42. Rajabi, F.; Arancon, R.A.D.; Luque, R. Oxidative Esterification of Alcohols and Aldehydes Using Supported Iron Oxide Nanoparticle Catalysts. *Catal. Commun.* **2015**, *59*, 101–103. [CrossRef]
43. Luo, Z.; Li, R.; Zhu, T.; Liu, C.-F.; Feng, N.; Nartey, K.A.; Liu, Q.; Xu, X. Iron-Catalyzed Oxidative Decabonylation/Radical Cyclization of Aliphatic Aldehydes with Biphenyl Isocyanides: A New Pathway For the Synthesis of 6-Alkylphenanthridines. *Asian J. Org. Chem.* **2021**, *10*, 926–930. [CrossRef]
44. Wang, W.D.; Wang, F.; Chang, Y.; Dong, Z. Biomass Chitosan-Derived Nitrogen-Doped Carbon Modified with Iron Oxide for the Catalytic Ammoxidation of Aromatic Aldehydes to Aromatic Nitriles. *Mol. Catal.* **2021**, *499*, 111293. [CrossRef]
45. Yeletsky, P.M.; Zaikina, O.O.; Sosnin, G.A.; Kukushkin, R.G.; Yakovlev, V.A. Heavy Oil Cracking in the Presence of Steam and Nanodispersed Catalysts Based on Different Metals. *Fuel Process. Technol.* **2020**, *199*, 106239. [CrossRef]
46. Ma, M.; Hou, P.; Zhang, P.; Cao, J.; Liu, H.; Yue, H.; Tian, G.; Feng, S. Magnetic Fe$_3$O$_4$ Nanoparticles as Easily Separable Catalysts for Efficient Catalytic Transfer Hydrogenation of Biomass-Derived Furfural to Furfuryl Alcohol. *Appl. Catal. A Gen.* **2020**, *602*, 117709. [CrossRef]
47. Liu, Y.-C.; Ko, B.-T.; Huang, B.-H.; Lin, C.-C. Reduction of Aldehydes and Ketones Catalyzed by a Novel Aluminum Alkoxide: Mechanistic Studies of Meerwein−Ponndorf−Verley Reaction. *Organometallics* **2002**, *21*, 2066–2069. [CrossRef]
48. Ermakova, M.A.; Ermakov, D.Y. High-Loaded Nickel–Silica Catalysts for Hydrogenation, Prepared by Sol–Gel: Route: Structure and Catalytic Behavior. *Appl. Catal. A Gen.* **2003**, *245*, 277–288. [CrossRef]
49. Ermakov, D.Y.; Bykova, M.V.; Selishcheva, S.A.; Khromova, S.A.; Yakovlev, V.A. Method of Preparing Hydrotreatment Catalyst. Patent RU2496580, 27 October 2013.
50. Asedegbega Nieto, E.; Ruiz, A.; Rodriguez-Ramos, I. Study of CO Chemisorption on Graphite-Supported Ru–Cu and Ni–Cu Bimetallic Catalysts. *Thermochim. Acta* **2005**, *434*, 113–118. [CrossRef]
51. Parris, G.E.; Klier, K. The Specific Copper Surface Areas in CuZnO Methanol Synthesis Catalysts by Oxygen and Carbon Monoxide Chemisorption: Evidence for Irreversible CO Chemisorption Induced by the Interaction of the Catalyst Components. *J. Catal.* **1986**, *97*, 374–384. [CrossRef]
52. Phillips, J.M.; Leibsle, F.M.; Holder, A.J.; Keith, T. A Comparative Study of Chemisorption by Density Functional Theory, Ab Initio, and Semiempirical Methods: Carbon Monoxide, Formate, and Acetate on Cu(110). *Surf. Sci.* **2003**, *545*, 1–7. [CrossRef]
53. Smirnov, A.A.; Khromova, S.A.; Bulavchenko, O.A.; Kaichev, V.V.; Saraev, A.A.; Reshetnikov, S.I.; Bykova, M.V.; Trusov, L.I.; Yakovlev, V.A. Effect of the Ni/Cu Ratio on the Composition and Catalytic Properties of Nickel-Copper Alloy in Anisole Hydrodeoxygenation. *Kinet Catal.* **2014**, *55*, 69–78. [CrossRef]
54. Gutowski, M.; Jaffe, J.E.; Liu, C.-L.; Stoker, M.; Hegde, R.I.; Rai, R.S.; Tobin, P.J. Thermodynamic Stability of High-K Dielectric Metal Oxides ZrO$_2$ and HfO$_2$ in Contact with Si and SiO$_2$. *Appl. Phys. Lett.* **2002**, *80*, 1897–1899. [CrossRef]
55. Khassin, A.A.; Yurieva, T.M.; Demeshkina, M.P.; Kustova, G.N.; Itenberg, I.S.; Kaichev, V.V.; Plyasova, L.M.; Anufrienko, V.F.; Molina, I.Y.; Larina, T.V.; et al. Characterization of the Nickel-Amesite-Chlorite-Vermiculite System. *Phys. Chem. Chem. Phys.* **2003**, *5*, 4025–4031. [CrossRef]
56. Kim, S.; Park, Y.M.; Choi, S.-H.; Kim, K.J.; Choi, D.H. Temperature-Dependent Carrier Recombination Processes in Nanocrystalline Si/SiO$_2$ Multilayers Studied by Continuous-Wave and Time-Resolved Photoluminescence. *J. Phys. D Appl. Phys.* **2007**, *40*, 1339. [CrossRef]
57. Batista, J.; Mandrino, D.; Jenko, M.; Martin, V. XPS and TPR Examinations of γ-Alumina-Supported Pd-Cu Catalysts. *Appl. Catal. A Gen.* **2001**, *206*, 113–124. [CrossRef]

58. Wöllner, A.; Lange, F.; Schmelz, H.; Knözinger, H. Characterization of Mixed Copper-Manganese Oxides Supported on Titania Catalysts for Selective Oxidation of Ammonia. *Appl. Catal. A Gen.* **1993**, *94*, 181–203. [CrossRef]
59. Moretti, G. Auger Parameter and Wagner Plot in the Characterization of Chemical States: Initial and Final State Effects. *J. Electron Spectrosc. Relat. Phenom.* **1995**, *76*, 365–370. [CrossRef]
60. Descostes, M.; Mercier, F.; Thromat, N.; Beaucaire, C.; Gautier-Soyer, M. Use of XPS in the Determination of Chemical Environment and Oxidation State of Iron and Sulfur Samples: Constitution of a Data Basis in Binding Energies for Fe and S Reference Compounds and Applications to the Evidence of Surface Species of an Oxidized Pyrite in a Carbonate Medium. *Appl. Surf. Sci.* **2000**, *165*, 288–302. [CrossRef]
61. Tan, B.J.; Klabunde, K.J.; Sherwood, P.M. X-ray Photoelectron Spectroscopy Studies of Solvated Metal Atom Dispersed Catalysts. Monometallic Iron and Bimetallic Iron-Cobalt Particles on Alumina. *Chem. Mater.* **1990**, *2*, 186–191. [CrossRef]
62. Selishchev, D.; Svintsitskiy, D.; Kovtunova, L.; Gerasimov, E.; Gladky, A.; Kozlov, D. Surface Modification of TiO_2 with Pd Nanoparticles for Enhanced Photocatalytic Oxidation of Benzene Micropollutants. *Colloids Surf. A Physicochem. Eng. Asp.* **2021**, *612*, 125959. [CrossRef]
63. Scofield, J.H. Hartree-Slater Subshell Photoionization Cross-Sections at 1254 and 1487 eV. *J. Electron Spectrosc. Relat. Phenom.* **1976**, *8*, 129–137. [CrossRef]
64. Copyright© 2005 Casa Software Ltd. Available online: http://www.casaxps.com/ (accessed on 28 February 2023).
65. Kovalevskiy, N.; Svintsitskiy, D.; Cherepanova, S.; Yakushkin, S.; Martyanov, O.; Selishcheva, S.; Gribov, E.; Kozlov, D.; Selishchev, D. Visible-Light-Active N-Doped TiO_2 Photocatalysts: Synthesis from $TiOSO_4$, Characterization, and Enhancement of Stability Via Surface Modification. *Nanomaterials* **2022**, *12*, 4146. [CrossRef] [PubMed]

Disclaimer/Publisher's Note: The statements, opinions and data contained in all publications are solely those of the individual author(s) and contributor(s) and not of MDPI and/or the editor(s). MDPI and/or the editor(s) disclaim responsibility for any injury to people or property resulting from any ideas, methods, instructions or products referred to in the content.

Article

Spatially Formed Tenacious Nickel-Supported Bimetallic Catalysts for CO_2 Methanation under Conventional and Induction Heating

Daniel Lach [1,*], Błażej Tomiczek [2], Tomasz Siudyga [1,*], Maciej Kapkowski [1], Rafał Sitko [1], Joanna Klimontko [3], Sylwia Golba [4], Grzegorz Dercz [4], Krzysztof Matus [5], Wojciech Borek [6] and Jaroslaw Polanski [1]

1. Centre for Materials and Drug Discovery, Institute of Chemistry, Faculty of Science and Technology, University of Silesia, Szkolna 9, 40-006 Katowice, Poland
2. Scientific and Didactic Laboratory of Nanotechnology and Material Technologies, Faculty of Mechanical Engineering, Silesian University of Technology, Konarskiego 18a, 44-100 Gliwice, Poland
3. Institute of Physics, Faculty of Science and Technology, University of Silesia, 75 Pułku Piechoty 1a, 41-500 Chorzów, Poland
4. Institute of Materials Engineering, Faculty of Science and Technology, University of Silesia, 75 Pułku Piechoty 1a, 41-500 Chorzów, Poland
5. Materials Research Laboratory, Faculty of Mechanical Engineering, Silesian University of Technology, Konarskiego 18a, 44-100 Gliwice, Poland
6. Department of Engineering Materials and Biomaterials, Faculty of Mechanical Engineering, Silesian University of Technology, Konarskiego 18a, 44-100 Gliwice, Poland
* Correspondence: daniel.lach@us.edu.pl (D.L.); tomasz.siudyga@us.edu.pl (T.S.)

Abstract: The paper introduces spatially stable Ni-supported bimetallic catalysts for CO_2 methanation. The catalysts are a combination of sintered nickel mesh or wool fibers and nanometal particles, such as Au, Pd, Re, or Ru. The preparation involves the nickel wool or mesh forming and sintering into a stable shape and then impregnating them with metal nanoparticles generated by a silica matrix digestion method. This procedure can be scaled up for commercial use. The catalyst candidates were analyzed using SEM, XRD, and EDXRF and tested in a fixed-bed flow reactor. The best results were obtained with the Ru/Ni-wool combination, which yields nearly 100% conversion at 248 °C, with the onset of reaction at 186 °C. When we tested this catalyst under inductive heating, the highest conversion was observed already at 194 °C.

Keywords: CO_2 methanation; bimetallic catalyst; Ni-wool support; Ni-mesh support; Au; Pd; Re; Ru nanoparticles; spatial and tenacious form; induction heating

1. Introduction

Excess anthropogenic CO_2 emission gave rise to novel sustainable chemistry and engineering ideas. Power-to-gas is an example of such a concept that targets CO_2 mitigation by using surplus energy, particularly renewable energy, to generate hydrogen, for example, from the hydrolysis of water and a further reaction of this hydrogen with carbon dioxide [1]. The main product is methane, which we can use as a synthetic natural gas (SNG). The rising prices of natural gas additionally make the concept highly attractive. The crucial reaction of the process is CO_2 methanation ($CO_2 + 4H_2 \rightleftarrows CH_4 + 2H_2O$). This reaction is discussed in detail in [2,3]. However, the practical course of CO_2 hydrogenation to CH_4 is impeded by many side processes, which depend on the reaction conditions. Therefore, to improve the selectivity and yield of this reaction and reduce the costs of the process, it is necessary to search for new high-performance and low-temperature catalysts.

Nickel-based catalysts are an essential class for CO_2 methanation [3–6]. The nickel catalyst was already used in the pioneering research on the hydrogenation of carbon oxides to methane by Paul Sabatier and Jean B. Senderens in 1902 [7]. It is characterized by a high selectivity to methane and is often a good compromise between high catalytic activity and low price. The methanation mechanism on the surface of Ni catalyst [8–10], the influence of support [11], and the synergies between Ni and other metals or promoters [3–6] were broadly investigated. It was also noted that the Ni catalyst in CO_2 methanation may be deactivated as a result of the formation of mobile nickel subcarbonyls due to the interaction of metal particles with the formed or temporarily present CO [9]. Therefore, one of the critical treatments to improve catalytic activity is surface modifications that allow for rapid removal of the surface nickel carbonyl species by surface-dissociated hydrogen. This process is promoted by defecting the Ni surface, which can act as the trap for hydrogen surface transport, reducing the activation energy of hydrogen dissociation [10]. For example, such a mechanism was proved by the high-activity CO_2 methanation with the sponge Ni-catalyst, which has many fcc-Ni crystal defects [12]. Due to their excellent mass and heat transfer efficiency, nickel foams are attractive as a substrate for microstructural catalysts, especially for highly exothermic reactions such as methanation. The desirable mechanical strength, high surface area to volume, and low flow pressure drop in the fixed bed reactor are also advantages. This fact was used by the authors of the composite catalyst Ni-Al_2O_3/Ni-foam [13] and Ru/CeO_2/Ni-foam [14]. tenacious catalyst produced by the wash-coating method with the Ni/CeO_2 component on an aluminum honeycomb bed was also presented in [15]. Another example of the Ni-based formed catalysts is a quaternary disc-shaped system (made of Ni, Ti, Ce, and yttria-stabilized zirconia (YSZ)) [16]. For the above examples, the conversion to methane at 250 °C varies between 15% and 65%. The preparation method is often multi-stage, energy- and time-consuming, and requires several constituent materials. There are already high-efficiency and low-temperature catalysts for CO_2 methanation, but in the form of grains or nickel nanowires. For example, we presented such materials in [17,18]. However, there have been no reported attempts to prepare tenacious and compact nickel-based spatial catalysts with satisfactory results for potential commercialization, maintaining high-performance and low-temperature catalysis in CO_2 methanation.

This article presents a novel approach to preparing bimetallic catalysts with spatially formed tenacious Ni-support based on mesh or wool. The supports were combined with nano -Au, -Pd, -Re, or -Ru. Nanometals for impregnation were generated using our recently developed method for powder catalysts [17–21]. This method minimizes the use of an expensive catalyst component in bimetallic conjugation. In addition, our new catalyst support formation procedure could be easily scaled to a commercial product. The Ru/Ni-wool combination achieved the best result, which provides almost 100% conversion in CO_2 methanation at 248 °C with the onset of reaction at 186 °C. The best sample was also tested in a methanation reactor with induction heating. For such a system, the highest conversion was noted already at 194 °C.

2. Results and Discussion
2.1. The Catalysts Design, Preparation, and Structure

Multi-component materials are commonly used in engineering to improve catalyst performance [22,23]. A typical representative of a heterogeneous catalyst consists of a metal and a support in the form of oxides (e.g., SiO_2, Al_2O_3, TiO_2), zeolites, carbon, or metaloorganic compounds [6,11,24,25]. Maximizing the metal surface area for a specific metal weight is essential in optimizing the catalyst [26]. Therefore, the small metal particles (typically less than 1–10 nm) are synthesized, and anchored to a thermally stable, high-surface-area support. However, the final catalytic material is often in powder or non-solid form. This form is not particularly commercially valuable. Scaling up is also a typical problem for such a catalyst form. Relatedly, we formulated a tenacious spatial catalyst consisting of commercially available nickel wool or mesh and enriched with selected metal nanoparticles: Au, Pd, Re, Ru. A scheme of the preparation procedure is shown in Figure 1. The obtained materials were tested as candidates for CO_2 methanation catalysts.

The morphology and composition of the resulting bimetallic system were studied using scanning electron microscopy (SEM) (Figures 2–4), specific surface area (SSA) (Table 1), X-ray diffraction spectroscopy (XRD) (Table 2 and Figure 5), and energy-dispersive X-ray fluorescence spectrometry (EDXRF) (Table 3). Additional materials from the analyses are included in the Supplementary Materials.

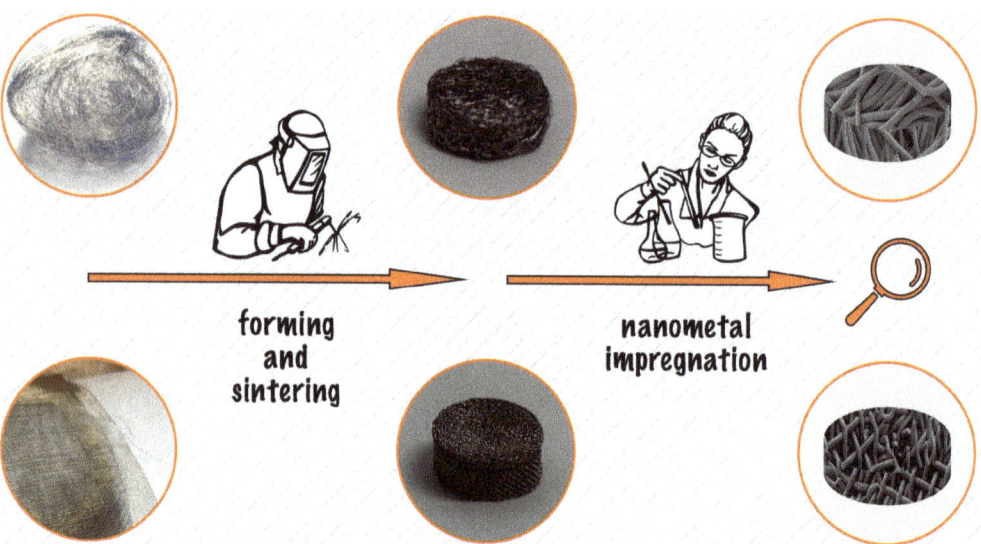

Figure 1. Scheme of the preparation procedure for Ni-wool or Ni-mesh supported catalysts.

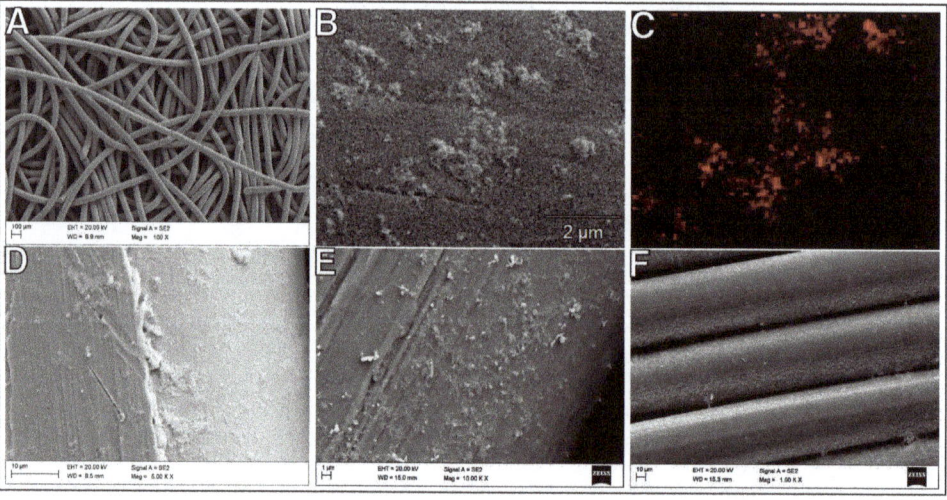

Figure 2. Scanning electron microscopy (SEM) images for nanometal/Ni-wool catalysts; (**A**) Ni-wool support fibers with Ru nanoparticles, (**B**) Ru nanoparticles on the support surface, (**C**) Ru nanoparticles (red spots) after EDS mapping, (**D**) Re nanoparticles on the fiber surface, (**E**) Pd nanoparticles on the fiber edge surface, (**F**) support fibers covered with Au nanoparticles.

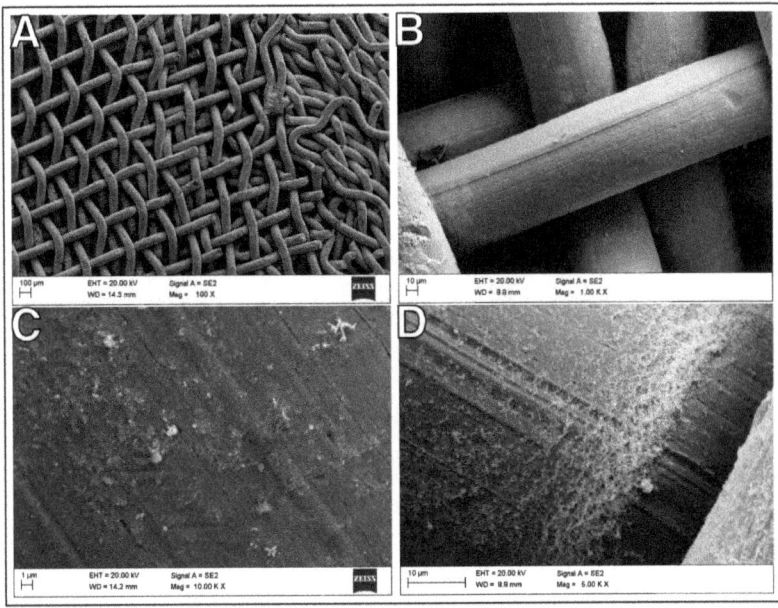

Figure 3. Scanning electron microscopy (SEM) images for nanometal/Ni-mesh catalysts, (**A**) Ni-mesh support fibers with Ru nanoparticles, (**B**) Re nanoparticles on the fibers surface, (**C**) Pd nanoparticles on the fiber surface, (**D**) Au nanoparticles on the surface of the fibers in contact.

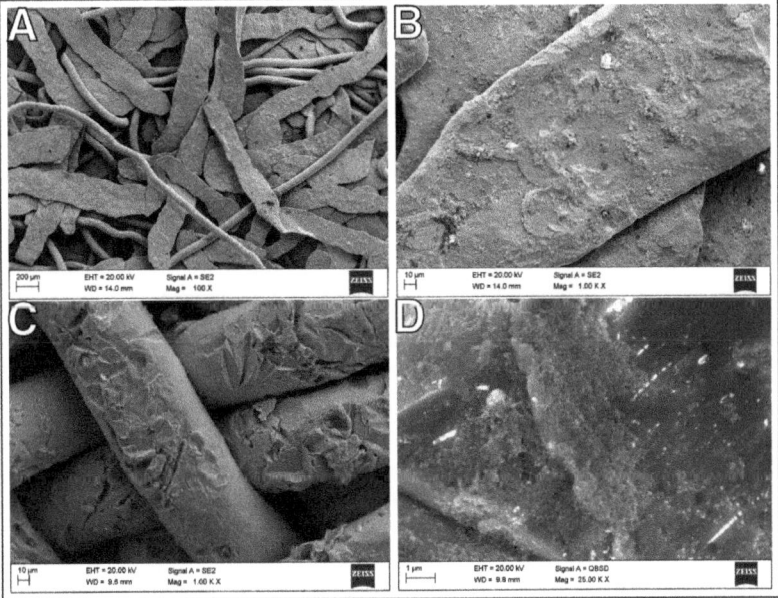

Figure 4. Scanning electron microscopy (SEM) images for nanoRu/Ni-ground_wool and nanoRu/Ni-blasted_mesh catalysts, (**A**) support fibers of ground nickel wool ornamented with naonRu, (**B**) Ru nanoparticles on the surface of ground fibers, (**C**) support fibers of sandblasted nickel mesh ornamented with naonRu, (**D**) Ru nanoparticles on the surface and in the pits of Ni-blasted_mesh fibers.

Table 1. Specific surface area (SSA) for the tested supports.

No	Support	S Bet [m²/g]
1	Ni-wool	0.104
2	Ni-ground_wool	0.338
3	Ni-mesh	0.280
4	Ni-blasted_mesh	0.097

Table 2. The average crystallite size and lattice parameters of the investigated catalysts as determined by the X-ray diffraction technique (XRD) method.

No.	Catalyst	Lattice Parameters [Å]	D [nm]				
			Ni	Pd	Au	Ru	Re
1	1%Ru/Ni-wool	a = 3.516 (±0.005) for Ni	40	-	-	6	-
2	1.5%Ru/Ni-wool	a = 3.528 (±0.003) for Ni	50	-	-	10	-
3	1%Ru/Ni-ground_wool	a = 3.516 (±0.004) for Ni	20	-	-	7	-
4	1%Ru/Ni-mesh	a = 3.519 (±0.006) for Ni	40	-	-	8	-
5	1%Ru/Ni-blasted_mesh	a = 3.530 (±0.004) for Ni	40	-	-	9	-
6	1%Re/Ni-wool	a = 3.520 (±0.003) for Ni	25	-	-	-	6
7	1%Re/Ni-mesh	a = 3.528 (±0.004) for Ni	95	-	-	-	7
8	1%Pd/Ni-wool	a = 3.524 (±0.003) for Ni a = 3.886 (±0.006) for Pd	60	12	-	-	-
9	1%Pd/Ni-mesh	a = 3.516 (±0.004) for Ni a = 3.880 (±0.005) for Pd	70	8	-	-	-
10	1%Au/Ni-wool	a = 3.523 (±0.004) for Ni a = 4.071 (±0.006) for Au	55	-	6	-	-
11	1%Au/Ni-mesh	a = 3.533 (±0.004) for Ni a = 4.079 (±0.005) for Au	60	-	6	-	-

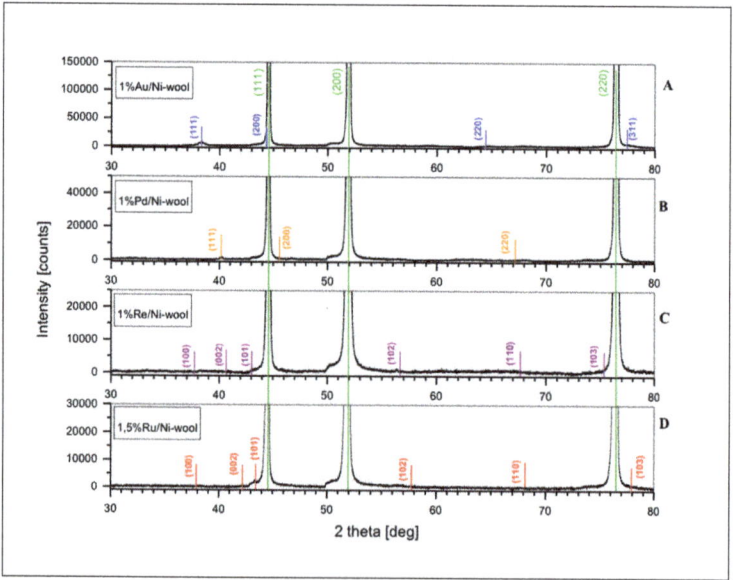

Figure 5. X-ray diffraction patterns at 2θ: 30°–80° for 1.0% Au/Ni-wool (**A**), 1.0% Pd/Ni-wool (**B**), 1.0% Re/Ni-wool (**C**), 1.5% Ru/Ni-wool (**D**) samples. Miller indices for experimental peaks of Ni (green) wool fibers and Au (blue), Pd (orange), Re (purple), Ru (red) metals are marked.

Table 3. Mass of support and nanometals, and EDXRF analysis of Ru, Re, Pd, Au, and Ni for the tested catalytic materials.

No.	Catalyst	[1] Support Mass [mg]	[2] Nanometal Mass [mg]	Weight Percentage of a Chemical Element [wt%]				
				Ru	Re	Pd	Au	Ni
1	1%Ru/Ni-wool	733.42	0.211	0.94	-	-	-	97.03
2	1.5%Ru/Ni-wool	726.10	0.428	1.50	-	-	-	96.50
3	1%Ru/Ni-ground_wool	726.38	0.209	0.91	-	-	-	91.70
4	1%Ru/Ni-mesh	1034.12	0.318	0.95	-	-	-	96.50
5	1%Ru/Ni-blasted_mesh	991.11	0.132	0.69	-	-	-	96.34
6	1%Re/Ni-wool	731.78	1.727	-	0.60	-	-	97.80
7	1%Re/Ni-mesh	842.10	2.502	-	0.51	-	-	98.20
8	1%Pd/Ni-wool	716.37	0.231	-	-	0.76	-	97.00
9	1%Pd/Ni-mesh	968.40	0.214	-	-	0.76	-	96.60
10	1%Au/Ni-wool	702.60	0.202	-	-	-	0.88	96.09
11	1%Au/Ni-mesh	836.00	0.239	-	-	-	0.78	96.29

[1] Mass after degreasing and drying of the support material. [2] Mass of individual nanometals deposited on the silica carrier and used after digesting for support impregnation.

Two types of nickel support were made. The first was made of rolled up nickel mesh and impulse sintered. The second was formed from nickel wool, which was compressed and impulse sintered. Pictures of these materials are shown in Figures 1–3, and in the Supplementary Materials (Figures S1–S3). In our previous study of a CO_2 methanation catalyst supported by nickel nanowires, we emphasized the significant effect of the extended surface area of the catalytic material [18]. Here, we also tried to improve the specific surface area of the presented supports. Modifications were made by ball milling wool and in the case of mesh by sandblasting. The specific surface area (SSA) of support materials is given in Table 1. The specific surface area of the Ni-mesh support is larger than that of Ni-wool. The difference in the diameter of the nickel wire in both cases decides this result. For the mesh wire, the lateral surface of the cylinder (with the same height compared) is almost 50,000 nm^2 larger. The milling process increased the SSA of the wool-type threefold. In the second modification, sandblasting did not improve the SSA of Ni-mesh and even lowered it according to the S bet analysis. Although SEM images (Figure 4) show furrows, pits, and roughness, we hypothesize that the walls of the grooves and irregularities have been smoothed out, hence, the failing of increased SSA. Nevertheless, SSA improvement research still needs to continue.

As we have already mentioned, in the preparation of a heterogeneous multi-component catalyst, in this case a bimetallic one, it is important to obtain a narrow size distribution of metallic nanoparticles and their large dispersion on the support. This feature was achieved using our proprietary method of synthesis of metal nanoparticles on a silica matrix, then digestion of the matrix with sodium hydroxide and uniform suspension of nanoparticles in the impregnation solution for support coverage. We described this method in [17]. Nanoparticles on silica with an average size of 4.1 nm for Ru, 4.4 nm for Pd, 5.1 nm for Au, and 1.8 nm for Re were used. Transmission electron microscope (TEM) images of the nanoparticles and their size distribution are given in the Supplementary Materials (Figures S4–S8). The structure of the catalyst material after ornamentation with nanoparticles of selected metals is shown in Figures 2–4, and in the Supplementary Materials (Figures S1–S3). The material forms a conglomerate of mesh or wool fibers with metal particles Au, Pd, Re, Ru, respectively. The nanometal coating on the substrate fibers is distributed over the entire surface in a non-uniform manner. Metal aggregation on the fibers is visible, in particular at the crossing of the support wires. The concentration of metal nanoparticles can also be seen in any imperfections or scratches on the surface of the nickel fibers. Microscopic examinations proved that the distribution of particles strongly depends on the surface roughness of the nickel support. Different shape and size of the nanoparticles are observed depending on the selected metal. The size and lattice parameters of the nanometal particles were determined using the XRD technique. We used the Scherrer equation to estimate the average crystalline particle size from the highest intensity diffraction peaks. The measured values of metal nanoparticles range from about 6 nm to 12 nm. Lattice parameters and average crystallite dimensions (D) are listed in Table 2. For compositions with the best-performing support (Ni-wool), the XRD spectra are shown in Figure 5.

The X-ray diffraction patterns of the 1%Au/Ni-wool, 1%Pd/Ni-wool, 1%Re/Ni-wool, and 1.5%Ru/Ni-wool are given in the range of the 2θ angle from 30 to 80 degrees. They clearly show the diffraction lines that correspond to the face-centered cubic (Fm3m) phase of Ni (JCPDS 01-077-8341), whereas only the most intense peaks of the cubic (Fm3m) phases of Au ($2\theta_{111} \sim 38°$) and Pd ($2\theta_{111} \sim 40°$) are identified. The overlapping diffraction lines were observed. The strongest diffraction lines of the hexagonal (P63/mmc) phases of Ru and Re ($2\theta_{101} \sim 43° - 44°$) overlap Ni (111) diffraction line, whereas the less intensive peaks of Ru and Re were not detected. The qualitative and quantitative elemental analysis was performed by EDXRF spectrometry. The results of the quantitative analysis calculated by the fundamental parameter method are presented in Table 3. The content of nanometal in the sample was up to 1%. This percentage is the optimal support load as studied in [27] and is consistent with our experience and testing Above this concentration, we observed either a complete coverage of the support fibers or an agglomeration, which increased the size of the nanoparticles. These effects reduced the number of nanometal-support connections (synergy centers between materials), decreasing catalyst activity.

2.2. The Catalysts in CO_2 Methanation

Kinetic limitations affecting the hydrogenation of carbon dioxide to methane with an acceptable rate and selectivity necessitate the use of a catalyst [3]. The set of catalytic materials presented above was tested in relation to methane conversion during a temperature increase, as shown in Figure 6. The best Ru/Ni-wool composition was determined. It achieves almost 100% conversion at 248 °C, with the onset of reaction at 186 °C. This composition is consistent with our previous research [17–19] and confirms the privilege of the Ru/Ni connection in CO_2 methanation catalysis, which we wrote about in [28]. Approximately 100% conversion of the best composition in relation to the reference sample-pure nickel wool support is possible at a temperature lower by as much as 289 °C. Compared to the previously studied CO_2 methanation catalysts, such as Ru/Ni-nanowires, Ru/Ni-grains, almost complete conversion of a mixture of 20% CO_2 and 80% H_2 to methane at a flow rate 3 dm^3/h is for temperatures as follows: Ru/Ni-nanowires 179 °C, Ru/Ni-grains 204 °C, Ru/Ni-wool 248 °C. For powder catalysts, weight hourly space velocity (WHSV) was equal to 6.5 h^{-1}, and for the present sample it was 1.8 h^{-1}. In turn, for example, for the most similar, tenacious, and spatial materials, the conversion at 250 °C is Ni-sponge 83% [12] and Ru/CeO$_2$/Ni-foam disc ca. 15% [14]. The gas hourly space velocity (GHSV) values, calculated with inlet flow rate of CO_2, were Ni-sponge 4200 h^{-1}, Ru/CeO$_2$/Ni-foam disc approx. 714 h^{-1}, and Ru/Ni-wool 3612 h^{-1}. The difference in performance in the case of the first comparison can be explained by the specific surface area, the number of active centers or diffusion, which advantage grains and nanowires. However, in the second case, we see a significant superiority of the obtained material over previous commensurable materials, probably thanks to the Ru-Ni synergy and differences in the adsorption of surface forms of reactants. The presented material does not use typical oxides (CeO_2, ZrO_2, Al_2O_3, SiO_2, TiO_2) as the support construction, and the reaction path runs only through the area of Ru and Ni atoms.

In this research, we also attempted to modify the support surface. Results in CO_2 methanation for the best compositions in comparison with ground or sandblasted nano-Ru support samples are shown in Figure 7. Supports crafted of Ni-wool fare much better than those of Ni-mesh. The difference in the morphology of the material can explain the observed phenomenon. The Ni-wool support is a highly irregular arrangement of fibers, which may impact a more turbulent flow of gases and a longer contact time of the reactants with the catalyst. There can be a difference in diffusion effects for both types of supports [29,30]. No significant catalytic improvement was observed for the ground nickel wool support. However, the earlier sandblasting of the mesh and the formation an irregularly layered support from its pieces increases the activity of Ru/Ni-blasted_mesh relative to Ru/Ni-mesh at about 280 °C by as much as 90%. The difference in favor of the Ru/Ni-wool at around 250 °C is 65%. The improvement of the catalyst mesh benchmark can be explained by the hydrogen traps in pits after sandblasting and structural changes in the support (see Figure 4c). These changes probably translate into increased hydrogen uptake, improved hydrogen spillover, and transport of the species adsorbed or formed on the surface of the catalytic material.

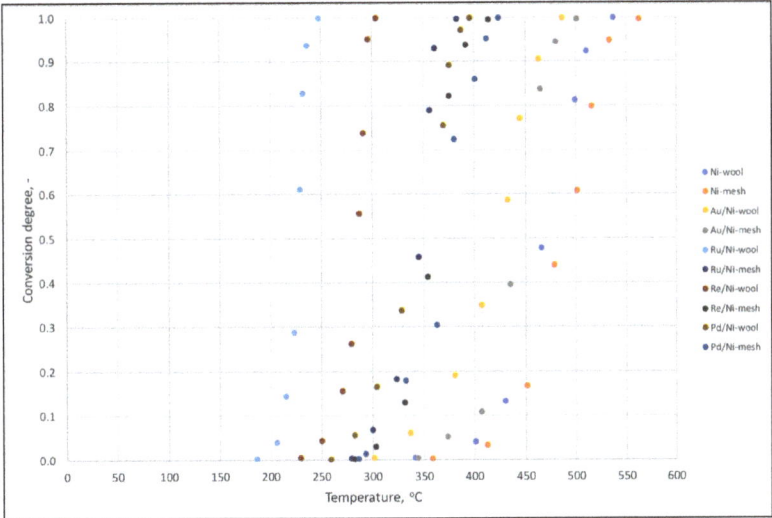

Figure 6. CO_2 conversion of catalysts made of nickel wool or nickel mesh and ornamented with ca. 1% nano-Ru, -Re, -Pd, or -Au.

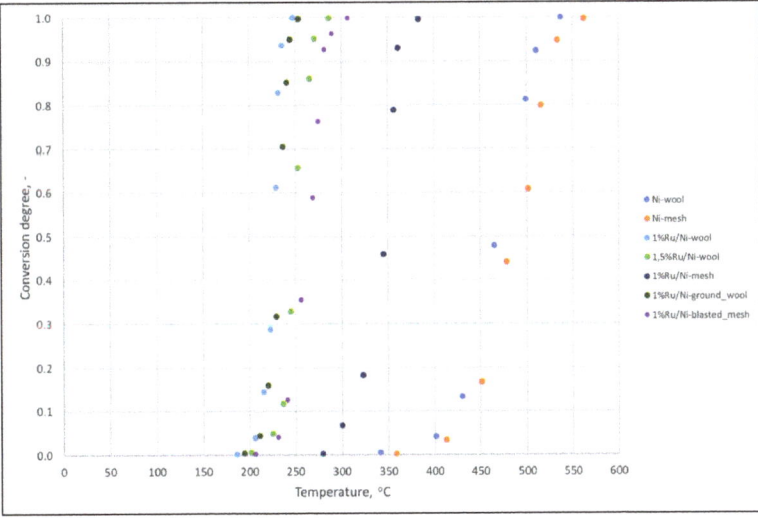

Figure 7. CO_2 conversion of catalysts made of modified nickel wool, nickel mesh, and ornamented with nano-Ru.

The best Ru/Ni-wool composition was tested in CO_2 methanation for 24 h at the highest conversion temperature (248 °C). No significant decrease in catalyst efficiency was observed during this time. XRD analysis of the sample also showed no destabilization. A slight difference in the values of the lattice constants Ni (0.002 Å) was noted; however, it is within the limits of the measurement error. We performed XPS (X-ray photoelectron spectroscopy) to profile the sample before and after the reaction. Analysis of chemical states indicated the presence of oxidized forms of composition metals. Carbon species have also been detected on the surface. After methanation, a significant share of carbon bonding to oxygen was observed, mainly corresponding to the C=O bond. Spectra, measurement details, and additional descriptions are given in the Supplementary Materials (Figure S9). Long-term catalyst deactivation tests (over 24 h) have not yet been performed (subject to further research). However, we assume that the behavior of this catalytic material will be analogous to the ones we studied earlier [17–19], and reactivation will be feasible by hydrogen treatment. For comparative purposes, we additionally tested the best sample in a reactor with direct bed induction heating. Induced heating eliminates limitations in heat transfer in the catalyst bed and improves energy efficiency, which we wrote about in [19], and it was broadly described in [31,32]. In such a system, it was possible to decrease the initial reaction temperature to Ti = 172 °C. The conversion degree of 99.9% was reached at 194 °C as opposed to 248 °C for the conventional heating system. The exact explanation of the reason for the improvement may be a topic for a separate publication, but our hypothesis assumes the generation of eddy currents in the support filaments, which can affect the electron modification of atoms and thus the potential differences and the energy barrier to overcome by intermediates and surface moiety in the mechanism of CO_2 methanation. The theory of hot electrons may also play a role here [33,34]. Research into scaling up and potential commercialization of the presented material is still in progress.

3. Materials and Methods

3.1. Catalysts Preparation

The catalysts were made in two steps: (1) preparation of the nickel support, (2) generation of Au, Pd, Re, or Ru nanoparticles and their subsequent ornamentation on the support surface.

3.1.1. Ni Support Preparation

Two types of nickel support were made from commercially available materials. The first one uses the nickel wool brand "Elemental Microanalysis". The thickness of the nickel wool wire was 0.065 mm. The second one was "Speorl KG" mesh with a wire thickness of 0.08 mm and a mesh size of 125 × 224 µm. The rolled mesh or wool was placed in a cylindrical graphite matrix and closed on both sides with copper stamps, which act as electrodes in the impulse resistance welding process. For sample formation, a thermo-mechanical simulator Gleeble-3800 from Dynamic System Inc. was used. A particular set of tools has been developed for impulse sintering of porous nickel skeletons, whose scheme is shown in Figure 8.

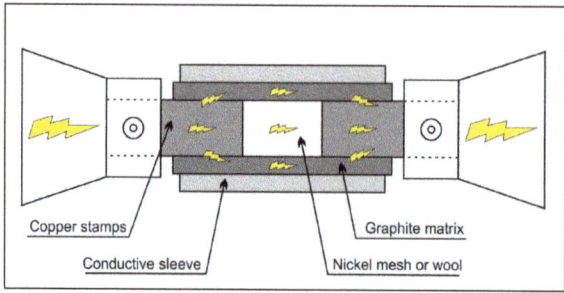

Figure 8. Scheme of preparation of the nickel support.

The die filled with nickel wool or mesh sample was placed into the Gleeble-3800 simulator. In the first step, the nickel mesh or wool were compressed to a distance of 6 mm between the copper stamps. During the initial pressing, compressive stresses of about 10–15 MPa were generated in the sample. When a vacuum of about 3×10^{-1} mBar was created in the Gleeble chamber, a program was started which consisted of heating the sample to a temperature of 700 °C at 10 s with further compression of the sample to a thickness value of, respectively, 3.3 mm for mesh or 3 mm for wool and with a disc diameter of 8.5 mm. During the experiment, a high electric current passes through the sample, simultaneously heating it to the pre-set temperature with at a predetermined heating rate 70 °C/s. The mass of the tested supports for each catalytic material is given in Table 3. Modified supports were prepared similarly, but the wool had been ground previously for 15 min in a planetary ball mill with 20 mm size zirconia balls. In turn, the mesh was sandblasted, cut into fragments, irregularly layered, and formed into the target disc.

3.1.2. Impregnation of Ni Support with Nanometal

Metal nanoparticles were prepared according to our method described in [17]. Metal nanoparticles digested from the silica precursor with 40% NaOH solution were washed to neutral pH and centrifuged. Then the nanoparticles were suspended in 0.7 mL of isopropyl alcohol in a sonic bath. The solution was taken into a 1 mL syringe fitted with a needle. A nanometal solution was spotted onto the previously degreased and dried nickel support. The soaked material was dried at 110 °C in an oven. The application and drying procedure were repeated until the solution was exhausted. Each time the application side of the nickel support was changed. The mass of individual nanometals deposited on the silica carrier and used after digesting for impregnation is given in Table 3.

3.2. Method of Catalysts Characterization

Images of the surface morphology of the studied materials were obtained with a scanning electron microscope SUPRA 35 Zeiss with EDS detector for microanalysis of chemical composition.

The quantitative and qualitative chemical composition was confirmed by energy dispersive X-ray fluorescence spectrometry (EDXRF), performed on an Epsilon 3 spectrometer (Panalytical, Almelo, The Netherlands) with an Rh target X-ray tube with 50 µm Be window and max. power of 9 W. The spectrometer was equipped with a thermoelectrically cooled silicon drift detector (SDD) with an 8 µm Be window and a resolution of 135 eV at 5.9 keV. The quantitative analysis was performed using Omnian software and was based on the fundamental parameter method and following measurement conditions: 12 kV, 50 µm Al primary beam filter, 300 s counting time, helium atmosphere for Pd and Ru determination; 30 kV, 100 µm Ag primary beam filter, 120 s counting time, air atmosphere for Ni, Re, and Au determination. The current of the X-ray tube was fixed so that it would not exceed a dead-time loss of ca. 50%.

The X-ray diffraction experiments were performed on a PANalytical Empyrean diffractometer with Cu Kα radiation (40 kV, 30 mA) equipped with a PIXcel detector. Data were collected in the 20°–100° 2θ range with 0.0131° step. A qualitative phase analysis employed the "X'Pert High Score Plus" computer program and the data from ICDD PDF-4 database. Crystal lattice parameters were calculated using the Chekcell V4 program.

The specific surface area (SSA) was determined using a Gemini VII 2390 a analyzer (Micromeritics Instruments Corp., Norcross, GA, USA) at the boiling point of nitrogen (−196 °C) using the Brunauer–Emmet–Teller (BET) method. Samples before the measurements were thermal-treated at 300 °C for 1 h to remove gases and vapors that may have adsorbed on the surface during the synthesis. This was performed with a VacPrep 061 degassing system (Micromeritics Instruments Corp., Norcross, GA, USA). Samples not analyzed immediately after the degassing procedure were kept at 60 °C. Correctness of the instrument was verified by analyzing a Carbon Black reference material of known surface area (P/N 004-16833-00 from Micromeritics, Norcross, GA, USA).

3.3. Methanation

The catalysts were tested in an 8 mm diameter fixed bed quartz flow reactor under atmospheric pressure. The feed mix was 20% CO_2 + 80% H_2 and was fed continuously at a flow rate of 3 dm^3/h. The conversion of CO_2 to CH_4 was investigated by exhaust gas analysis using an on-site gas analyzer GX-6000 RIKEN and a gas chromatograph SRI 310 C equipped with a thermal conductivity detector (1/8 inch diameter, 3 m long column; micropacked with active carbon 80–100 mesh; 80 °C temperature of column with argon as the carrier gas with flow rate of 10 dm^3/h^{-1}). The methane detection limit was 1 ppm for the GX-6000 and 10 ppm for the gas chromatography SRI 310 C.

For a selected catalytic system with high activity, comparative tests were carried out by replacing thermal heating with induction heating (according to the methodology used in [19]). A 100 W induction heater was used for this purpose, keeping the other parameters unchanged (size and dimensions of the catalyst bed, substrate flows). Temperature of the gases flowing out of the catalytic bed was measured.

4. Conclusions

The search for new methods of carbon dioxide management increases the interest in catalytic methanation of CO_2. In search of novel bimetallic catalyst candidates for this reaction, we developed a spatially formed tenacious Ni-support based on mesh or wool. Au, Pd, Re, Ru nanometals were selected to ornamentation the support of the tested catalysts. We developed a new method for the catalyst support formation and its impregnation with nanometals. The catalyst preparation could be easily scaled up to a commercial procedure. The obtained catalyst candidates were analyzed by SEM, XRD, and EDXRF. The best combination appeared to be the Ru/Ni-wool achieving almost 100% conversion at 248 °C, with the reaction onset at 186 °C. We also tried to modify the support surface by milling wool or sandblasting and irregular layering of the mesh. There was a 90% improvement in conversion at 280 °C for Ru/Ni-blasted_mesh compared to Ru/Ni-mesh. However, this combination had a 65% lower conversion than the best Ru/Ni-wool at 250 °C. Comparatively, we tested the activity of Ru/Ni-wool in a reactor with induction heating, which significantly improves the efficiency. For such a system, the initial reaction temperature was Ti = 172 °C and a conversion degree of 99.9% was reached at 194 °C.

Supplementary Materials: The following supporting information can be downloaded at: https://www.mdpi.com/article/10.3390/ijms24054729/s1.

Author Contributions: Conceptualization, D.L., B.T., M.K., T.S. and J.P.; methodology, D.L., B.T., G.D., T.S. and W.B.; validation, T.S., R.S., J.K., S.G., G.D., K.M. and W.B.; formal analysis, D.L., T.S., R.S., J.K., S.G., G.D., K.M. and W.B.; investigation, D.L., B.T., M.K., T.S., R.S., J.K., S.G., K.M. and W.B.; resources, D.L., B.T., T.S. and M.K.; data curation, D.L., B.T., T.S., R.S., J.K. and S.G.; writing—original draft preparation, D.L. and J.P.; writing—review and editing, B.T., T.S., R.S., J.K., S.G. and W.B.; visualization, D.L., B.T., M.K., T.S. and M.K.; supervision, D.L., B.T. and J.P.; project administration, D.L.; funding acquisition, J.P. All authors have read and agreed to the published version of the manuscript.

Funding: This research was funded by the National Science Center OPUS 2018/29/B/ST8/02303. The research activities co-financed by the funds granted under the Research Excellence Initiative of the University of Silesia in Katowice, Poland.

Institutional Review Board Statement: Not applicable.

Informed Consent Statement: Not applicable.

Data Availability Statement: Data is contained within the article or Supplementary Materials.

Acknowledgments: The authors would like to thank the University of Silesia in Katowice for including their research in the Research Excellence Initiative program.

Conflicts of Interest: The authors declare no conflict of interest. The funders had no role in the design of the study; in the collection, analyses, or interpretation of data; in the writing of the manuscript; or in the decision to publish the results.

References

1. Wulf, C.; Linßen, J.; Zapp, P. Review of Power-to-Gas Projects in Europe. *Energy Procedia* **2018**, *155*, 367–378. [CrossRef]
2. Gao, J.; Wang, Y.; Ping, Y.; Hu, D.; Xu, G.; Gu, F.; Su, F. A Thermodynamic Analysis of Methanation Reactions of Carbon Oxides for the Production of Synthetic Natural Gas. *RSC Adv.* **2012**, *2*, 2358. [CrossRef]
3. Lee, W.J.; Li, C.; Prajitno, H.; Yoo, J.; Patel, J.; Yang, Y.; Lim, S. Recent Trend in Thermal Catalytic Low Temperature CO_2 Methanation: A Critical Review. *Catal. Today* **2021**, *368*, 2–19. [CrossRef]
4. Mills, G.A.; Steffgen, F.W. Catalytic Methanation. *Catal. Rev.* **1974**, *8*, 159–210. [CrossRef]
5. Aziz, M.A.A.; Jalil, A.A.; Triwahyono, S.; Ahmad, A. CO_2 Methanation over Heterogeneous Catalysts: Recent Progress and Future Prospects. *Green Chem.* **2015**, *17*, 2647–2663. [CrossRef]
6. Tsiotsias, A.I.; Charisiou, N.D.; Yentekakis, I.V.; Goula, M.A. Bimetallic Ni-Based Catalysts for CO_2 Methanation: A Review. *Nanomaterials* **2020**, *11*, 28. [CrossRef]
7. Sabatier, P.; Senderens, J.-B. Nouvelles Synthèses Du Méthane. *Comptes Rendus Académie Sci.* **1902**, *134*, 514–516.
8. Choe, S.J.; Kang, H.J.; Park, D.H.; Huh, D.S.; Park, J. Adsorption and Dissociation Reaction of Carbon Dioxide on Ni(111) Surface: Molecular Orbital Study. *Appl. Surf. Sci.* **2001**, *181*, 265–276. [CrossRef]
9. Choe, S.J.; Kang, H.J.; Kim, S.J.; Park, S.B.; Park, D.H.; Huh, D.S. Adsorbed Carbon Formation and Carbon Hydrogenation for CO_2 Methanation on the Ni(111) Surface: ASED-MO Study. *Bull. Korean Chem. Soc.* **2005**, *26*, 1682–1688. [CrossRef]
10. Weng, M.H.; Chen, H.-T.; Wang, Y.-C.; Ju, S.-P.; Chang, J.-G.; Lin, M.C. Kinetics and Mechanisms for the Adsorption, Dissociation, and Diffusion of Hydrogen in Ni and Ni/YSZ Slabs: A DFT Study. *Langmuir* **2012**, *28*, 5596–5605. [CrossRef]
11. Le, T.A.; Kim, M.S.; Lee, S.H.; Kim, T.W.; Park, E.D. CO and CO_2 Methanation over Supported Ni Catalysts. *Catal. Today* **2017**, *293–294*, 89–96. [CrossRef]
12. Tada, S.; Ikeda, S.; Shimoda, N.; Honma, T.; Takahashi, M.; Nariyuki, A.; Satokawa, S. Sponge Ni Catalyst with High Activity in CO_2 Methanation. *Int. J. Hydro. Energy* **2017**, *42*, 30126–30134. [CrossRef]
13. Li, Y.; Zhang, Q.; Chai, R.; Zhao, G.; Liu, Y.; Lu, Y.; Cao, F. Ni-Al$_2$O$_3$/Ni-Foam Catalyst with Enhanced Heat Transfer for Hydrogenation of CO_2 to Methane. *AIChE J.* **2015**, *61*, 4323–4331. [CrossRef]
14. Cimino, S.; Cepollaro, E.M.; Lisi, L.; Fasolin, S.; Musiani, M.; Vázquez-Gómez, L. Ru/Ce/Ni Metal Foams as Structured Catalysts for the Methanation of CO_2. *Catalysts* **2020**, *11*, 13. [CrossRef]
15. Fukuhara, C.; Hayakawa, K.; Suzuki, Y.; Kawasaki, W.; Watanabe, R. A Novel Nickel-Based Structured Catalyst for CO_2 Methanation: A Honeycomb-Type Ni/CeO$_2$ Catalyst to Transform Greenhouse Gas into Useful Resources. *Appl. Catal. Gen.* **2017**, *532*, 12–18. [CrossRef]
16. Moon, D.H.; Lee, S.M.; Ahn, J.Y.; Nguyen, D.D.; Kim, S.S.; Chang, S.W. New Ni-Based Quaternary Disk-Shaped Catalysts for Low-Temperature CO_2 Methanation: Fabrication, Characterization, and Performance. *J. Environ. Manag.* **2018**, *218*, 88–94. [CrossRef]
17. Polanski, J.; Siudyga, T.; Bartczak, P.; Kapkowski, M.; Ambrozkiewicz, W.; Nobis, A.; Sitko, R.; Klimontko, J.; Szade, J.; Lelątko, J. Oxide Passivated Ni-Supported Ru Nanoparticles in Silica: A New Catalyst for Low-Temperature Carbon Dioxide Methanation. *Appl. Catal. B Environ.* **2017**, *206*, 16–23. [CrossRef]
18. Siudyga, T.; Kapkowski, M.; Janas, D.; Wasiak, T.; Sitko, R.; Zubko, M.; Szade, J.; Balin, K.; Klimontko, J.; Lach, D.; et al. Nano-Ru Supported on Ni Nanowires for Low-Temperature Carbon Dioxide Methanation. *Catalysts* **2020**, *10*, 513. [CrossRef]
19. Siudyga, T.; Kapkowski, M.; Bartczak, P.; Zubko, M.; Szade, J.; Balin, K.; Antoniotti, S.; Polanski, J. Ultra-Low Temperature Carbon (Di)Oxide Hydrogenation Catalyzed by Hybrid Ruthenium–Nickel Nanocatalysts: Towards Sustainable Methane Production. *Green Chem.* **2020**, *22*, 5143–5150. [CrossRef]
20. Kapkowski, M.; Ambrożkiewicz, W.; Siudyga, T.; Sitko, R.; Szade, J.; Klimontko, J.; Balin, K.; Lelątko, J.; Polanski, J. Nano Silica and Molybdenum Supported Re, Rh, Ru or Ir Nanoparticles for Selective Solvent-Free Glycerol Conversion to Cyclic Acetals with Propanone and Butanone under Mild Conditions. *Appl. Catal. B Environ.* **2017**, *202*, 335–345. [CrossRef]
21. Kapkowski, M.; Popiel, J.; Siudyga, T.; Dzida, M.; Zorębski, E.; Musiał, M.; Sitko, R.; Szade, J.; Balin, K.; Klimontko, J.; et al. Mono- and Bimetallic Nano-Re Systems Doped Os, Mo, Ru, Ir as Nanocatalytic Platforms for the Acetalization of Polyalcohols into Cyclic Acetals and Their Applications as Fuel Additives. *Appl. Catal. B Environ.* **2018**, *239*, 154–167. [CrossRef]
22. Kumar, G.; Nikolla, E.; Linic, S.; Medlin, J.W.; Janik, M.J. Multicomponent Catalysts: Limitations and Prospects. *ACS Catal.* **2018**, *8*, 3202–3208. [CrossRef]
23. Mizuno, N.; Misono, M. Heterogeneous Catalysis. *Chem. Rev.* **1998**, *98*, 199–218. [CrossRef] [PubMed]
24. Dai, H.; Xiong, S.; Zhu, Y.; Zheng, J.; Huang, L.; Zhou, C.; Deng, J.; Zhang, X. NiCe Bimetallic Nanoparticles Embedded in Hexagonal Mesoporous Silica (HMS) for Reverse Water Gas Shift Reaction. *Chin. Chem. Lett.* **2022**, *33*, 2590–2594. [CrossRef]
25. Guo, J.; Duan, Y.; Liu, Y.; Li, H.; Zhang, Y.; Long, C.; Wang, Z.; Yang, Y.; Zhao, S. The Biomimetic Engineering of Metal–Organic Frameworks with Single-Chiral-Site Precision for Asymmetric Hydrogenation. *J. Mater. Chem. A* **2022**, *10*, 6463–6469. [CrossRef]
26. Ross, J.R.H. *Contemporary Catalysis: Fundamentals and Current Applications*; Elsevier: Amsterdam, The Netherlands, 2019; ISBN 978-0-444-63474-0.
27. Lange, F.; Armbruster, U.; Martin, A. Heterogeneously-Catalyzed Hydrogenation of Carbon Dioxide to Methane Using RuNi Bimetallic Catalysts. *Energy Technol.* **2015**, *3*, 55–62. [CrossRef]
28. Polanski, J.; Lach, D.; Kapkowski, M.; Bartczak, P.; Siudyga, T.; Smolinski, A. Ru and Ni—Privileged Metal Combination for Environmental Nanocatalysis. *Catalysts* **2020**, *10*, 992. [CrossRef]

29. Kärger, J.; Goepel, M.; Gläser, R. Diffusion in Nanocatalysis. In *Nanotechnology in Catalysis*; Van de Voorde, M., Sels, B., Eds.; Wiley-VCH Verlag GmbH & Co. KGaA: Weinheim, Germany, 2017; pp. 293–334; ISBN 978-3-527-69982-7.
30. Tesser, R.; Santacesaria, E. Revisiting the Role of Mass and Heat Transfer in Gas–Solid Catalytic Reactions. *Processes* **2020**, *8*, 1599. [CrossRef]
31. Zheng, Y.; Wang, Y.; Yuan, Y.; Huang, H. Metal-based Heterogeneous Electrocatalysts for Electrochemical Reduction of Carbon Dioxide to Methane: Progress and Challenges. *ChemNanoMat* **2021**, *7*, 502–514. [CrossRef]
32. Ghosh, S.; Ourlin, T.; Fazzini, P.; Lacroix, L.; Tricard, S.; Esvan, J.; Cayez, S.; Chaudret, B. Magnetically Induced CO_2 Methanation In Continuous Flow over Supported Nickel Catalysts with Improved Energy Efficiency. *ChemSusChem* **2023**, *16*, e202201724. [CrossRef]
33. Wang, C.; Liu, Y. Ultrafast Optical Manipulation of Magnetic Order in Ferromagnetic Materials. *Nano Converg.* **2020**, *7*, 35. [CrossRef]
34. Guo, J.; Zhang, Y.; Shi, L.; Zhu, Y.; Mideksa, M.F.; Hou, K.; Zhao, W.; Wang, D.; Zhao, M.; Zhang, X.; et al. Boosting Hot Electrons in Hetero-Superstructures for Plasmon-Enhanced Catalysis. *J. Am. Chem. Soc.* **2017**, *139*, 17964–17972. [CrossRef]

Disclaimer/Publisher's Note: The statements, opinions and data contained in all publications are solely those of the individual author(s) and contributor(s) and not of MDPI and/or the editor(s). MDPI and/or the editor(s) disclaim responsibility for any injury to people or property resulting from any ideas, methods, instructions or products referred to in the content.

Article

Biomimetic Catalysts Based on Au@TiO$_2$-MoS$_2$-CeO$_2$ Composites for the Production of Hydrogen by Water Splitting

Kenneth Fontánez [1], Diego García [2], Dayna Ortiz [3], Paola Sampayo [3], Luis Hernández [3], María Cotto [3], José Ducongé [3], Francisco Díaz [3], Carmen Morant [4], Florian Petrescu [3], Abniel Machín [5,*] and Francisco Márquez [3,*]

1. Department of Chemistry, University of Puerto Rico, Rio Piedras Campus, San Juan 00925, Puerto Rico
2. Department of Biochemistry, School of Medicine, University of Puerto Rico, Medical Sciences Campus, San Juan 00936, Puerto Rico
3. Nanomaterials Research Group, Department of Natural Sciences and Technology, Division of Natural Sciences, Technology and Environment, Universidad Ana G. Méndez-Gurabo Campus, Gurabo 00778, Puerto Rico
4. Department of Applied Physics, Autonomous University of Madrid, Instituto de Ciencia de Materiales Nicolás Cabrera, 28049 Madrid, Spain
5. Department of Natural Sciences and Technology, Division of Natural Sciences, Technology and Environment, Universidad Ana G. Méndez-Cupey Campus, San Juan 00926, Puerto Rico
* Correspondence: machina1@uagm.edu (A.M.); fmarquez@uagm.edu (F.M.)

Abstract: The photocatalytic hydrogen evolution reaction (HER) by water splitting has been studied, using catalysts based on crystalline TiO$_2$ nanowires (TiO$_2$NWs), which were synthesized by a hydrothermal procedure. This nanomaterial was subsequently modified by incorporating different loadings (1%, 3% and 5%) of gold nanoparticles (AuNPs) on the surface, previously exfoliated MoS$_2$ nanosheets, and CeO$_2$ nanoparticles (CeO$_2$NPs). These nanomaterials, as well as the different synthesized catalysts, were characterized by electron microscopy (HR-SEM and HR-TEM), XPS, XRD, Raman, Reflectance and BET surface area. HER studies were performed in aqueous solution, under irradiation at different wavelengths (UV-visible), which were selected through the appropriate use of optical filters. The results obtained show that there is a synergistic effect between the different nanomaterials of the catalysts. The specific area of the catalyst, and especially the increased loading of MoS$_2$ and CeO$_2$NPs in the catalyst substantially improved the H$_2$ production, with values of ca. 1114 μm/hg for the catalyst that had the best efficiency. Recyclability studies showed only a decrease in activity of approx. 7% after 15 cycles of use, possibly due to partial leaching of gold nanoparticles during catalyst use cycles. The results obtained in this research are certainly relevant and open many possibilities regarding the potential use and scaling of these heterostructures in the photocatalytic production of H$_2$ from water.

Keywords: hydrogen production; TiO$_2$; gold nanoparticles; MoS$_2$; CeO$_2$; water splitting

Citation: Fontánez, K.; García, D.; Ortiz, D.; Sampayo, P.; Hernández, L.; Cotto, M.; Ducongé, J.; Díaz, F.; Morant, C.; Petrescu, F.; et al. Biomimetic Catalysts Based on Au@TiO$_2$-MoS$_2$-CeO$_2$ Composites for the Production of Hydrogen by Water Splitting. *Int. J. Mol. Sci.* **2023**, *24*, 363. https://doi.org/10.3390/ijms24010363

Academic Editor: Shaodong Zhou

Received: 27 November 2022
Revised: 16 December 2022
Accepted: 22 December 2022
Published: 26 December 2022

Copyright: © 2022 by the authors. Licensee MDPI, Basel, Switzerland. This article is an open access article distributed under the terms and conditions of the Creative Commons Attribution (CC BY) license (https://creativecommons.org/licenses/by/4.0/).

1. Introduction

There is a global concern about the present and future consequences of climate change. One of the main focuses in the last decade has been cutting or reducing the dependence of fossil fuels to meet our energy requirements [1,2]. Hydrogen, as an energy vector [3], is a promising candidate because it can be obtained from renewable sources like water, its combustion products are mainly water or water vapor, is less toxic than gasoline or any other usual fuel, among others [4,5].

Photosynthesis is considered the best and most efficient model that allows the conversion of solar energy for the generation of clean fuel. In nature, photosynthesis operates by supplying electrons to the active center of photosystem-II. This process is carried out through four consecutive steps of proton-coupled electron transfer, generating, as a final result of the process, products derived from reduced carbon that are the basis of life and biological activity. Considering the inspiration of this natural process, continuous efforts have been made for decades to implement and assemble some of these photosynthetic mechanisms in order to use solar energy to generate oxygen and hydrogen by splitting water [6–8].

One of the methods to produce hydrogen via water splitting is by photocatalysis [9–11]. Usually semiconductors including titanium oxide (TiO_2), zinc oxide (ZnO), iron (III) oxide (Fe_2O_3), zinc sulfide (ZnS), zirconium oxide (ZrO_2), cadmium sulfide (CdS), among others, are selected as photocatalysts due to their narrow bandgap and electronic structure [12–15]. TiO_2 is one of the most studied and used catalysts in photocatalysis for the reduction of water and degradation of organic pollutants [16], although it presents some disadvantages: (1) Recombination of photo-generated electron hole pairs [17]; (2) Fast backward reaction [18]; and (3) Inability to use visible light. The band gap of TiO_2 is 3.2 eV for anatase, 3.0 eV for rutile, and 3.4 eV for brookite, and with this band gap energy, only ultraviolet light (UV) can be used for hydrogen production [19]. To work with these limitations, multiple chemical modifications have been developed and implemented over the years.

One of them is the incorporation of noble metals, such as silver (Ag), gold (Au) and platinum (Pt), on the surface of titanium oxide due to the ability of the noble metal nanoparticles in reducing the fast recombination of the photogenerated charge carriers, enabling the use of visible light [20]. By reducing the photogenerated charge carriers, the UV activity is increased due to the electron transfer from the CB of TiO_2 to the noble metal nanoparticles [21]. The photoactivity in the visible range of the electromagnetic spectrum can be explained due to the surface plasmon resonance effect and charge separation by the transfer of photoexcited electrons from the metal nanoparticles to the CB of TiO_2 [22]. Obtaining heterostructures by coupling two or more materials with different properties makes it possible to improve the photocatalytic activity of the system [23,24]. Among these heterostructures, it is worth highlighting TiO_2-ZnO, TiO_2-WO_3, TiO_2-CdS, TiO_2-SO_2, among others, which have shown considerable improvements compared to the materials used separately [23,24]. A semiconductor that has gained popularity in recent years has been cerium (IV) oxide, CeO_2. As with TiO_2, CeO_2 has a high bandgap energy (from 2.6 eV to 3.4 eV, depending on the synthesis process and the material obtained), and high thermal stability. CeO_2 can be synthesized with different morphologies, it can be doped with metal or non-metal ions, it can be combined with other materials to form more efficient heterostructures, and it can be used for a wide variety of catalytic processes [25]. The incorporation of metal dichalcogenides, such as molybdenum disulfide (MoS_2), has also been explored to replace the use of noble metal co-catalysts due to its high abundance, low cost, good stability, and high catalytic activity [26]. However, the use of MoS_2 could also reduce the charge transfer rate and, therefore, the efficiency in some catalytic hydrogen production processes [27].

As shown above, there is no perfect catalyst to produce hydrogen by water splitting. Therefore, the objective of this research has been to explore the capabilities of different materials to produce hydrogen under visible and ultraviolet light and to combine these to obtain catalytically active heterostructures. To achieve this, different cocatalysts were incorporated onto TiO_2. The fifteen photocatalysts synthesized are made up of gold nanoparticles (Au NPs; 1%, 3%, 5% by weight), cerium (IV) oxide nanoparticles (CeO_2 NPs; 1%, 3%, 5% by weight), and molybdenum disulfide (MoS_2; 1%, 3%, 5% by weight), and have been fully characterized and evaluated in the photocatalytic reaction of hydrogen production by water splitting.

2. Results and Discussion

2.1. Characterization of Catalysts

Fifteen catalysts were synthesized, based on Au nanoparticles deposited on TiO_2 nanowires (TiO_2NWs), MoS_2, and CeO_2 nanoparticles (CeO_2NPs). The proportion of Au nanoparticles, as well as MoS_2 and CeO_2NPs, were conveniently varied, considering TiO$_2$NWs as the base component. These catalysts were used for the hydrogen evolution reaction (HER) from the photocatalytic decomposition of water, and the most efficient catalyst (3%Au@TiO$_2$NWs-5%MoS$_2$-5%CeO$_2$NPs) was fully characterized by different techniques.

Table S1 shows the BET surface area of the different components and of the synthesized catalysts. As can be seen, TiO$_2$NWs shows a high surface area of 236 m^2/g that increases with the incorporation of Au nanoparticles on the surface, going from 242 m^2/g (1%Au@TiO$_2$NWs) to 263 m^2/g (3%Au@TiO$_2$NWs), and to 275 m^2/g (5%Au@TiO$_2$NWs). This effect of increasing the area by incorporating nanoparticles has been previously described [28]. The other two components of the synthesized catalysts (MoS$_2$ and CeO$_2$NPs) also show high areas that justify the high area values observed in the catalysts, which range between 248 and 396 m^2/g. As can be seen in Table S1, in general, the surface area increases with the addition of Au, although this trend is not so clear with the increase of the other two components.

The catalyst precursors were characterized by electron microscopy. Figure 1a shows the HR-SEM image of TiO$_2$NWs, characterized by being formed by square-section wires, with diameters ranging from ca. 200 to 300 nm and lengths of up to 10 μm. MoS$_2$, previously delaminated by prolonged ultrasound treatment, shows a layered structure with variable lengths from 1 to 2 μm (see Figure 1b). The effect of exfoliation of MoS$_2$, by high power ultrasound, can be seen in the HR-SEM micrograph of Figure S1, taken at low magnification. Figure S1 shows a MoS$_2$ particle in an intermediate stage of delamination, and before the layers have dispersed. CeO$_2$NPs are characterized by presenting spherical aggregates of more than 400 nm which, in turn, are formed by very homogeneous nanoparticles with sizes ranging from 4 to 6 nm (Figure 1c). Figure 1d shows the HR-SEM image of the catalyst that showed the highest efficiency (3%Au@TiO$_2$NWs-5%MoS$_2$-5%CeO$_2$NPs). As can be seen in Figure 1d, the different components of the catalyst show a good dispersion. The components of the studied catalysts were characterized by HR-TEM. Figure 2a shows the atomic resolution image of TiO$_2$NWs. The material is highly crystalline, showing the distinct lattice fringes with an interplanar spacing of 0.33 nm, indexed to (110) crystal plane which corresponds to the rutile phase [29]. On the other hand, the growth of TiO$_2$NWs takes place along (001) direction determined by HR-TEM image, which is consistent with XRD results that will be discussed later. MoS$_2$ shows a high level of exfoliation (Figure 2b), which allows us to observe the detail of the atomic structure of a monolayer. As can be seen in the inset of Figure 2b, corresponding to the selected area electron diffraction (SAED), the material is highly crystalline. Apparently, and although it is still to be confirmed, the HR-TEM analyses seem to indicate the presence of structural defects generated by the appearance of vacancies in the two-dimensional structure of the material. These defects could be due to the intense exfoliation process produced by high intensity ultrasound, and could be related to the high activity of the catalysts. Figure 2c shows the HR-TEM image of CeO$_2$NPs. As seen in the SAED, the material is crystalline. Part of the image has been further magnified to show detail of the lattice fringes, with an interplanar spacing of 0.31 nm, indexed to (111) crystal plane corresponding to the characteristic face-centered cubic fluorite-type structure [30].

Figure 3 shows the results obtained by X-ray diffraction (XRD) for the most efficient catalyst (3%Au@TiO$_2$NWs-5%MoS$_2$-5%CeO$_2$NPs), along with that of MoS$_2$, CeO$_2$NPs and TiO$_2$NWs for comparison purposes. As shown in Figure 3a, well-defined diffraction peaks are observed at ca. 14°, 32°, 39°, 49°, and 59° that have been ascribed to (002), (100), (103), (105), and (110) planes of 2H-type MoS$_2$ hexagonal phase (JCPDS # 75–1539), respectively [31–33]. Figure 3b shows the diffraction pattern of CeO$_2$NPs. The most intense peaks are observed at ca. 28°, 33°, 47° and 56°, and correspond to (111), (200), (220) and (311) crystal planes, respectively [34–36]. These peaks are characteristic of CeO$_2$ with face-centered cubic fluorite-type structure. Figure 3c shows the XRD pattern of TiO$_2$NWs. The diffraction peaks at ca. 27°, 36°, 41°, y 54° were ascribed to (110), (101), (111), y (211) TiO$_2$ crystalline planes in rutile phase (JCPDS 75-1750) [37,38]. The XRD of the most efficient catalyst is shown in Figure 3d. As can be seen there, the main peaks of all three components are present. However, the presence of Au, which should be shown as a very low intensity peak at ca. 38° [39], corresponding to Au (111), is not observed in the catalyst, possibly due to the high dispersion of the metal.

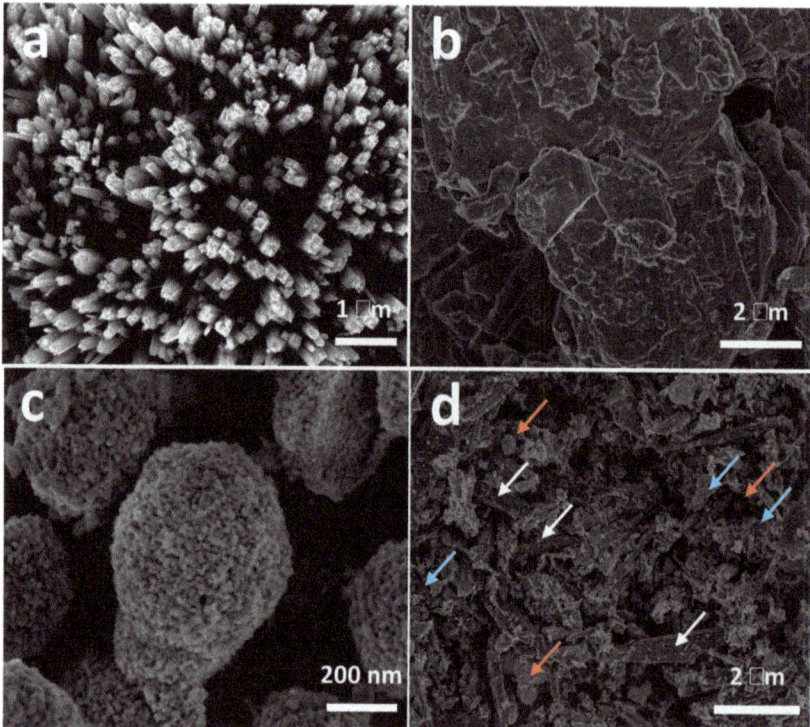

Figure 1. HR-SEM micrographs of TiO$_2$NWs (**a**), MoS$_2$ nanosheets (**b**), CeO$_2$NPs (**c**) and the 3%Au@TiO$_2$NWs-5%MoS$_2$-5%CeO$_2$NPs catalyst (**d**). The arrows in (**d**) indicate the different components of the catalyst: TiO$_2$NWs (white), MoS$_2$ (blue), and CeO$_2$NPs (red).

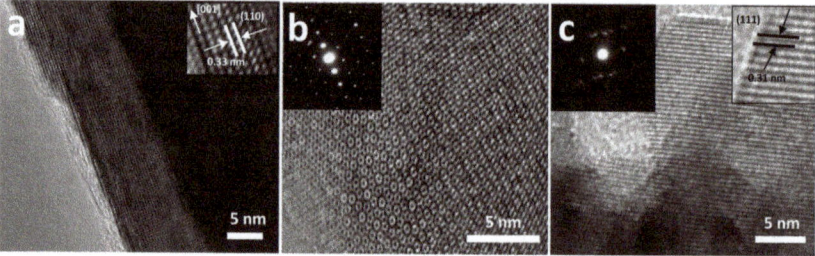

Figure 2. HR-TEM micrographs of the different components of the catalysts: TiO$_2$NWs and inset at atomic resolution showing the direction of growth and the lattice fringes (**a**); MoS$_2$ single layer and inset corresponding to the selected area electron diffraction, SAED (**b**); and CeO$_2$NPs and insets corresponding to SAED and micrograph at higher magnification showing the lattice fringes (**c**).

The different materials, as well as the most efficient catalyst, were characterized by Raman spectroscopy (Figure 4). MoS$_2$ shows two very characteristic bands at 383 cm^{-1} and 407 cm^{-1} (Figure 4a), which have been assigned to the E$^1_{2g}$ and A$_{1g}$ modes, respectively [40].

The position of these bands has been correlated with the number of layers of the material, so the results suggest that the exfoliation process was very efficient, generating MoS_2 flakes with few layers [41,42]. CeO_2NPs (Figure 4b) shows an intense band at ca. 457 cm^{-1}, and a much less pronounced one at 607 cm^{-1} that have been assigned to a cubic fluorite structure, as already evidenced from the XRD results. The main band at 457 cm^{-1} corresponds to a triply degenerate F_{2g} mode of symmetric stretching vibrations of oxygen ions around Ce^{4+} ions in octahedral CeO_8 [43]. The asymmetry of the band at 457 cm^{-1} has been associated with structural defects due to the presence of oxygen vacancies in the oxide [44,45], which could also be correlated with the reactivity of the material. Figure 4c shows the Raman spectrum of TiO_2NWs, whose bands at ca. 448 cm^{-1} and 610 cm^{-1} have been assigned to the vibration modes E_g and A_{1g} of TiO_2 in the rutile phase, as already evidenced by XRD. The Raman spectrum of the most efficient catalyst (Figure 4d) shows two pronounced bands at 448 cm^{-1} and 610 cm^{-1}, and a shoulder at ca. 687 cm^{-1}. These three bands come from rutile, which is the major component of the catalyst. Additionally, two small peaks are observed at ca. 384 cm^{-1} and 407 cm^{-1} assigned to MoS_2. Due to the position of the CeO_2NPs bands, the signal of this material is masked under the strong contribution of rutile.

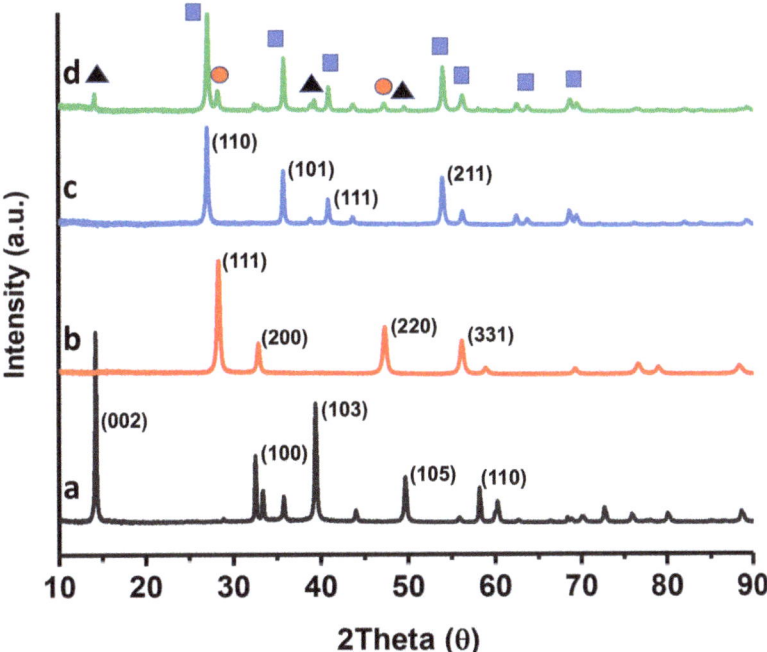

Figure 3. XRD patterns of MoS_2 (**a**); CeO_2NPs (**b**); TiO_2NWs (**c**); and 3%Au@TiO_2NWs-5%MoS_2-5%CeO_2NPs (**d**). The most intense peaks in the catalyst have been associated with the different components (black triangles: MoS_2, red circles: CeO_2NPs, blue squares: TiO_2NWs).

The most efficient catalyst (3%Au@TiO_2NWs-5%MoS_2-5%CeO_2NPs) was also characterized by X-ray photoelectron spectroscopy (XPS). Ti2p (Figure 5a) shows two components at 464.3 eV and 458.7 eV that were ascribed to the $Ti2p_{1/2}$ and $Ti2p_{3/2}$ transitions, respectively [3,9]. These transitions are quite symmetrical, so any additional contribution was ruled out. Figure 5b shows the transition corresponding to O1s. As can be seen, the transition is clearly asymmetric and has been deconvoluted into two components at ca. 530.3 eV and 532.3 eV. The most intense peak (530.3 eV) has been assigned to oxygen in the TiO_2 lattice [3,46], which also masks the possible contribution of oxygen in the CeO_2 lattice, while the component observed at 532.3 eV has been assigned to oxygen vacancies in CeO_2 [47] (as suggested by the asymmetry of the main peak of CeO_2NPs in Raman spectroscopy), or to non-lattice oxygen [46]. Figure 5c shows the Au4f transition, with peaks at 84.1 eV and 87.7 eV and a characteristic spin-orbit splitting of ca. 3.6 eV, which have been clearly assigned to the presence of metallic Au [48]. The transition corresponding to Ce3d is shown in Figure 5d. This transition, which is very complex due to a state hybridization process, evidences two distinguishable series of peaks corresponding to the Ce^{4+} and Ce^{3+} species. The different peaks were labeled u, u′, u″, v, v′ and v″ to represent the different electronic states of Ce^{4+} and Ce^{3+} [47,49]. The presence of Ce^{3+} ions gives rise to a charge imbalance, responsible for oxygen vacancies and the presence of defects and unsaturated chemical bonds in the nanomaterial. These defects in CeO_2NPs support the results previously shown by Raman and XPS.

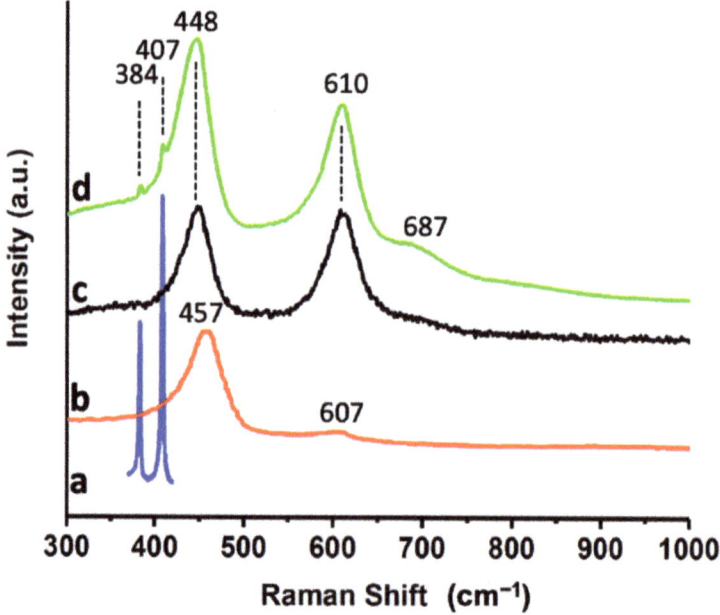

Figure 4. Raman spectra of MoS_2 (**a**); CeO_2NPs (**b**); TiO_2NWs (**c**); and 3%Au@TiO_2NWs-5%MoS_2-5%CeO_2NPs (**d**).

Figure 5d shows the Mo3d and S2s transitions. The Mo3d shows two peaks at 232.4 eV and 229.2 eV, which have been attributed to the $Mo3d_{3/2}$ and $Mo3d_{5/2}$ doublet, respectively, characteristic of the Mo^{4+} state in MoS_2 [48,50]. The observed peak at ca. 226.4 eV was assigned to S2s [48], typical of MoS_2. The slight asymmetry of the Mo3d peaks could point to a possible mixture of oxidation states, which could also be correlated with the potential presence of defects in the material lattice and reactivity.

The efficiency of radiation absorption by the catalysts is a critical factor for their activity, so the different catalysts and nanomaterials used were analyzed using Tauc diagrams [51]. As shown in Figure 6, TiO_2NWs showed a bandgap in the border region between UV and visible (2.95 eV), slightly different from the expected value for TiO_2 in rutile phase (3.05 eV) [52]. CeO_2NPs and MoS_2 show bandgaps at 2.61 eV and 2.46 eV, respectively, clearly in the visible region. The most efficient catalyst, formed as a heterostructure of these components in addition to the presence of Au nanoparticles, shows a bandgap at 2.23 eV (ca. 555 nm), which clearly justifies the activity of the 3%Au@TiO_2NWs-5%MoS_2-5%CeO_2NPs heterostructure under irradiation with visible light, as will be described in the section corresponding to catalytic results.

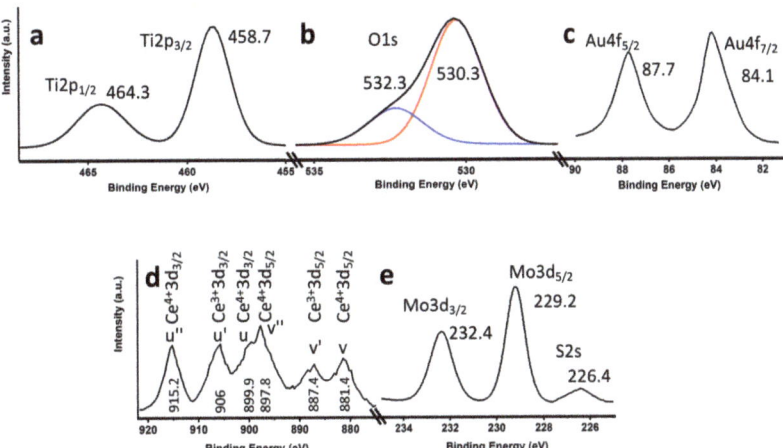

Figure 5. XPS core level spectra for Ti2p (**a**); O1s (**b**); Au4f (**c**); Ce3d (**d**); and Mo3d/S2s (**e**).

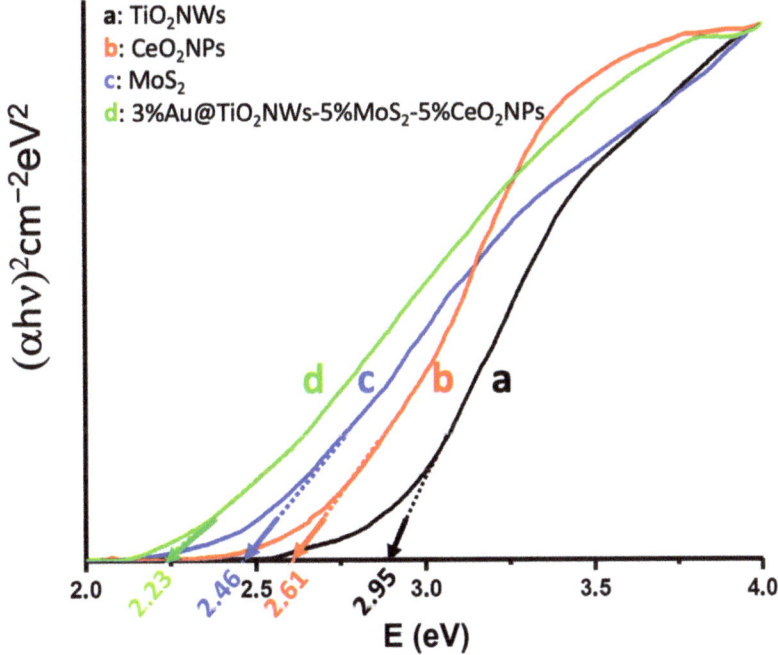

Figure 6. Tauc plots of $(\alpha h\nu)^2$ versus energy (eV), and determination of the bandgap energy of TiO_2NWs (**a**); CeO_2NPs (**b**); MoS_2 (**c**); and 3%Au@TiO_2NWs-5%MoS_2-5%CeO_2NPs (**d**).

2.2. Photocatalytic Hydrogen Production

Before proceeding to evaluate the activity of the synthesized catalysts, several preliminary studies were carried out to establish the optimal reaction conditions. To do this, we started from the heterostructure with the highest proportion of each of the components

(5%Au@TiO$_2$NWs-5%MoS$_2$-5%CeO$_2$NPs). Initially, a study of the optimum pH was carried out (see Figure S2a), and it was established that the most suitable was pH = 7. Another of the preliminary studies that was carried out allowed establishing the optimum amount of catalyst in the reaction medium. As can be seen in Figure S2b, there is a clear correlation between the amount of catalyst and the HER efficiency, so a loading of 50 mg of catalyst (for a total reaction volume of 100 mL) was established as the optimal amount.

The photocatalytic efficiency of the catalysts for the evolution of H$_2$ was evaluated in the presence of Na$_2$SO$_3$ (0.02 M) and Na$_2$S (0.4 M) as sacrificial reagents. In all cases, it was found that there is no evolution of H$_2$ without either photocatalyst or irradiation (see Figure S2c). Figure 7 shows the results of the photocatalytic hydrogen production from the fifteen synthesized catalysts. In all cases, the activity under irradiation with different wavelengths (220, 280, 320, 400, 500, 600 and 700 nm) was evaluated. As can be seen, the maximum hydrogen production is observed under irradiation at 400 nm, this wavelength being the one corresponding to the bandgap of TiO$_2$NWs (see Table S2). At more energetic wavelengths the behavior varies from one system to another. Figure 7a,d, corresponding to the heterostructures Au@TiO$_2$NWs-5%MoS$_2$-3%CeO$_2$NPs and Au@TiO$_2$NWs-5%MoS$_2$-1%CeO$_2$NPs, respectively, show a similar trend under irradiation at 220, 280 and 320 nm, with stable values of hydrogen production in catalysts with 3% and 5% Au. Catalysts with lower Au loading (1%) show much lower activities. In all the catalysts a decrease in activity is observed under irradiation in the visible range (λ > 400 nm), although surprisingly hydrogen production is observed even at very low energy wavelengths (i.e., 700 nm). This behavior clearly points to a synergy between the different nanomaterials, and specifically to the gold load, which, as observed, has a significant effect on the levels of hydrogen production. The highest catalytic activity is observed with 3%Au@TiO$_2$NWs-5%MoS$_2$-5%CeO$_2$NPs, showing a hydrogen production of 1114 µm/hg under irradiation at 400 nm. In contrast, the catalyst with the lowest efficiency was the Au@TiO$_2$NWs-5%MoS$_2$-1%CeO$_2$NPs heterostructure, and specifically the one with 1%Au, with only a hydrogen production of 606 µm/hg (Figure 7d). This behavior is evidence of the relevance of CeO$_2$NPs in the heterostructure. As previously discussed in Section 2.1, CeO$_2$NPs present oxygen vacancies and defects that may be responsible for the reactivity of the material. These defects, as already described [47], might originate from the effective separation of photogenerated electron–hole pairs within the composite endorsing the charge transfer efficiency. The effect of MoS$_2$ on the catalyst activity is indisputable, although it is not as relevant as for CeO$_2$NPs. When comparing the catalysts Au@TiO$_2$NWs-1%MoS$_2$-5%CeO$_2$NPs (Figure 7e), Au@TiO$_2$NWs-3%MoS$_2$-5%CeO$_2$NPs (Figure 7b) and Au@TiO$_2$NWs-5%MoS$_2$-5%CeO$_2$NPs (Figure 7c) it is observed that by increasing the amount of MoS$_2$ the production of H$_2$ also increases substantially. This effect is more notable when going from 3% to 5% of MoS$_2$ in the heterostructure. MoS$_2$ has high surface area values, as shown in Table S1. The increase in MoS$_2$ loading produces an increase in the specific area of the heterostructure, together with the improvement in the conductivity properties of the catalyst [53], which could justify the effect of this nanomaterial on the reactivity observed in Figure 7. On the other hand, the presence of Au, and especially the generation of surface plasmons generated by the Au nanoparticles on the surface of the heterostructure [54], represents an adjuvant factor on the photocatalytic HER.

The recyclability of the most efficient catalyst (3%Au@TiO$_2$NWs-5%MoS$_2$-5%CeO$_2$NPs) was also evaluated. For this, 15 consecutive reactions were carried out with the same catalyst (Figure S3) using a larger reaction volume (200 mL). After each reaction, the catalyst was recovered by centrifugation (3000 rpm, 15 min), washed with deionized water, and dried overnight in an oven at 50 °C. The recovered catalyst was used again in the next reaction, using the same experimental conditions, and keeping the reaction temperature constant at 20 °C. After 15 cycles of use, the results obtained (Figure S3) showed an efficiency of ca. 93% of the initial, which represents a loss of activity of around 7%. After cycle 15, the catalyst showed similar morphological characteristics although, according to EDX measurements, the gold loading was slightly reduced to approximately 2.4%, which could indicate a gold leaching effect during the use cycles, that would justify the loss of 7% of activity. This result is remarkable considering that the catalysts are heterostructures formed by four components whose synergistic behavior remains almost unchanged in each cycle of use.

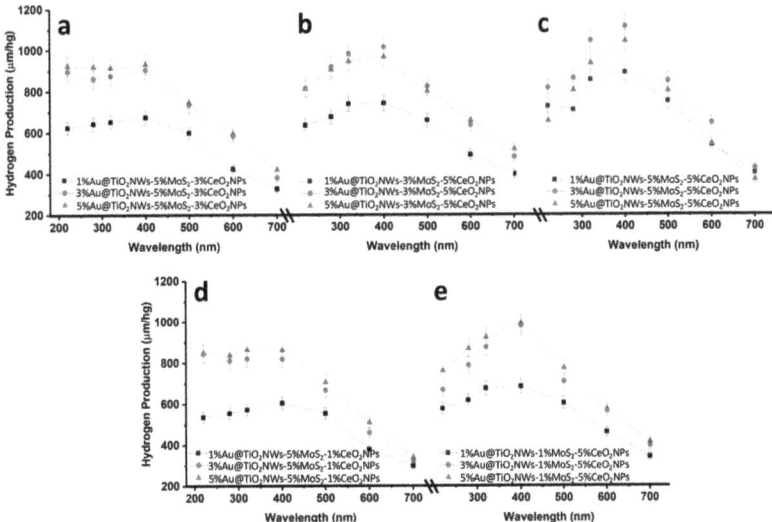

Figure 7. H_2 production profiles of the synthesized catalysts under irradiation at different wavelengths (the legends corresponding to the catalysts are shown in each of the figures). The estimated error bars for each of the values obtained are also shown.

Some results of recent studies on hydrogen production by water splitting using heterostructured catalysts are shown in Table S3 [55–62]. As can be seen there, the amount of hydrogen reported in our research is one of the highest, although a direct comparison with other research is not possible because the experimental conditions, reaction time, and even nanomaterials are not the same. Nonetheless, our hydrogen production results are certainly promising, and could potentially even be applicable to larger-scale processes.

2.3. Mechanism of the Photocatalytic Hydrogen Production

Figure S4 shows the effect of incorporating hole scavengers (methanol and EDTA-Na_2) on HER. Figure S4a to Figure S4e show the most efficient systems in the absence of scavengers and in the presence of each of them. As can be seen, the addition of methanol (5 mL) to the reaction medium clearly increases the production of hydrogen at all wavelengths. The effect of the addition of EDTA-Na_2 (0.1 M) produces an even greater effect, with pronounced increases in all the systems used and practically under any irradiation energy. The incorporation of larger amounts of methanol or EDTA-Na_2 did not produce significant changes, therefore, at least for these hole scavengers and for the reaction conditions used, the maximum possible H_2 production was reached. These results clearly suggest that electron-hole recombination occurs in the absence of scavengers, despite the effect of electron channeling towards the Au nanoparticles and MoS_2 nanosheets from the catalysts and the use of electron donors (Na_2SO_3, 0.02 M and Na_2S, 0.4 M) in the reaction mixture.

Considering these results, in addition to the determination of bandgaps (see Figure 6), a tentative mechanism for HER has been proposed (Figure 8). To do this, Mulliken's classical theory of electronegativity has been used [63,64], which makes it possible to establish the position of the edge of the band of the different nanomaterials that form the heterostructure and, based on it, to establish the direction of migration of the photogenerated charge carriers in the catalyst (see Equations (1) and (2)).

$$E_{CB} = \chi - E_C - 0.5E_g \quad (1)$$

$$E_{VB} = E_{CB} + E_g \quad (2)$$

where E_{CB} and E_{VB} are the edge potentials of the valence band and the conduction band, respectively, χ is the absolute electronegativity, E_C is the energy of free electrons on the hydrogen scale (4.50 eV) [65,66], and E_g is the experimentally determined bandgap (see Figure 6). The X values for TiO_2 (rutile) and CeO_2 are 5.81 and 5.56 eV, respectively [67,68]. The E_{CB} and E_{VB} edge positions for TiO_2NWs determined from Equations (1) and (2) are -0.165 and 2.785 eV, respectively, while for CeO_2NPs the calculated values were -0.245 and 2.365 eV (see Figure 8).

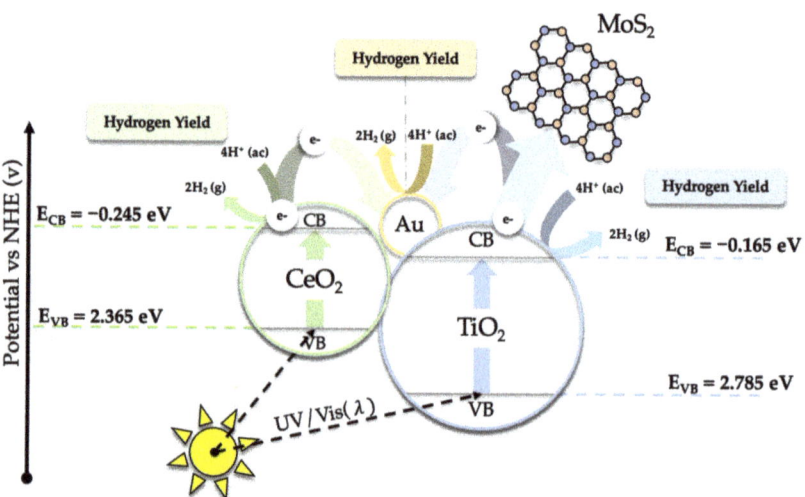

Figure 8. Mechanism proposed for the hydrogen production using Au@TiO_2NWs-MoS_2-CeO_2NPs catalysts under UV-visible irradiation.

As previously shown [58], the presence of CeO_2NPs and MoS_2 significantly broadens the range of light absorption, which increases the density of photogenerated electrons and the production of H_2. Under irradiation, the valence band electrons of both TiO_2NWs and CeO_2NPs are photoexcited to their corresponding conduction bands [9]. These electrons can reduce the water, generating H_2, or be transferred to the MoS_2 or Au nanoparticles. Both MoS_2 and Au act as sinks that channel electrons, prevent electron-hole recombination, and facilitate subsequent reactivity [3,11,69]. The holes that were created in the valence band of the heterostructure, as previously shown, undergo partial recombination with the photogenerated electrons, although this recombination is inhibited by adding hole scavengers (methanol or EDTA-Na_2) to the reaction mixture. As previously described, the presence of Ce^{3+} and Ce^{4+} species, identified by XPS (see Figure 5d), may play a relevant role in prolonging the lifetime of photoinduced charge carriers [58]. Ce^{4+} species can trap the electrons helping to avoid electron-hole recombination. On the other hand, Ce^{3+}/oxygen vacancies can provide abundant H_2O adsorption sites, which decreases the H_2O adsorption energy, increasing the efficiency of the water splitting reaction.

3. Materials and Methods

3.1. Reagents and Materials

All reagents were used as received without further purification. All solutions were prepared with deionized water (Milli-Q water, Burlington, MA, USA, 18.2 MΩcm^{-1} at 25 °C). $TiCl_4$ (99.9%) was provided by Fisher Scientific, Cayey, Puerto Rico. $HAuCl_4 \cdot 3H_2O$ (ACS Reagent, St. Louis, MO, USA, 49.0 + % Au basis), MoS_2 (Nanopowder, St. Louis,

MO, USA, 90 nm diameter, 99% trace metals basis), $Ce(NO_3)_3 \cdot 6H_2O$ (99.99%), Ethanol (200 proof, anhydrous, ≥99.5%), EDTA disodium salt dihydrate (OmniPur, St. Louis, MO, USA), and $NaBH_4$ (99.99% trace metals basis) were provided by Sigma Aldrich (Darmstadt, Germany). Methanol anhydrous for UHPLC-MS LiChrosolv (99.9%) was provided by Supelco (Bellefonte, PA, USA). Silicon p-type boron doped substrates (Si <100>), were provided by El-CAT (Ridgefield Park, NJ, USA). UHP N_2 (5.0), used for the photocatalytic reaction, was provided by Praxair, Gurabo, Puerto Rico.

3.2. Synthesis of Nanomaterials

The synthesis of titanium oxide nanowires (TiO_2NWs) has been previously described [70]. In a typical synthesis, a mixture of water and HCl (37% solution) (1:1, *v/v*) was prepared. Subsequently, $TiCl_4$ (3 mL) was added dropwise to 100 mL of the solution and allowed to mix for 30 min until the presence of suspended particles was not observed. The solution was then transferred to Teflon-lined autoclaves, and silicon substrates (Si <100>) with the polished surface facing the inside of the Teflon container, were incorporated into the solution. The autoclaves were then sealed and transferred to an oven. The autoclaves were treated at 180 °C for 24 h. Once the treatment time had elapsed, the autoclaves were left to cool for at least 12 h. As a result, the growth of a white deposit on the surface of the Si substrates was observed. The material obtained was washed with abundant deionized water, dried in an oven at 60 °C and stored in vials that were sealed until later use.

The deposition of gold nanoparticles (AuNPs) was carried out by dispersing 1 g of the support (TiO_2NWs), whose synthesis was previously described, in 100 mL of H_2O and the mixture was sonicated for 30 min. Next, the required amount of gold precursor ($HAuCl_4 \cdot 3H_2O$) was added to the reaction mixture and stirred for 1 h. Finally, the process continued with the reduction of gold by adding, dropwise, a $NaBH_4$ solution (10 mg in 10 mL of H_2O) under constant stirring. Once the 10 mL of $NaBH_4$ had been added, the resulting solution was kept under stirring for 1 h. The reaction product was separated by centrifugation (3000 rpm, 15 min), washed 4 times with deionized water, and dried overnight at 60 °C. The different Au@TiO_2NWs compounds were synthesized with 1%, 3% and 5% AuNPs on the surface and these materials were later used for the incorporation of the rest of the catalyst components.

CeO_2 nanoparticles (CeO_2NPs) were obtained through a coprecipitation process. For this, two solutions were prepared: (i) 250 mL of a solution of $Ce(NO_3)_3 \cdot 6H_2O$ (0.02 M) and (ii) 250 mL of a solution of K_2CO_3 (0.03 M). Both solutions were introduced dropwise into an Erlenmeyer flask containing 50 mL of water. During this process, the mixture was kept under constant stirring. As a result, a precipitate of $Ce_2(CO_3)_3$ was obtained, which was separated from the solution by centrifugation. The resulting solid was washed four times with deionized water and dried at 70 °C for 3 h. Next, the dry material was calcined in a muffle at 600 °C for 3 h, using an open crucible.

The commercial MoS_2 was subjected to an exfoliation process before being used. For this, 4 g of MoS_2 were mixed with 200 mL of deionized water. The resulting dispersion was sonicated using a Cole-Palmer Tip Sonicator (Cole-Parmer 750-Watt Ultrasonic Processor) for 6 h in pulsed mode (40% amplitude, pulse on 5 s, pulse off 10 s). Subsequently, the solution was kept static for sedimentation for 3 h. Next, the supernatant was extracted from the mixture and centrifuged for 30 min at 3000 rpm to remove the non-delaminated MoS_2. Next, the supernatant can be dried by evaporation at 50 °C to be used later, or manipulated directly as a suspension. The concentration of the suspension can be determined by measuring the absorbance at 672 nm, using the Beer-Lambert law, and considering ε as 3400 mL mg^{-1} m^{-1}.

The materials, whose synthesis has been described above, were used for the following stages of preparation of the catalysts. Thus, 300 mg of Au@TiO_2NWs were dispersed in a solution containing 20 mL of ethanol and 20 mL of deionized water, and the mixture was vigorously stirred for 1 h. Subsequently, cerium oxide nanoparticles (CeO_2NPs) were added, and the suspension was stirred for 2 h. The product was then separated from solution by centrifugation (3000 rpm, 15 min), washed 4 times with deionized water, and dried overnight at 60 °C. The incorporation of MoS_2 was carried out in a final synthesis step, and by a procedure similar to that previously described for CeO_2NPs. Once the synthesis process was finished, the product was recovered by centrifugation (3000 rpm, 15 min), washed four times with deionized water, dried overnight at 60 °C, and stored and sealed at room temperature until later use. The 15 synthesized catalysts, based on Au@TiO_2NWs-MoS_2-CeO_2NPs, were identified indicating the percentage of gold incorporated on the surface and the percentages of MoS_2 and CeO_2NPs in each case.

3.3. Characterization of the Catalysts

The surface morphology of the catalysts was evaluated using a FEI Verios 460 L High Resolution Scanning Electron Microscope (HR-SEM, Thermo Fisher Scientific, Hillsboro, OR, USA), equipped with a Quantax EDS Analyzer, and by High Resolution Transmission Electron Microscopy (HR-TEM), using a JEOL JEM 3000F (300 kV) microscope. XPS measurements were carried out using an ESCALAB 220i-XL spectrometer, using non-monochromatic Mg Kα (1253.6 eV) radiation from a twin anode, operating at 20 mA and 12 kV in the constant analyzer energy mode, with a PE of 50 eV. The crystallinity of the catalysts was studied by X-ray diffraction, using a Bruker D8-Advance diffractometer that operates at 40 kV and 40 mA in the range of 20–80°, using a Bragg-Brentano configuration, and at a scan speed of 1° min^{-1}. The catalysts were also characterized by Raman spectroscopy, using a DXR Thermo Raman Microscope, which uses a 532 nm laser source at 5 mW power and a 25 µm pinhole aperture with a 5 cm^{-1} nominal resolution. Bandgap measurements of the different materials were carried out using a Perkin Elmer Lambda 365 UV-Vis spectrophotometer (Perkin Elmer, Waltham, MA, USA), equipped with an integrating sphere. The bandgap value was obtained from the graph of the Kubelka-Munk function versus the absorbed light energy [51]. Brunauer Emmett Teller (BET) specific area measurements were carried out using a Micromeritics ASAP 2020 system, according to N_2 adsorption isotherms at 77 K.

3.4. Photocatalytic Hydrogen Production

The experimental setup for the characterization of the catalysts for the hydrogen evolution reaction (HER) by photocatalytic water splitting consisted of mixing 50 mg of the desired catalyst with 100 mL of deionized water in a 200 mL quartz reactor. Next, sacrificial electron donor solutions (Na_2SO_3, 0.02 M; Na_2S, 0.4 M) were added. In order to test the effect of adding additional hole scavengers to the reaction mixture, methanol (5 mL) and EDTA-Na_2 (0.1 M) were used. The reaction mixture was kept at 20 °C for 1 h before the start of the reaction, to guarantee temperature stability, and was purged with nitrogen (N_2, 5.0) during the pre-reaction process. Next, the reaction mixture was irradiated using a solar simulator, whose irradiation power in the absence of filters is 120 mW.cm^{-2}). To study the influence of irradiation energy on the water splitting reaction, different cut-off filters at 220, 280, 320, 400, 500, 600, and 700 nm were used, and the reaction was followed for two hours. The hydrogen produced was quantified using a gas chromatograph coupled to a thermal conductivity detector (GC-TCD, Perkin-Elmer Clarus 600).

4. Conclusions

A total of 15 catalysts with different amounts of Au, MoS_2, and CeO_2 (1%, 3%, and 5% by weight) incorporated onto TiO_2NWs were synthesized, and their photocatalytic activity was evaluated by the production of hydrogen via water splitting using visible and ultraviolet light. The highest hydrogen production was 1114 µm/hg, and was obtained with the 3%Au@TiO_2NWs-5%MoS_2-5%CeO_2NPs composite. The combination of the different materials caused a synergistic effect, increasing the catalytic activity and allowing the use of wavelengths ranging from 220 to the visible range, with remarkable efficiency even under irradiation at wavelengths as low in energy as 700 nm. The recyclability test showed an efficiency loss of ca. 7% after 15 cycles, suggesting a stable and suitable catalyst for the photocatalytic production of hydrogen by water splitting.

The results obtained in this research are certainly the starting point for further developments that allow us to delve into the mechanisms that control the HER. In this sense, the continuation of this research, already in progress, will analyze three factors that, in our opinion, are of great relevance for the catalytic systems studied: (i) the effect of increasing the loading of MoS_2 and CeO_2NPs in the heterostructure, plus beyond the 5% considered in the present investigation; (ii) analysis of how the use of cerium oxides in which the Ce^{3+}/Ce^{4+} ratio can be modulated influences HER; and (iii) characterization of possible leaching during catalyst use and regeneration cycles.

Supplementary Materials: The following supporting information can be downloaded at: https://www.mdpi.com/article/10.3390/ijms24010363/s1.

Author Contributions: Conceptualization, F.M., C.M. and A.M.; methodology, F.M.; formal analysis, A.M. and F.M.; investigation, K.F., D.G., D.O., P.S., F.D., L.H., C.M. and F.M.; resources, F.M., C.M. and F.P.; writing—original draft preparation, A.M. and F.M.; writing—review and editing, A.M., F.M., C.M., J.D. and M.C.; supervision, A.M. and F.M.; project administration, A.M., F.M. and M.C.; funding acquisition, F.M., A.M., C.M., M.C. and J.D. All authors have read and agreed to the published version of the manuscript.

Funding: Financial support from NSF Center for the Advancement of Wearable Technologies-CAWT (Grant 1849243), from the Consortium of Hybrid Resilient Energy Systems CHRES (DE-NA0003982), from The Puerto Rico-Louis Stokes Alliance for Minority Participation, PR-LSAMP (HRD-2008186), and from the Spanish Ministry of Economy and Competitiveness, under NanoCat-Com Project (PID2021-124667OB-I00), are gratefully acknowledged.

Institutional Review Board Statement: Not applicable for studies not involving humans or animals.

Informed Consent Statement: Not applicable.

Data Availability Statement: The data is contained in the article and is available from the corresponding authors on reasonable request.

Acknowledgments: The facilities provided by the National Center for Electron Microscopy at Complutense University of Madrid (Spain), by "Instituto de Micro y Nanotecnología IMN-CNM, CSIC, CEI UAM + CSIC" and by the Materials Characterization Center at University of Puerto Rico are gratefully acknowledged. K.F. thanks PR NASA Space Grant Consortium for a graduate fellowship (#80NSSC20M0052). D.O. thanks Consortium of Hybrid Resilient Energy Systems (CHRES) for a graduate fellowship. L.H. thanks PR-LSAMP for an undergraduate fellowship (HRD-2008186).

Conflicts of Interest: The authors declare no conflict of interest.

References

1. Gustavsson, L.; Nguyen, T.; Sathre, R.; Tettey, U.Y.A. Climate Effects of Forestry and Substitution of Concrete Buildings and Fossil Energy. *Renew. Sustain. Energy Rev.* **2021**, *136*, 110435. [CrossRef]
2. Hassan, A.; Ilyas, S.Z.; Jalil, A.; Ullah, Z. Monetization of the Environmental Damage Caused by Fossil Fuels. *Environ. Sci. Pollut. Res.* **2021**, *28*, 21204–21211. [CrossRef] [PubMed]
3. Machín, A.; Cotto, M.; Ducongé, J.; Arango, J.C.; Morant, C.; Márquez, F. Synthesis and Characterization of Au@TiO₂ NWs and Their Catalytic Activity by Water Splitting: A Comparative Study with Degussa P25. *Am. J. Eng. Appl. Sci.* **2017**, *10*, 298–311. [CrossRef]
4. Atilhan, S.; Park, S.; El-Halwagi, M.M.; Atilhan, M.; Moore, M.; Nielsen, R.B. Green Hydrogen as an Alternative Fuel for the Shipping Industry. *Curr. Opin. Chem. Eng.* **2021**, *31*, 100668. [CrossRef]
5. Qazi, U.Y. Future of Hydrogen as an Alternative Fuel for Next-Generation Industrial Applications; Challenges and Expected Opportunities. *Energies* **2022**, *15*, 4741. [CrossRef]
6. Valdés, Á.; Qu, Z.-W.; Kroes, G.-J.; Rossmeisl, J.; Nørskov, J.K. Oxidation and Photo-Oxidation of Water on TiO₂ Surface. *J. Phys. Chem. C* **2008**, *112*, 9872–9879. [CrossRef]
7. Schley, N.D.; Blakemore, J.D.; Subbaiyan, N.K.; Incarvito, C.D.; D'Souza, F.; Crabtree, R.H.; Brudvig, G.W. Distinguishing Homogeneous from Heterogeneous Catalysis in Electrode-Driven Water Oxidation with Molecular Iridium Complexes. *J. Am. Chem. Soc.* **2011**, *133*, 10473–10481. [CrossRef]

8. Etacheri, V.; Di Valentin, C.; Schneider, J.; Bahnemann, D.; Pillai, S.C. Visible-Light Activation of TiO_2 Photocatalysts: Advances in Theory and Experiments. *J. Photochem. Photobiol. C Photochem. Rev.* **2015**, *25*, 1–29. [CrossRef]
9. Pinilla, S.; Machín, A.; Park, S.-H.; Arango, J.C.; Nicolosi, V.; Márquez-Linares, F.; Morant, C. TiO_2-Based Nanomaterials for the Production of Hydrogen and the Development of Lithium-Ion Batteries. *J. Phys. Chem. B* **2018**, *122*, 972–983. [CrossRef]
10. Machín, A.; Cotto, M.; Duconge, J.; Arango, J.C.; Morant, C.; Pinilla, S.; Soto-Vázquez, L.; Resto, E.; Márquez, F. Hydrogen Production via Water Splitting Using Different Au@ZnO Catalysts under UV–Vis Irradiation. *J. Photochem. Photobiol. A Chem.* **2018**, *353*, 385–394. [CrossRef]
11. Machín, A.; Arango, J.C.; Fontánez, K.; Cotto, M.; Duconge, J.; Soto-Vázquez, L.; Resto, E.; Petrescu, F.I.T.; Morant, C.; Márquez, F. Biomimetic Catalysts Based on Au@ZnO–Graphene Composites for the Generation of Hydrogen by Water Splitting. *Biomimetics* **2020**, *5*, 39. [CrossRef]
12. Bisaria, K.; Sinha, S.; Singh, R.; Iqbal, H.M.N. Recent Advances in Structural Modifications of Photo-Catalysts for Organic Pollutants Degradation—A Comprehensive Review. *Chemosphere* **2021**, *284*, 131263. [CrossRef]
13. Wang, H.; Hu, P.; Zhou, J.; Roeffaers, M.B.J.; Weng, B.; Wang, Y.; Ji, H. Ultrathin 2D/2D $Ti_3C_2T_x$/Semiconductor Dual-Functional Photocatalysts for Simultaneous Imine Production and H_2 Evolution. *J. Mater. Chem. A* **2021**, *9*, 19984–19993. [CrossRef]
14. Qiu, Q.; Zhu, P.; Liu, Y.; Liang, T.; Xie, T.; Lin, Y. Highly Efficient In_2S_3/WO_3 Photocatalysts: Z-Scheme Photocatalytic Mechanism for Enhanced Photocatalytic Water Pollutant Degradation under Visible Light Irradiation. *RSC Adv.* **2021**, *11*, 3333–3341. [CrossRef]
15. Duan, X.; Yang, J.; Hu, G.; Yang, C.; Chen, Y.; Liu, Q.; Ren, S.; Li, J. Optimization of TiO_2/ZSM-5 Photocatalysts: Energy Band Engineering by Solid State Diffusion Method with Calcination. *J. Environ. Chem. Eng.* **2021**, *9*, 105563. [CrossRef]
16. Dharma, H.N.C.; Jaafar, J.; Widiastuti, N.; Matsuyama, H.; Rajabsadeh, S.; Othman, M.H.D.; Rahman, M.A.; Jafri, N.N.M.; Suhaimin, N.S.; Nasir, A.M.; et al. A Review of Titanium Dioxide (TiO_2)-Based Photocatalyst for Oilfield-Produced Water Treatment. *Membranes* **2022**, *12*, 345. [CrossRef]
17. Xia, C.; Nguyen, T.H.C.; Nguyen, X.C.; Kim, S.Y.; Nguyen, D.L.T.; Raizada, P.; Singh, P.; Nguyen, V.-H.; Nguyen, C.C.; Hoang, V.C.; et al. Emerging Cocatalysts in TiO_2-Based Photocatalysts for Light-Driven Catalytic Hydrogen Evolution: Progress and Perspectives. *Fuel* **2022**, *307*, 121745. [CrossRef]
18. Tang, R.; Gong, D.; Deng, Y.; Xiong, S.; Deng, J.; Li, L.; Zhou, Z.; Zheng, J.; Su, L.; Yang, L. π-π Stacked Step-Scheme PDI/g-C_3N_4/TiO_2@Ti_3C_2 Photocatalyst with Enhanced Visible Photocatalytic Degradation towards Atrazine via Peroxymonosulfate Activation. *Chem. Eng. J.* **2022**, *427*, 131809. [CrossRef]
19. Danfá, S.; Oliveira, C.; Santos, R.; Martins, R.C.; Quina, M.M.J.; Gomes, J. Development of TiO_2-Based Photocatalyst Supported on Ceramic Materials for Oxidation of Organic Pollutants in Liquid Phase. *Appl. Sci.* **2022**, *12*, 7941. [CrossRef]
20. Machín, A.; Soto-Vázquez, L.; Colón-Cruz, C.; Valentín-Cruz, C.A.; Claudio-Serrano, G.J.; Fontánez, K.; Resto, E.; Petrescu, F.I.; Morant, C.; Márquez, F. Photocatalytic Activity of Silver-Based Biomimetics Composites. *Biomimetics* **2021**, *6*, 4. [CrossRef]
21. Al Jitan, S.; Li, Y.; Bahamon, D.; Žerjav, G.; Tatiparthi, V.S.; Aubry, C.; Sinnokrot, M.; Matouk, Z.; Rajput, N.; Gutierrez, M.; et al. Unprecedented Photocatalytic Conversion of Gaseous and Liquid CO_2 on Graphene-Impregnated Pt/Cu-TiO_2. *SSRN J.* **2022**. [CrossRef]
22. Li, B.; Ding, Y.; Li, Q.; Guan, Z.; Zhang, M.; Yang, J. The Photothermal Effect Enhance Visible Light-Driven Hydrogen Evolution Using Urchin-like Hollow RuO_2/TiO_2/Pt/C Nanomaterial. *J. Alloys Compd.* **2022**, *890*, 161722. [CrossRef]
23. Rashid, M.M.; Simončič, B.; Tomšič, B. Recent Advances in TiO_2-Functionalized Textile Surfaces. *Surf. Interfaces* **2021**, *22*, 100890. [CrossRef]
24. Nguyen, T.T.; Cao, T.M.; Balayeva, N.O.; Pham, V.V. Thermal Treatment of Polyvinyl Alcohol for Coupling MoS_2 and TiO_2 Nanotube Arrays toward Enhancing Photoelectrochemical Water Splitting Performance. *Catalysts* **2021**, *11*, 857. [CrossRef]
25. Maver, K.; Arčon, I.; Fanetti, M.; Al Jitan, S.; Palmisano, G.; Valant, M.; Štangar, U.L. Improved Photocatalytic Activity of SnO_2-TiO_2 Nanocomposite Thin Films Prepared by Low-Temperature Sol-Gel Method. *Catal. Today* **2022**, *397–399*, 540–549. [CrossRef]
26. Li, Z.; Li, H.; Wang, S.; Yang, F.; Zhou, W. Mesoporous Black TiO_2/MoS_2/Cu_2S Hierarchical Tandem Heterojunctions toward Optimized Photothermal-Photocatalytic Fuel Production. *Chem. Eng. J.* **2022**, *427*, 131830. [CrossRef]
27. Wang, P.; Yuan, Y.; Liu, Q.; Cheng, Q.; Shen, Z.; Yu, Z.; Zou, Z. Solar-Driven Lignocellulose-to-H_2 Conversion in Water Using 2D-2D MoS_2/TiO_2 Photocatalysts. *ChemSusChem* **2021**, *14*, 2860–2865. [CrossRef]
28. Machín, A.; Fontánez, K.; Duconge, J.; Cotto, M.C.; Petrescu, F.I.; Morant, C.; Márquez, F. Photocatalytic Degradation of Fluoroquinolone Antibiotics in Solution by Au@ZnO-rGO-GC_3N_4 Composites. *Catalysts* **2022**, *12*, 166. [CrossRef]
29. He, Z.; Cai, Q.; Fang, H.; Situ, G.; Qiu, J.; Song, S.; Chen, J. Photocatalytic Activity of TiO_2 Containing Anatase Nanoparticles and Rutile Nanoflower Structure Consisting of Nanorods. *J. Environ. Sci.* **2013**, *25*, 2460–2468. [CrossRef]
30. Soni, S.; Chouhan, N.; Meena, R.K.; Kumar, S.; Dalela, B.; Mishra, M.; Meena, R.S.; Gupta, G.; Kumar, S.; Alvi, P.A.; et al. Electronic Structure and Room Temperature Ferromagnetism in Gd-doped Cerium Oxide Nanoparticles for Hydrogen Generation via Photocatalytic Water Splitting. *Glob. Chall.* **2019**, *3*, 1800090. [CrossRef]
31. Kong, D.; He, H.; Song, Q.; Wang, B.; Lv, W.; Yang, Q.-H.; Zhi, L. Rational Design of MoS_2@graphene Nanocables: Towards High Performance Electrode Materials for Lithium Ion Batteries. *Energy Environ. Sci.* **2014**, *7*, 3320–3325. [CrossRef]

32. Wang, L.; Li, J.; Zhou, H.; Huang, Z.; Zhai, B.; Liu, L.; Hu, L. Three-Dimensionally Layers Nanosheets of MoS$_2$ with Enhanced Electrochemical Performance Using as Free-Standing Anodes of Lithium Ion Batteries. *J. Mater. Sci. Mater. Electron.* **2018**, *29*, 3110–3119. [CrossRef]
33. Vattikuti, S.V.P.; Shim, J. Synthesis, Characterization and Photocatalytic Performance of Chemically Exfoliated MoS$_2$. *IOP Conf. Ser. Mater. Sci. Eng.* **2018**, *317*, 012025. [CrossRef]
34. Jayakumar, G.; Irudayaraj, A.A.; Raj, A.D. Investigation on the Synthesis and Photocatalytic Activity of Activated Carbon–Cerium Oxide (AC–CeO$_2$) Nanocomposite. *Appl. Phys. A* **2019**, *125*, 742. [CrossRef]
35. Ederer, J.; Šťastný, M.; Došek, M.; Henych, J.; Janoš, P. Mesoporous Cerium Oxide for Fast Degradation of Aryl Organophosphate Flame Retardant Triphenyl Phosphate. *RSC Adv.* **2019**, *9*, 32058–32065. [CrossRef]
36. Tamizhdurai, P.; Sakthinathan, S.; Chen, S.-M.; Shanthi, K.; Sivasanker, S.; Sangeetha, P. Environmentally Friendly Synthesis of CeO$_2$ Nanoparticles for the Catalytic Oxidation of Benzyl Alcohol to Benzaldehyde and Selective Detection of Nitrite. *Sci. Rep.* **2017**, *7*, 46372. [CrossRef]
37. Zhou, M.; Liu, Y.; Wu, B.; Zhang, X. Different Crystalline Phases of Aligned TiO$_2$ Nanowires and Their Ethanol Gas Sensing Properties. *Phys. E Low-Dimens. Syst. Nanostruct.* **2019**, *114*, 113601. [CrossRef]
38. Akita, A.; Kobayashi, H.; Tada, H. Action of Chloride Ions as a Habit Modifier in the Hydrothermal Crystal Growth of Rutile TiO$_2$ Nanorod from SnO$_2$ Seed Crystal. *Chem. Phys. Lett.* **2020**, *761*, 138003. [CrossRef]
39. *The International Center for Diffraction Data (ICDD) No. 00-004-0784*; ICDD: Newtown Square, PA, USA, 2008.
40. Wieting, T.J.; Verble, J.L. Infrared and Raman Studies of Long-Wavelength Optical Phonons in Hexagonal MoS$_2$. *Phys. Rev. B* **1971**, *3*, 4286–4292. [CrossRef]
41. Li, H.; Zhang, Q.; Yap, C.C.R.; Tay, B.K.; Edwin, T.H.T.; Olivier, A.; Baillargeat, D. From Bulk to Monolayer MoS$_2$: Evolution of Raman Scattering. *Adv. Funct. Mater.* **2012**, *22*, 1385–1390. [CrossRef]
42. Castellanos-Gomez, A.; Quereda, J.; van der Meulen, H.P.; Agraït, N.; Rubio-Bollinger, G. Spatially Resolved Optical Absorption Spectroscopy of Single- and Few-Layer MoS$_2$ by Hyperspectral Imaging. *Nanotechnology* **2016**, *27*, 115705. [CrossRef] [PubMed]
43. Jayakumar, G.; Albert Irudayaraj, A.; Dhayal Raj, A. A Comprehensive Investigation on the Properties of Nanostructured Cerium Oxide. *Opt. Quant. Electron.* **2019**, *51*, 312. [CrossRef]
44. Dos Santos, M.L.; Lima, R.C.; Riccardi, C.S.; Tranquilin, R.L.; Bueno, P.R.; Varela, J.A.; Longo, E. Preparation and Characterization of Ceria Nanospheres by Microwave-Hydrothermal Method. *Mater. Lett.* **2008**, *62*, 4509–4511. [CrossRef]
45. Cui, J.; Hope, G.A. Raman and Fluorescence Spectroscopy of CeO$_2$, Er$_2$O$_3$, Nd$_2$O$_3$, Tm$_2$O$_3$, Yb$_2$O$_3$, La$_2$O$_3$, and Tb$_4$O$_7$. *J. Spectrosc.* **2015**, *2015*, 940172. [CrossRef]
46. Bharti, B.; Kumar, S.; Lee, H.-N.; Kumar, R. Formation of Oxygen Vacancies and Ti^{3+} State in TiO$_2$ Thin Film and Enhanced Optical Properties by Air Plasma Treatment. *Sci. Rep.* **2016**, *6*, 32355. [CrossRef]
47. Ma, R.; Islam, M.J.; Reddy, D.A.; Kim, T.K. Transformation of CeO$_2$ into a Mixed Phase CeO$_2$/Ce$_2$O$_3$ Nanohybrid by Liquid Phase Pulsed Laser Ablation for Enhanced Photocatalytic Activity through Z-Scheme Pattern. *Ceram. Int.* **2016**, *42*, 18495–18502. [CrossRef]
48. Briggs, D.; Seah, M. *Practical Surface Analysis*; Wiley: New York, NY, USA, 1994.
49. Pal, P.; Pahari, S.K.; Sinhamahapatra, A.; Jayachandran, M.; Kiruthika, G.V.M.; Bajaj, H.C.; Panda, A.B. CeO$_2$ Nanowires with High Aspect Ratio and Excellent Catalytic Activity for Selective Oxidation of Styrene by Molecular Oxygen. *RSC Adv.* **2013**, *3*, 10837. [CrossRef]
50. Ma, J.; Xing, M.; Yin, L.; San Hui, K.; Hui, K.N. Porous Hierarchical TiO$_2$/MoS$_2$/rGO Nanoflowers as Anode Material for Sodium Ion Batteries with High Capacity and Stability. *Appl. Surf. Sci.* **2021**, *536*, 147735. [CrossRef]
51. Makuła, P.; Pacia, M.; Macyk, W. How To Correctly Determine the Band Gap Energy of Modified Semiconductor Photocatalysts Based on UV–Vis Spectra. *J. Phys. Chem. Lett.* **2018**, *9*, 6814–6817. [CrossRef]
52. Fonseca-Cervantes, O.R.; Pérez-Larios, A.; Romero Arellano, V.H.; Sulbaran-Rangel, B.; Guzmán González, C.A. Effects in Band Gap for Photocatalysis in TiO$_2$ Support by Adding Gold and Ruthenium. *Processes* **2020**, *8*, 1032. [CrossRef]
53. Saha, D.; Kruse, P. Editors' Choice—Review—Conductive Forms of MoS$_2$ and Their Applications in Energy Storage and Conversion. *J. Electrochem. Soc.* **2020**, *167*, 126517. [CrossRef]
54. Amendola, V.; Pilot, R.; Frasconi, M.; Maragò, O.M.; Iatì, M.A. Surface Plasmon Resonance in Gold Nanoparticles: A Review. *J. Phys. Condens. Matter* **2017**, *29*, 203002. [CrossRef] [PubMed]
55. Yonggang, L.; Yingzhen, Z.; Zengxing, L.; Shen, X.; Jianying, H.; Kim, H.N.; Yuekun, L. Molybdenum Sulfide Cocatalyst Activation upon Photodeposition of Cobalt for Improved Photocatalytic Hydrogen Production Activity of ZnCdS. *Chem. Eng. J.* **2021**, *425*, 131478. [CrossRef]
56. Dung, V.D.; Thuy, T.D.N.; Periyayya, U.; Yeong-Hoon, C.; Gyu-Cheol, K.; Jin-Kyu, Y.; Duy-Thanh, T.; Thanh Duc, L.; Hyuk, C.; Hyun You, K.; et al. Insightful Understanding of Hot-carrier Generation and Transfer in Plasmonic Au@CeO$_2$ Core–Shell Photocatalysts for Light-Driven Hydrogen Evolution Improvement. *App. Catal. B Envirom.* **2021**, *286*, 119947. [CrossRef]
57. Rusinque, B.; Escobedo, S.; de Lasa, H. Hydrogen Production via Pd-TiO$_2$ Photocatalytic Water Splitting under Near-UV and Visible Light: Analysis of the Reaction Mechanism. *Catalysts* **2021**, *11*, 405. [CrossRef]

58. Chengzhang, Z.; Yuting, W.; Zhifeng, J.; Fanchao, X.; Qiming, X.; Cheng, S.; Qing, T.; Weixin, Z.; Xiaoguang, D.; Shaobin, W. CeO_2 Nanocrystal-modified Layered $MoS_2/g-C_3N_4$ as 0D/2D Ternary Composite for Visible-Light Photocatalytic Hydrogen Evolution: Interfacial Consecutive Multi-Step Electron Transfer and Enhanced H_2O Reactant Adsorption. *Appl. Catal. B Envirom.* **2019**, *259*, 118072. [CrossRef]
59. Donghyung, K.; Kijung, Y. Boron Doping Induced Charge Transfer Switching of a C_3N_4/ZnO Photocatalyst from Z-Scheme to Type II to Enhance Photocatalytic Hydrogen Production. *Appl. Catal. B Envirom.* **2021**, *282*, 119538. [CrossRef]
60. Bashiri, R.; Irfan, M.S.; Mohamed, N.M.; Sufian, S.; Ling, L.Y.; Suhaimi, N.A.; Samsudin, M.F.R. Hierarchically $SrTiO_3$@TiO_2@Fe_2O_3 Nanorod Heterostructures for Enhanced Photoelectrochemical Water Splitting. *Int. J. Hydrogen Energy* **2021**, *46*, 24607–24619. [CrossRef]
61. Tian, Y.; Yang, X.; Li, L.; Zhu, Y.; Wu, Q.; Li, Y.; Ma, F.; Yu, Y. A Direct Dual Z-scheme 3DOM SnS_2–ZnS/ZrO_2 Composite with Excellent Photocatalytic Degradation and Hydrogen Production Performance. *Chemosphere* **2021**, *279*, 130882. [CrossRef]
62. Ma, B.; Bi, J.; Lv, J.; Kong, C.; Yan, P.; Zhao, X.; Zhang, X.; Yang, T.; Yang, Z. Inter-Embedded Au-Cu_2O Heterostructure for the Enhanced Hydrogen Production from Water Splitting under the Visible Light. *Chem. Eng. J.* **2021**, *41*, 126709. [CrossRef]
63. Jourshabani, M.; Shariatinia, Z.; Badiei, A. Synthesis and Characterization of Novel Sm_2O_3/S-Doped $g-C_3N_4$ Nanocomposites with Enhanced Photocatalytic Activities under Visible Light Irradiation. *Appl. Surf. Sci.* **2018**, *427*, 375–387. [CrossRef]
64. Prabavathi, S.L.; Saravanakumar, K.; Nkambule, T.T.I.; Muthuraj, V.; Mamba, G. Enhanced Photoactivity of Cerium Tungstate-Modified Graphitic Carbon Nitride Heterojunction Photocatalyst for the Photodegradation of Moxifloxacin. *J. Mater. Sci. Mater. Electron.* **2020**, *31*, 11434–11447. [CrossRef]
65. Cao, J.; Li, X.; Lin, H.; Chen, S.; Fu, X. In Situ Preparation of Novel p–n Junction Photocatalyst BiOI/$(BiO)_2CO_3$ with Enhanced Visible Light Photocatalytic Activity. *J. Hazard. Mater.* **2012**, *239–240*, 316–324. [CrossRef]
66. Nethercot, A.H. Prediction of Fermi Energies and Photoelectric Thresholds Based on Electronegativity Concepts. *Phys. Rev. Lett.* **1974**, *33*, 1088–1091. [CrossRef]
67. Kutchinsky, J.; Taboryski, R.; Sørensen, C.B.; Hansen, J.B.; Lindelof, P.E. Experimental Investigation of Supercurrent Enhancement in S–N–S Junctions by Non-Equilibrium Injection into Supercurrent-Carrying Bound Andreev States. *Phys. C Supercond.* **2001**, *352*, 4–10. [CrossRef]
68. Elaziouti, A.; Laouedj, N.; Bekka, A.; Vannier, R.-N. Preparation and Characterization of p–n Heterojunction $CuBi_2O_4$/CeO_2 and Its Photocatalytic Activities under UVA Light Irradiation. *J. King Saud Univ. Sci.* **2015**, *27*, 120–135. [CrossRef]
69. Lin, Y.; Ren, P.; Wei, C. Fabrication of MoS_2/TiO_2 Heterostructures with Enhanced Photocatalytic Activity. *CrystEngComm* **2019**, *21*, 3439–3450. [CrossRef]
70. Soto-Vázquez, L.; Rolón-Delgado, F.; Rivera, K.; Cotto, M.C.; Ducongé, J.; Morant, C.; Pinilla, S.; Márquez-Linares, F.M. Catalytic Use of TiO_2 Nanowires in the Photodegradation of Benzophenone-4 as an Active Ingredient in Sunscreens. *J. Environ. Manag.* **2019**, *247*, 822–828. [CrossRef]

Disclaimer/Publisher's Note: The statements, opinions and data contained in all publications are solely those of the individual author(s) and contributor(s) and not of MDPI and/or the editor(s). MDPI and/or the editor(s) disclaim responsibility for any injury to people or property resulting from any ideas, methods, instructions or products referred to in the content.

Article

Zr-Based Metal–Organic Frameworks with Phosphoric Acids for the Photo-Oxidation of Sulfides

Zhenghua Zhao [1,2], Mingjie Liu [1,2], Kai Zhou [1,2], Hantao Gong [1,2], Yajing Shen [2], Zongbi Bao [1,2], Qiwei Yang [1,2], Qilong Ren [1,2] and Zhiguo Zhang [1,2,*]

[1] Key Laboratory of Biomass Chemical Engineering of Ministry of Education, College of Chemical and Biological Engineering, Zhejiang University, Hangzhou 310058, China
[2] Institute of Zhejiang University—Quzhou, Quzhou 324000, China
* Correspondence: zhiguo.zhang@zju.edu.cn

Abstract: Heterogeneous Brønsted acidic catalysts such as phosphoric acids are the conventional activators for organic transformations. However, the photocatalytic performance of these catalysts is still rarely explored. Herein, a novel Zr-based metal–organic framework **Zr-MOF-P** with phosphoric acids as a heterogeneous photocatalyst has been fabricated, which shows high selectivity and reactivity towards the photo-oxidation of sulfides under white light illumination. A mechanism study indicates that the selective oxygenation of sulfides occurs with triplet oxygen rather than common reactive oxygen species (ROS). When **Zr-MOF-P** is irradiated, the hydroxyl group of phosphoric acid is converted into oxygen radical, which takes an electron from the sulfides, and then the activated substrates react with the triplet oxygen to form sulfoxides, avoiding the destruction of the catalysts and endowing the reaction with high substrate compatibility and fine recyclability.

Keywords: metal–organic frameworks; phosphoric acid; photo-oxidation; oxygen radical

1. Introduction

Porous solid Brønsted acids such as phosphoric acids are important heterogeneous catalysts for diverse chemical reactions [1–3]. An effective way to construct these catalysts is to introduce them into highly porous and stable metal–organic frameworks (MOFs) [4–6]. Phosphoric acid-based MOFs have received extensive attention in recent years due to their high reactivity and selectivity towards many types of reactions, especially asymmetric transformations [7–11]. Diverse phosphoric acid ligands are synthesized to construct different MOFs; among these, the coordination of carboxyl groups with Zr(IV) show the best stability [10,11]. However, in view of the large conjugate structure associated with phosphoric acid-based MOFs, their photocatalytic performance is rarely explored.

MOFs with tailorable structures and high porosities have been demonstrated to be efficient photocatalysts towards various types of reactions, such as H_2O splitting [12–15], CO_2 reduction [16–19], organic pollutant degradation [20–23], and organic transformations [24–28]. Ligand structure in MOFs is essential to the regulation of their photocatalytic performance, including the photoresponsivity and active site. Compared to other ligands, phosphoric acids have versatile sites towards different substrates [29,30], and the moderate acidity of phosphoric acid makes it an ideal candidate for the construction of materials to undergo excited-state proton transfer (ESIPT) under photo-excitation, which is a common reaction in molecules with acidic hydrogen atoms and conjugate systems [31–33]. ESIPT products are reactive and the reaction is in general wholly reversible, so it would be an effective strategy to construct novel photocatalytic MOFs through introducing phosphoric acids into photo responsive ligands. Thereinto, binaphthol (BINOL)-based phosphoric acids with a large conjugated structure, ESIPT propriety, fine visible-light-response, and good modifiability could be functioned as effective photocatalysts [34–39]. In this context, BINOL is chosen as a potential skeleton to fabricate photo responsive phosphoric acid ligands for MOF construction.

Photocatalyzed selective oxidation of sulfides using oxygen as the oxidizer is an environment-friendly way to produce sulfoxides, which are key intermediates for bioactive ingredients in the pharmaceutical industry [40–42]. Generally, oxygen is activated by the photocatalyst through energy transfer or electron transfer to generate reactive oxygen species (ROS) such as $^1O_2$ and $O_2^{-•}$, followed by the reaction with sulfide to produce sulfoxide [43–48]. Due to the high activity of ROS, the generated sulfoxide and some substituent groups with low stability would also be oxidized, causing low selectivity and poor substrate compatibility. Moreover, the photocatalysts might also be affected by hyperactive ROS and the lifetime of catalysts is shortened, which are unfavorable for the application of photocatalysts, especially the recycling of heterogeneous photocatalysts. Avoiding the production of ROS, making the photocatalyst directly activate the sulfide and react with triplet oxygen, is an effective way to improve the selectivity and substrate compatibility [49]. Considering that the phosphoric acid-based materials may follow the ESIPT route, we believe phosphoric acids containing Zr-based MOFs would be a good candidate material for the selective photocatalytic oxygenation of sulfides to sulfoxides and may have better selectivity and substrate compatibility. In this view, a BINOL-based phosphoric acid ligand and corresponding Zr-MOFs were fabricated and employed in this reaction; their catalytic performance and reaction mechanism were also investigated.

2. Results and Discussion

2.1. Synthesis and Characterization

The ligand 3,3′,6,6′-tetrakis(4-benzoic acid)-1,1′-binaphthyl phosphate (L_1H_4) was synthesized with an optimized route according to the literature [7]; the new route was shortened to seven steps and the total yield was increased to 31%. To study the effect of the phosphate hydroxyl group, a phosphate-hydroxyl-protected ligand 3,3′,6,6′-tetrakis(4-methyl benzoate)-1,1′-binaphthyl methyl phosphate (L_2Me_4) was also synthesized by an additional methylation reaction for the precursor of L_1H_4 (L_1Me_4). Zr-MOF-P was prepared through a solvothermal reaction with L_1H_4, $ZrCl_4$, formic acid, and trifluoroacetic acid (TFA) in N,N-dimethylformamide (DMF) at 120 °C as light yellow octahedral crystals (Scheme 1).

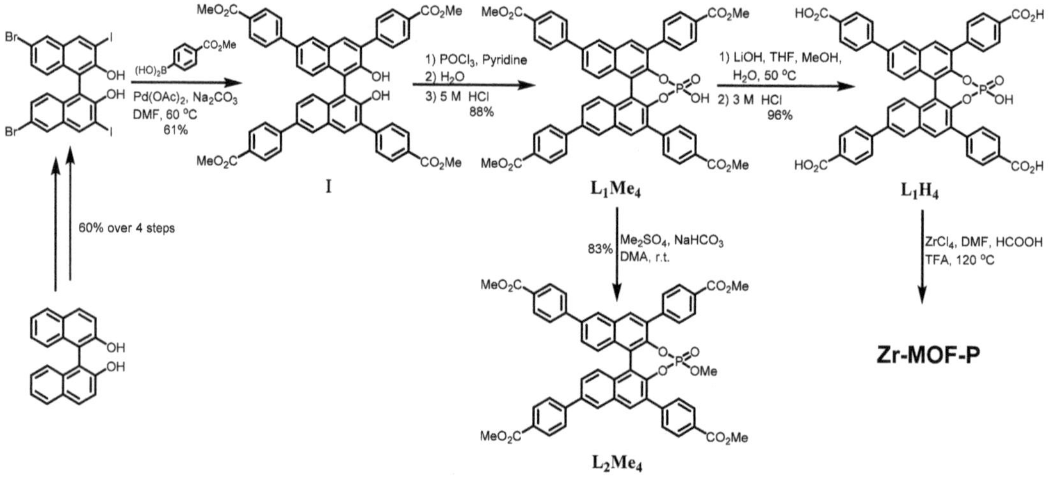

Scheme 1. Synthesis of ligand and Zr-MOF-P.

Scanning electron microscope (SEM) images (Figure 1) and powder X-ray diffraction (PXRD) patterns (Figure 2a) confirmed that Zr-MOF-P was successfully synthesized. Single crystal X-ray diffraction showed that **Zr-MOF-P** crystallizes in the monoclinic space group (a = 21.112 Å, b = 38.991 Å, c = 19.209 Å, and β = 120.902°) and is similar to most of the reported Zr-MOFs composed of carboxylate-based tetrahedral linkers [50–53]; it also exhibited the **flu** topology (Figure 3, Table S1, and Figure S1). The phosphoric acid ligand presents a distorted tetrahedron structure, with a 55.70° dihedral angle between the two naphthalene groups, constructing a cavity with a diameter of 12.2 Å. The total solvent accessible volume of **Zr-MOF-P** is estimated to be 73.6%, calculated by the PLATON routine [54]. The simulated PXRD pattern is similar to the experimental data, demonstrating the phase purity of **Zr-MOF-P**. Due to the large ligand and high porosity of **Zr-MOF-P**, its crystallinity was destroyed after removing the solvent molecules in the channels by vacuum-drying. Therefore, we tried the supercritical CO_2-drying method to keep its structural integrity, and the results showed that the crystal structure of the dried MOF was still partly destroyed (Figure 2a). Considering that the removal of the solvent would break the structure of **Zr-MOF-P**, it was directly used in the photocatalytic reactions after being washed with DMF several times through suction filtration. Unsurprisingly, **Zr-MOF-P** showed high thermal stability, thermogravimetric analysis (TGA) indicated that the guest molecules are removed before the temperature reaches 160 °C, and the frameworks remained stable below 300 °C (Figure 2b).

Figure 1. SEM images of the synthesized **Zr-MOF-P**.

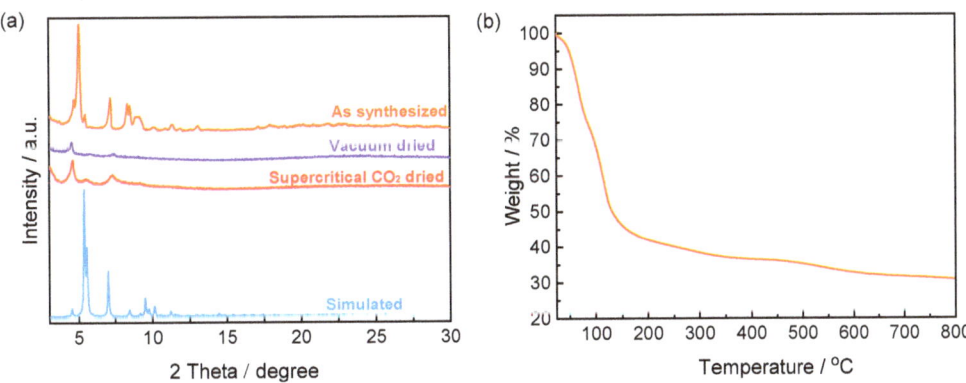

Figure 2. (**a**) PXRD patterns of **Zr-MOF-P** under different conditions. (**b**) TGA profile of **Zr-MOF-P**.

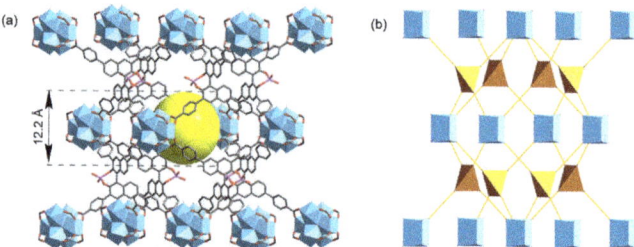

Figure 3. (a) Structure of **Zr-MOF-P**. C, gray; O, red; P, pink; Zr, cyan polyhedra; the yellow sphere is the cavity; H atoms are omitted for clarity. (b) The **flu** topology of **Zr-MOF-P**. The orange and cyan polyhedra represent 4- and 8-connected nodes, respectively.

To identify the possible photocatalytic application of **Zr-MOF-P**, its photo-electrochemical properties were tested. As shown in Figure 4a, UV–Vis spectra indicated that this MOF could be excited by visible light for its obvious absorption below 600 nm, and the band gap was estimated to be 2.83 eV. Mott–Schottky measurements were performed at the frequency of 1000, 1500, and 2000 Hz to identify the semiconductor characteristics of **Zr-MOF-P**, the flat band position determined from the same intersection is about −1.26 V vs. Ag/AgCl (−1.04 V vs. NHE), and the positive slope of the C^{-2} values indicates the character of n-type semiconductors [55–57]. Thus, the conduction band (CB) is −1.04 V vs. NHE, and the valence band (VB) is 1.79 V vs. NHE (Figure 4b). The VB of **Zr-MOF-P** is higher than the oxidation potentials of sulfides, but lower than that of sulfoxides [49], indicating that it could be used in the photo-oxidation of sulfides to sulfoxides. Photo-electrochemical measurements showed that **Zr-MOF-P** had an obvious photocurrent response, illustrating that the hole−electron pair could be separated under visible light irradiation (Figure 4c). The weak fluorescence emission indicated the low electron−hole recombination rate in **Zr-MOF-P**, which favored the electron transfer between photocatalysts and substrates (Figure 4d). All the photo-electrochemical measurements clearly demonstrated that **Zr-MOF-P** would be an ideal photocatalyst for the photo-oxidation of sulfides to sulfoxides.

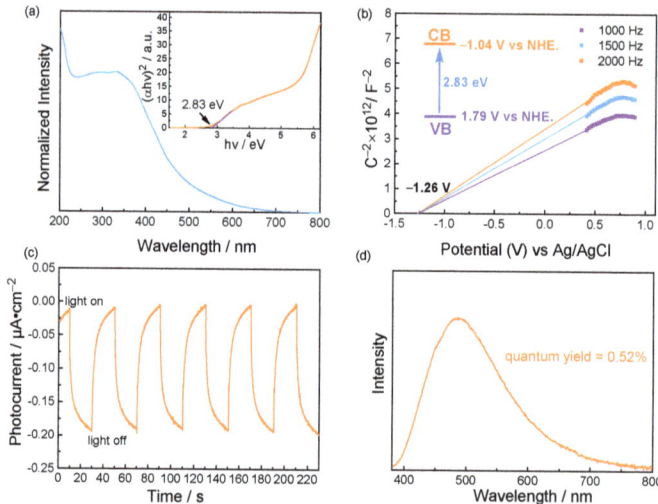

Figure 4. (a) UV–Vis diffuse reflectance spectrum of **Zr-MOF-P**; insets are the Tauc plots. (b) Mott–Schottky plots for **Zr-MOF-P**. (c) Photocurrent responses of **Zr-MOF-P**. (d) Fluorescence spectrum and quantum yield of **Zr-MOF-P**, λ_{ex} = 320 nm.

2.2. Photo-Oxidation of Thioanisol

Considering the excellent photo-electric performance of **Zr-MOF-P**, we investigated its photocatalytic activity towards the photo-oxidation of sulfide into sulfoxide, and thioanisole was selected as the model substrate. The reaction was initially carried out in acetonitrile with **Zr-MOF-P** under white light irradiation and an O_2 atmosphere at room temperature. As shown in Table 1, after 9 h, 19% of sulfide was oxidized into sulfoxide and no overoxidized product (sulfone) was produced. Based on the reported works, the yield is obviously affected by the type of solvents. Therefore, different solvents were explored, and the protic solvent trifluoroethanol (TFEA) was found to be the optimal solvent with a yield of 97% (Table 1, entry 5). Comparative experiments indicated that the photocatalyst, white light irradiation, and O_2 are all indispensable (Table 1, entries 8–10). Moreover, the reaction still went smoothly on the gram scale with a yield as high as 95%.

Table 1. Photo-oxidation of thioanisol [1].

Entry	Solvent	Change in Other Conditions	Yield [2]/%
1	MeCN	none	19
2	MeOH	none	58
3	EtOH	none	74
4	EtOAc	none	trace
5	CF_3CH_2OH	none	97 (95) [3]
6	$CHCl_3$	none	56
7	DMF	none	3
8	CF_3CH_2OH	no photocatalyst	1
9	CF_3CH_2OH	dark	n.d. [4]
10	CF_3CH_2OH	N_2 atmosphere	n.d. [4]

[1] Conditions: thioanisol (0.1 mmol), **Zr-MOF-P** (4 mg), O_2 (1 atm), solvent (2 mL), white LED (5 W), room temperature, 9 h. [2] Determined by gas chromatography (GC) (anisole as internal standard) and GC-MS. [3] Isolated yield of gram-scale experiment. [4] n.d. = not detected.

2.3. Photocatalytic Mechanism

Most of the research reported that ROS such as $^1O_2$, $O_2^{-\bullet}$, or $\cdot OH$ originating from oxygen under the activation of a photocatalyst were important active species in the photocatalytic oxidation of sulfide. Therefore, we carried out a series of experiments to confirm whether ROS participate in the **Zr-MOF-P**-catalyzed reaction. Quenching experiments through adding different scavengers of ROS was firstly performed. Diazabicyclo[2.2.2]octane (DABCO, TCI, Tokyo, Japan) is a scavenger for $^1O_2$; its addition showed no effect on the yield of sulfoxide, excluding the participation of $^1O_2$ in the reaction (Table 2, entry 2). $O_2^{-\bullet}$ is another common ROS participating in the photocatalytic oxidation reaction, while the reaction still went smoothly with the addition of benzoquinone (BQ, J&K Scientific, Beijing, China) as an $O_2^{-\bullet}$ scavenger (Table 2, entry 3). Other ROS such as $\cdot OH$ and H_2O_2 are also capable of oxidizing sulfide, which were excluded through the quenching experiments with the addition of i-PrOH and catalase (TCI, Tokyo, Japan), respectively (Table 2, entries 4 and 5). The radical scavenger hydroquinone (HQ, J&K Scientific, Beijing, China) and electro trapper $CuSO_4$ showed significant inhibition to the reaction, demonstrating that the reaction might have undergone an electron-transfer-induced free radical pathway (Table 2, entries 6 and 7). Moreover, the addition of the sulfide radical cation scavenger 1,4-dimethoxybenzene (DMB, Aladdin, Shanghai, China) also repressed the reaction with a decreased yield of 78% (Table 2, entry 8). Considering that the optimal solvent TFEA is conducive to maintaining the stability of cations [58,59], the sulfide radical cation should be a critical intermediate in the catalytic process. Quenching experiments using **L$_1$Me$_4$** as the catalyst also showed similar results (Table 2, entries 9–15), which indicated that the ligand in **Zr-MOF-P** was the active component.

Table 2. Quenching experiments [1].

Entry	Photocatalyst	Additive	Yield [2]/%
1	Zr-MOF-P	none	97
2	Zr-MOF-P	DABCO	95
3	Zr-MOF-P	BQ	96
4	Zr-MOF-P	i-PrOH	94
5	Zr-MOF-P	catalase [3]	93
6	Zr-MOF-P	HQ	10
7	Zr-MOF-P	$CuSO_4$	18
8	Zr-MOF-P	DMB	78
9 [4]	L_1Me_4	none	99
10 [4]	L_1Me_4	DABCO	91
11 [4]	L_1Me_4	BQ	98
12 [4]	L_1Me_4	i-PrOH	92
13 [4]	L_1Me_4	HQ	11
14 [4]	L_1Me_4	$CuSO_4$	26
15 [4]	L_1Me_4	DMB	15
16 [4]	L_2Me_4	none	2

[1] Conditions: thioanisol (0.1 mmol), photocatalyst (4 mg), scavengers (0.2 mmol), O_2 (1 atm), solvent (2 mL), white LED (5 W), room temperature, 9 h. [2] Determined by GC, anisole as internal standard. [3] 0.1 g of catalase (>200,000 unit/g) [4] 4 h.

To further exclude the participation of ROS, electron paramagnetic resonance (EPR) tests were performed by adopting 2,2,6,6-tetramethylpiperidine (TEMP) and 5,5-dimethyl-1-pryyoline-Noxide (DMPO) as trappers. As shown in Figure 5a, no EPR signal was detected under white light irradiation, which means that no $^1O_2$, $O_2^{-•}$, or ·OH was produced by the photocatalyst. Moreover, the probe molecules Singlet Oxygen Sensor Green, nitrotetrazolium blue chloride, and coumarin-3-carboxylic acid for $^1O_2$, $O_2^{-•}$, and ·OH were added to the suspension of Zr-MOF-P under white light irradiation; the results still showed that no ROS was produced (Figure 5b–d). Based on the above mechanism research experiments, the possibility of ROS participating in the reaction was ruled out; sulfide was directly activated by Zr-MOF-P, and then reacted with $^3O_2$.

The active site of Zr-MOF-P was found through another controlled experiment. L_1Me_4 showed fairly good activity towards the reaction with a yield of 99% (Table 2, entry 9), while the phosphate-hydroxyl-protected L_2Me_4 almost had no catalytic activity with a yield as low as 2% (Table 2, entry 16). EPR spectra showed a single and unstructured signal of Zr-MOF-P with a g value of 2.0033 after illumination, and the solid L_1Me_4 also had the same signal while L_2Me_4 did not, demonstrating the existence of photo-induced oxygen radicals in Zr-MOF-P and L_1Me_4 [60]. Moreover, the EPR signal of Zr-MOF-P after illumination was significantly decreased after the addition of thioanisole, which indicated an electron transfer process between thioanisole and the photo-induced oxygen radical (Figure 6). Therefore, a proposed mechanism was shown in Scheme 2. The ligand was firstly excited by visible light, followed with an ESIPT process, producing the oxygen radical A. The photo-induced oxygen radical takes an electron from sulfide, generating the reduced ligand B and sulfide radical cation C. Then C reacts with $^3O_2$, which converts into the persulfoxide radical D. The reduced ligand B donates an electron to D, which is recovered, and D is transformed into persulfoxide E. Finally, E reacts with another sulfide molecule and two molecules of sulfoxide are produced.

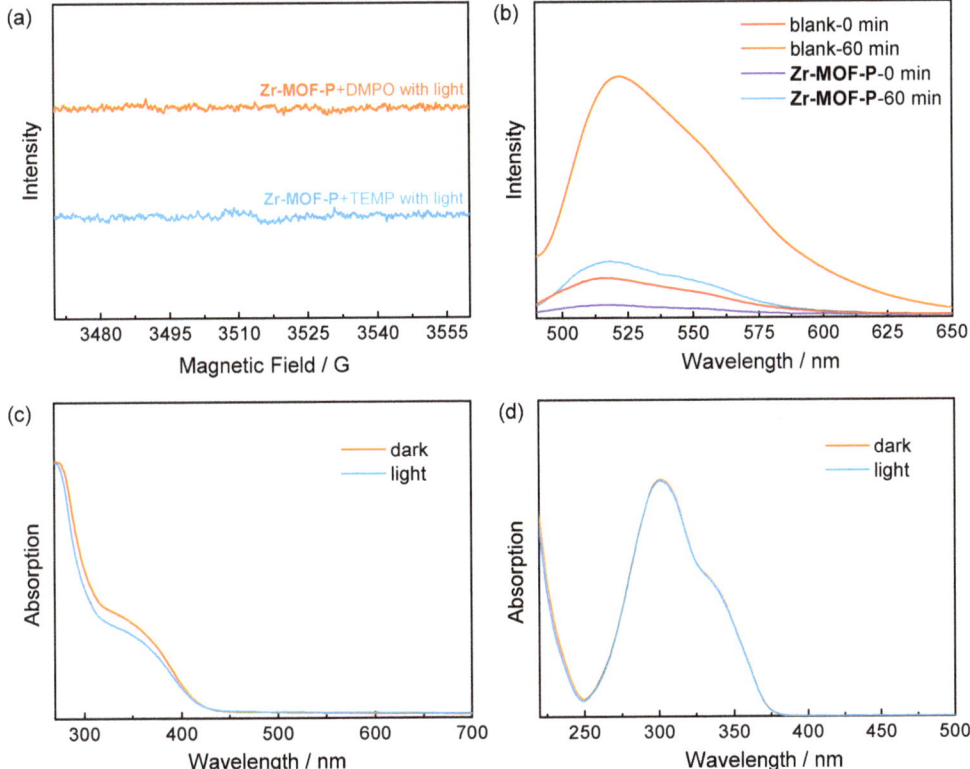

Figure 5. (**a**) EPR spectra of **Zr-MOF-P** using TEMP and DMPO as trapping agents under irradiation. (**b**) Fluorescence spectra of Singlet Oxygen Sensor Green in the suspension of **Zr-MOF-P** with and without illumination (λ_{ex} = 480 nm). (**c**) UV–Vis spectra of nitrotetrazolium blue chloride in the suspension of **Zr-MOF-P** with and without illumination. (**d**) UV–Vis spectra of coumarin-3-carboxylic acid in the suspension of **Zr-MOF-P** with and without illumination.

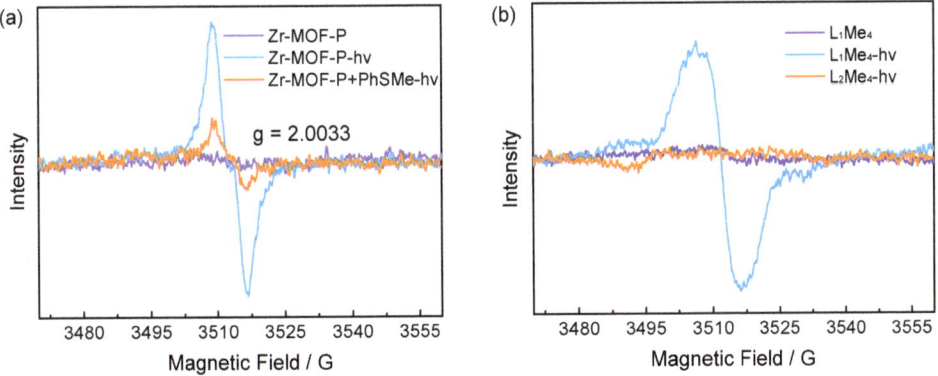

Figure 6. (**a**) EPR spectra of **Zr-MOF-P** under different conditions. (**b**) EPR spectra of **L$_1$Me$_4$** and **L$_2$Me$_4$** under different conditions.

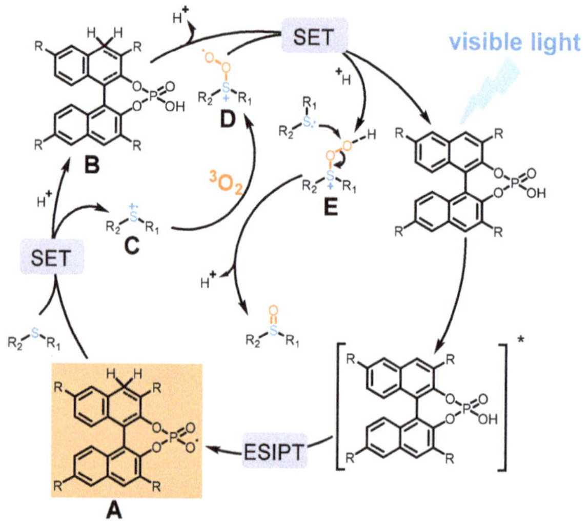

Scheme 2. Proposed mechanism of **Zr-MOF-P** for photocatalytic oxidation of sulfides. * excited state.

2.4. Substrate Compatibility and Recyclability

Encouraged by the unusual photocatalytic mechanism of **Zr-MOF-P**, various sulfides with different substituents were employed in the reaction (Scheme 3). Methylphenyl sulfide derivatives with halogen at the *ortho*- or *para*- positions of phenyl rings were all completely transformed into corresponding sulfoxides, and the conversion of *ortho*-substituted sulfide were slower than the *para*-substituted sulfide due to the steric hindrance (**2–7**). Other substituents such as nitro (**8**), methyl (**9**), and methoxy (**10, 11**) were all tolerated in the reaction. Without the participation of ROS, amino (**12**)- and hydroxy (**13**)-substituted sulfides could be oxidized to sulfoxides and no side reaction was observed. The photosensitive iodine was also well tolerated, and almost quantitatively corresponding sulfoxide was obtained (**14**). Moreover, even the diphenyl sulfide which was difficult to be oxidized by most photocatalysts could be successfully transformed into sulfoxide (**15**) with excellent yields. All of the above results indicated that the avoidance of producing ROS would endow **Zr-MOF-P** with high substrate compatibility and selectivity.

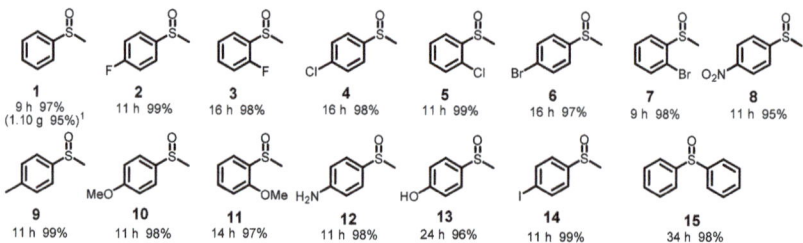

Scheme 3. Photocatalytic oxidation of various sulfides. Conditions: sulfides (0.1 mmol), photocatalyst (4 mg), O_2 (1 atm), solvent (2 mL), white LED (5 W), and room temperature; the yields and products were determined by GC (anisole as internal standard) and GC-MS. [1] Isolated yield of gram-scale experiment.

As a heterogenous photocatalyst, recyclability is an advantage compared with homogeneous catalysts. **Zr-MOF-P** can be easily separated through centrifugation when the reaction finished, and it can be directly used for subsequent runs without additional processing. The photocatalytic activity of **Zr-MOF-P** shows no noticeable change after five cycles of the photo-oxidation of thioanisole (Figure 7a). PXRD spectra and SEM images of the recycled photocatalyst also show little change compared with the pristine **Zr-MOF-P** (Figure 7b,c). Therefore, a photocatalyst with high stability and recyclability for the photo-oxidation of sulfides has been constructed.

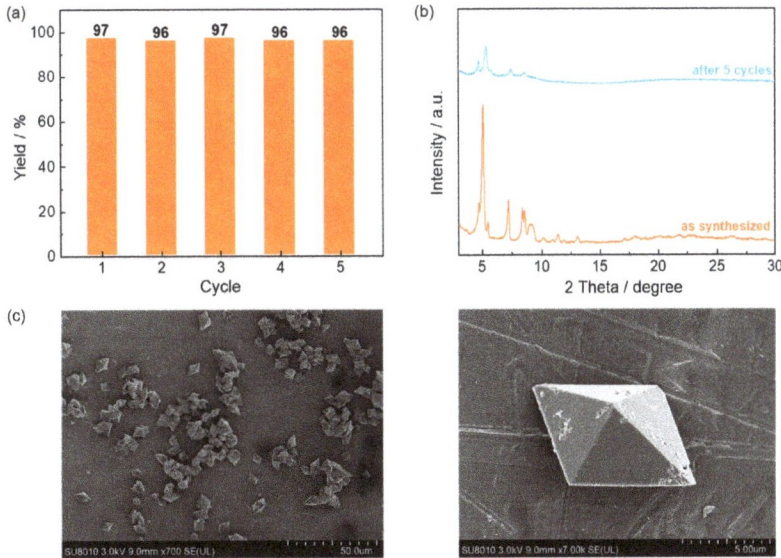

Figure 7. (**a**) The yield of methylphenyl sulfoxide in five cycles with **Zr-MOF-P** as photocatalyst. (**b**) PXRD patterns of **Zr-MOF-P** before and after five cycles. (**c**) SEM images of **Zr-MOF-P** after five cycles.

3. Materials and Methods

3.1. Instruments

Powder X-ray diffraction (PXRD) was carried out with a PANalytical X'Pert3 Powder-17005730 X-ray Powder Diffractometer equipped with two Cu anodes (λ_1 = 1.540598 Å, λ_2 = 1.544426 Å, ratio $K_{\alpha 2}/K_{\alpha 1}$ = 0.5) at 40 kV and 40 mA. Thermogravimetric analysis (TGA) was performed using a TA Discovery SDT 650 heated from room temperature to 800 °C under N_2 atmosphere at the heating rate of 10 °C·min^{-1}. Scanning electron microscopy (SEM) images were obtained using a Hitachi SU-8010 microscope (Tokyo, Japan). UV–Vis diffuse reflectance spectra were obtained on a Shimadzu UV-2600i (Kyoto, Japan) spectrophotometer equipped with an integrated sphere and a white standard of $BaSO_4$ was used as a reference. UV–Vis spectra were obtained on a Shimadzu UV-2600i spectrophotometer. Fluorescence spectra and quantum yield were obtained on an Edinburgh Instruments FLS1000 fluorescence spectrophotometer (Livingston, UK). Nuclear magnetic resonance (NMR) data were collected on a Bruker Avance III 500 spectrometer (Berlin, Germany). HRMS was recorded on an Agilent G6545 Q-TOF (Santa Clara, CA, USA). Electrochemical characterizations were carried out with a CH Instruments CHI660E workstation (Shanghai, China). The photocatalytic reactions were performed in a PerfectLight PCX50C photoreactor (Beijing, China) with 5 W white light LED. Gas chromatographic (GC) analyses were performed using a Shimadzu 2010 gas chromatograph (Kyoto, Japan) equipped with an HP-5MS capillary column (30 m × 0.25 mm × 0.25 µm) and a flame ionization detector. Gas chromatography–mass spectrometry (GC-MS) was recorded on a Waters GCT Premier mass spectrometer (Milford, MA, USA). Electron paramagnetic resonance (EPR) measurements were carried out on a Bruker model A300 spectrometer (Berlin, Germany).

3.2. Synthesis

All the reagents in experiments are commercially available and used without further purification. L_1H_4 was synthesized from BINOL according to literature [7], and we optimized the reported synthesis route. Considering **I** was prepared via several protection and deprotection reactions from 6,6'-dibromo-3,3'-diiodo-1,1'-binaphthyl-2,2'-diol in literature, we tried to synthesize **I** directly through a Suzuki-Miyaura coupling reaction between 6,6'-dibromo-3,3'-diiodo-1,1'-binaphthyl-2,2'-diol and 4-(methoxycarbonyl) benzeneboronic acid.

The synthesis of **I** is as follows: 6,6'-dibromo-3,3'-diiodo-1,1'-binaphthyl-2,2'-diol (2.00 g, 2.87 mmol), 4-(methoxycarbonyl)benzeneboronic acid (5.17 g, 28.74 mmol), Pd(OAc)$_2$ (129 mg, 0.57 mmol), Na$_2$CO$_3$ (2.13 g, 20.09 mmol), DMF (32 mL), and H$_2$O (32 mL) were added into a 350 mL Schlenck tube under Ar atmosphere. The reaction was stirred for 24 h in a 60 °C oil bath. After the reaction finished, it was cooled to room temperature, the mixture was extracted with CH$_2$Cl$_2$, and the organic phase was washed with H$_2$O three times. Then, the organic phase was dried over anhydrous Na$_2$SO$_4$, and the solvent was filtered and concentrated. Crude product was purified by column chromatography on silica gel (2/1 petroleum ether/ethyl acetate, R_f = 0.55) to afford 1.44 g (1.75 mmol, 61% yield) of **I** as a white solid.

The synthesis of **L$_2$Me$_4$** is as follows: **L$_1$Me$_4$** (50.0 mg, 0.056 mmol), dimethyl sulfate (14.3 mg, 10.7 µL, 0.113 mmol), NaHCO$_3$ (10.4 mg, 0.124 mmol), and *N,N*-Dimethylacetamide (0.7 mL) were added into a 10 mL Schlenck tube under Ar atmosphere. The reaction was stirred for 24 h at room temperature. After the reaction finished, H$_2$O (2 mL) was added to quench the reaction, the mixture was extracted with CH$_2$Cl$_2$, and the organic phase was washed with H$_2$O three times. Then, the organic phase was dried over anhydrous Na$_2$SO$_4$, and the solvent was filtered and concentrated. Crude product was purified by column chromatography on silica gel (2/1 petroleum ether/ethyl acetate, R_f = 0.30) to afford 38.6 mg (0.043 mmol, 76% yield) of **L$_2$Me$_4$** as a white solid. ^{1}H NMR (500 MHz, CDCl$_3$) δ 8.26–8.16 (m, 12H), 7.84 (d, *J* = 7.7 Hz, 4H), 7.80 (t, *J* = 8.1 Hz, 4H), 7.69–7.65 (m, 2H), 7.56 (d, *J* = 8.8 Hz, 1H), 7.49 (d, *J* = 8.9 Hz, 1H), 4.01–3.91 (m, 12H), 3.14 (d, *J* = 9.9 Hz, 3H) (Figure S2). ^{13}C NMR (126 MHz, CDCl$_3$) δ 167.00, 166.89, 166.86, 166.74, 144.51, 144.44, 141.10, 141.05, 137.99, 137.95, 133.95, 133.11, 132.34, 131.95, 131.90, 131.86, 131.71, 131.67, 130.35, 130.34, 130.03, 130.00, 129.89, 129.77, 129.69, 129.63, 129.47, 129.43, 127.79, 127.27, 127.23, 126.91, 126.84, 126.69, 126.64, 122.55, 122.51, 77.29, 77.03, 76.78, 55.13, 52.35, 52.26, 52.25, 52.21 (Figure S3). HRMS (ESI): [M + H]$^+$ Calcd for C$_{53}$H$_{40}$O$_{12}$P$^+$ 899.2252; Found 899.2252.

The synthesis of **Zr-MOF-P** is as follows: **L$_1$H$_4$** (200 mg, 0.242 mmol), ZrCl$_4$ (169 mg, 0.0725 mmol), anhydrous formic acid (10 mL), and trifluoroacetic acid (2 mL) were added in DMF (40 mL). After 10 min of ultrasonic vibration, the mixture was heated in a 100 mL Teflon-sealed autoclave at 120 °C for 3 days. Then, the mixture was cooled to room temperature, light yellow powders (310 mg) were collected through centrifugation, and washed with DMF. Because the removing of solvent molecules from MOF channels will distort the framework, **Zr-MOF-P** was dipped in DMF and was collected through suction filtration before use.

3.3. Electrochemical Characterization

Electrochemical characterizations were carried out using a CH Instruments CHI660E workstation through a three-electrode system in 0.2 M Na$_2$SO$_4$ aqueous solution.

Mott–Schottky plots of **Zr-MOF-P** were measured using the photocatalyst-coated glassy carbon as working electrode, Ag/AgCl as reference electrode, and Pt plate as counter electrode at frequencies of 1000, 1500, and 2000 Hz, respectively. Preparation of the working electrode is as follows: 5 mg **Zr-MOF-P** was dispersed in 1 mL ethanol, and 10 µL 5 wt% Nafion was added as binder. Then, 20 µL of the solution was coated on the surface of the glassy carbon electrode and dried at room temperature. This process was repeated until the electrode was completely covered.

Photocurrent measurements of **Zr-MOF-P** were measured using the photocatalyst-coated Pt plate as working electrode, Ag/AgCl as reference electrode, and Pt plate as counter electrode, and a 40 W White light LED was used as light source. Preparation of the working electrode is as follows: 5 mg **Zr-MOF-P** was dispersed in 1 mL ethanol, and 10 μL 5 wt% Nafion was added as binder. Then, 50 μL of the solution was coated on the Pt plate and dried at room temperature. This process was repeated until 1 cm^2 of the Pt plate was completely covered.

3.4. Photocatalytic Reaction

The photocatalytic reactions were performed on a PerfectLight PCX50C photoreactor (Beijing, China) equipped with 5 W white LEDs. In addition, the reaction was carried out at 25 °C by circulating refrigeration equipment. For the photo-oxidation of sulfides to sulfoxides, 4 mg photocatalyst, 0.1 mmol substrate, and 2 mL solvent were added into a 10 mL Schlenck tube under O_2 atmosphere. The reaction mixture was magnetically stirred at 150 rpm and illuminated with 5 W white LEDs. After the reaction finished, 20 μL of anisole was added as the internal standard and stirred for 10 min. Then, the photocatalyst was separated through centrifugation and washed with solvent. The products were analyzed by GC and GC-MS.

For gram-scale reaction, thioanisol (8.86 mmol, 1.10 g), TFEA (100 mL), and **Zr-MOF-P** (20 mg) were stirred at room temperature for 7 days in oxygen atmosphere (1 atm) under the irradiation of white LEDs. After the reaction finished, photocatalyst was separated through centrifugation and washed with ethyl acetate several times. The combined organic phase was concentrated over a rotary evaporator, and **1** (1.18 g, 95%) was obtained through column chromatography as a colorless oil.

3.5. EPR Measurements

EPR spectra were obtained on a Bruker model A300 spectrometer (Berlin, Germany) at room temperature. The spectrometer parameters are shown as follows: sweep width, 100 G; center field, 3510.890 G; microwave bridge frequency, 9.839 GHz; power, 20.37 mW; modulation frequency, 100 kHz; modulation amplitude, 1 G; conversion time, 42.00 s; sweep time 42.00 s; receiver gain, 2.00×10^4. The preparation of the liquid samples was similar to the photocatalyst reaction. The signal after irradiation was measured after 5 min of irradiation with a 50 W Xe lamp with stirring, and the mixture was transferred to 3 mm diameter glass tubes as soon as possible to record the signals. Furthermore, for solid samples, about 2 mg of target compound was put into a 3 mm diameter glass tube, and the signal after irradiation was also measured after 5 min of irradiation with a 50 W Xe lamp.

3.6. ROS Detection with Probe Molecules

$^1O_2$ detection: **Zr-MOF-P** (2 mg) and TFEA (1.5 mL) containing Singlet Oxygen Sensor Green (10 μM) were added into a 10 mL Schlenck tube under air atmosphere. The reaction mixture was magnetically stirred for 30 min in the dark before illuminated with white LEDs for 1 h. After the reaction finished, photocatalyst was separated through centrifugation and the supernatant was examined with fluorescence spectrophotometer. The result was compared with the blank group and unilluminated control group, showing that no $^1O_2$ produced.

$O_2^{-\bullet}$ detection: **Zr-MOF-P** (2 mg) and TFEA (1.5 mL) containing nitrotetrazolium blue chloride (0.1 mM) were added into a 10 mL Schlenck tube under air atmosphere. The reaction mixture was magnetically stirred for 30 min in the dark before illuminated with white LEDs for 2 h. After the reaction finished, photocatalyst was separated through centrifugation and the supernatant was examined with UV–Vis spectrophotometer. The result was compared with the unilluminated control group, showing that no $O_2^{-\bullet}$ produced.

·OH detection: **Zr-MOF-P** (2 mg) and TFEA (1.5 mL) containing coumarin-3-carboxylic acid (0.1 mM) were added into a 10 mL Schlenck tube under air atmosphere. The reaction mixture was magnetically stirred for 30 min in the dark before illuminated with white LEDs for 2 h. After the reaction finished, photocatalyst was separated through centrifugation and the supernatant was examined with UV–Vis spectrophotometer. The result was compared with the unilluminated control group, showing that no ·OH produced.

4. Conclusions

A photocatalyst **Zr-MOF-P** based on a BINOL-derived phosphoric acid ligand for the selective oxidation of sulfides under white light irradiation was prepared. Comprehensive mechanistic studies indicated that **Zr-MOF-P** had appropriate photo-electrochemical properties for this reaction, and the ESIPT process produced the reactive oxygen radical, which would take an electron from the sulfides. Thus, the sulfides were activated and, subsequently, react with ground state oxygen, producing sulfoxides. The unique mechanism without the participation of ROS ensured the high selectivity and substrate compatibility of the reaction. Moreover, as a heterogeneous photocatalyst, **Zr-MOF-P** had sufficient stability, as it can be easily separated and re-used at least five times without any noticeable change in reactivity. This study demonstrates that phosphoric acids with a large conjugate structure can be used as photocatalysts, and they might have potential applications in more kinds of photocatalytic reactions. Further applications for **Zr-MOF-P** are under study in our group.

Supplementary Materials: The following supporting information can be downloaded at: https://www.mdpi.com/article/10.3390/ijms232416121/s1.

Author Contributions: Methodology, Z.Z. (Zhenghua Zhao) and Z.Z. (Zhiguo Zhang); validation, Z.Z. (Zhenghua Zhao), M.L., K.Z. and H.G.; formal analysis, Z.B., Q.Y. and Q.R.; investigation, Z.Z. (Zhenghua Zhao); writing—original draft preparation, Z.Z. (Zhenghua Zhao), Y.S. and Z.Z. (Zhiguo Zhang); writing—review and editing, Z.Z. (Zhenghua Zhao), Y.S. and Z.Z. (Zhiguo Zhang); supervision, Z.Z. (Zhiguo Zhang); All authors have read and agreed to the published version of the manuscript.

Funding: This research was funded by the National Key R&D Program of China (grant number 2021YFC2103704) and National Natural Science Foundation of China (grant number 21878266, 22078288, and U21A20301).

Institutional Review Board Statement: Not applicable.

Informed Consent Statement: Not applicable.

Data Availability Statement: CCDC 2218003 contains the supplementary crystallographic data of **Zr-MOF-P**: these data can be obtained free of charge through www.ccdc.cam.ac.uk/data_request/cif (accessed on 26 November 2022), or by emailing data_request@ccdc.cam.ac.uk, or by contacting The Cambridge Crystallographic Data Centre, 12 Union Road, Cambridge CB2 1EZ, UK; fax: +44-1223-336033.

Acknowledgments: We gratefully acknowledge Jianyang Pan (Research and Service Center, College of Pharmaceutical Sciences, Zhejiang University) for NMR characterization.

Conflicts of Interest: The authors declare no conflict of interest.

References

1. Wang, S.S.; Yang, G.Y. Recent Advances in Polyoxometalate-Catalyzed Reactions. *Chem. Rev.* **2015**, *115*, 4893–4962. [CrossRef] [PubMed]
2. Mansir, N.; Taufiq-Yap, Y.H.; Rashid, U.; Lokman, I.M. Investigation of heterogeneous solid acid catalyst performance on low grade feedstocks for biodiesel production: A review. *Energy Convers. Manag.* **2017**, *141*, 171–182. [CrossRef]
3. Doustkhah, E.; Lin, J.; Rostamnia, S.; Len, C.; Luque, R.; Luo, X.; Bando, Y.; Wu, K.C.; Kim, J.; Yamauchi, Y.; et al. Development of Sulfonic-Acid-Functionalized Mesoporous Materials: Synthesis and Catalytic Applications. *Chem. Eur. J.* **2019**, *25*, 1614–1635. [CrossRef] [PubMed]
4. Buru, C.T.; Farha, O.K. Strategies for Incorporating Catalytically Active Polyoxometalates in Metal−Organic Frameworks for Organic Transformations. *ACS Appl. Mater. Interfaces* **2020**, *12*, 5345–5360. [CrossRef]
5. Gong, W.; Liu, Y.; Li, H.; Cui, Y. Metal-organic frameworks as solid Brønsted acid catalysts for advanced organic transformations. *Coord. Chem. Rev.* **2020**, *420*, 213400. [CrossRef]

6. Chen, Y.; Guerin, S.; Yuan, H.; O'Donnell, J.; Xue, B.; Cazade, P.A.; Haq, E.U.; Shimon, L.J.W.; Rencus-Lazar, S.; Tofail, S.A.M.; et al. Guest Molecule-Mediated Energy Harvesting in a Conformationally Sensitive Peptide-Metal Organic Framework. *J. Am. Chem. Soc.* **2022**, *144*, 3468–3476. [CrossRef]
7. Zheng, M.; Liu, Y.; Wang, C.; Liu, S.; Lin, W. Cavity-induced enantioselectivity reversal in a chiral metal–organic framework Brønsted acid catalyst. *Chem. Sci.* **2012**, *3*, 2623–2627. [CrossRef]
8. Zhang, Z.; Ji, Y.R.; Wojtas, L.; Gao, W.Y.; Ma, S.; Zaworotko, M.J.; Antilla, J.C. Two homochiral organocatalytic metal organic materials with nanoscopic channels. *Chem. Commun.* **2013**, *49*, 7693–7695. [CrossRef]
9. Chen, X.; Jiang, H.; Li, X.; Hou, B.; Gong, W.; Wu, X.; Han, X.; Zheng, F.; Liu, Y.; Jiang, J.; et al. Chiral Phosphoric Acids in Metal–Organic Frameworks with Enhanced Acidity and Tunable Catalytic Selectivity. *Angew. Chem. Int. Ed.* **2019**, *58*, 14748–14757. [CrossRef]
10. Gong, W.; Chen, X.; Jiang, H.; Chu, D.; Cui, Y.; Liu, Y. Highly Stable Zr(IV)-Based Metal–Organic Frameworks with Chiral Phosphoric Acids for Catalytic Asymmetric Tandem Reactions. *J. Am. Chem. Soc.* **2019**, *141*, 7498–7508. [CrossRef] [PubMed]
11. Dorneles de Mello, M.; Kumar, G.; Tabassum, T.; Jain, S.K.; Chen, T.H.; Caratzoulas, S.; Li, X.; Vlachos, D.G.; Han, S.I.; Scott, S.L.; et al. Phosphonate-Modified UiO-66 Brønsted Acid Catalyst and Its Use in Dehydra-Decyclization of 2-Methyltetrahydrofuran to Pentadienes. *Angew. Chem. Int. Ed.* **2020**, *59*, 13260–13266. [CrossRef] [PubMed]
12. Fang, X.; Shang, Q.; Wang, Y.; Jiao, L.; Yao, T.; Li, Y.; Zhang, Q.; Luo, Y.; Jiang, H.L. Single Pt Atoms Confined into a Metal–Organic Framework for Efficient Photocatalysis. *Adv. Mater.* **2018**, *30*, 1705112. [CrossRef] [PubMed]
13. Wu, X.P.; Gagliardi, L.; Truhlar, D.G. Cerium Metal–Organic Framework for Photocatalysis. *J. Am. Chem. Soc.* **2018**, *140*, 7904–7912. [CrossRef] [PubMed]
14. Xiao, Y.; Qi, Y.; Wang, X.; Wang, X.; Zhang, F.; Li, C. Visible-Light-Responsive 2D Cadmium–Organic Framework Single Crystals with Dual Functions of Water Reduction and Oxidation. *Adv. Mater.* **2018**, *30*, 1803401. [CrossRef] [PubMed]
15. Zuo, Q.; Liu, T.; Chen, C.; Ji, Y.; Gong, X.; Mai, Y.; Zhou, Y. Ultrathin Metal-Organic Framework Nanosheets with Ultrahigh Loading of Single Pt Atoms for Efficient Visible-Light-Driven Photocatalytic H2 Evolution. *Angew. Chem. Int. Ed. Engl.* **2019**, *58*, 10198–10203. [CrossRef]
16. Xiao, J.D.; Han, L.; Luo, J.; Yu, S.H.; Jiang, H.L. Integration of Plasmonic Effects and Schottky Junctions into Metal-Organic Framework Composites: Steering Charge Flow for Enhanced Visible-Light Photocatalysis. *Angew. Chem. Int. Ed.* **2018**, *57*, 1103–1107. [CrossRef]
17. Qin, J.S.; Yuan, S.; Zhang, L.; Li, B.; Du, D.Y.; Huang, N.; Guan, W.; Drake, H.F.; Pang, J.; Lan, Y.Q.; et al. Creating Well-Defined Hexabenzocoronene in Zirconium Metal–Organic Framework by Postsynthetic Annulation. *J. Am. Chem. Soc.* **2019**, *141*, 2054–2060. [CrossRef]
18. Wang, X.-K.; Liu, J.; Zhang, L.; Dong, L.-Z.; Li, S.-L.; Kan, Y.-H.; Li, D.-S.; Lan, Y.-Q. Monometallic Catalytic Models Hosted in Stable Metal-Organic Frameworks for Tunable CO_2 Photoreduction. *ACS Catal.* **2019**, *9*, 1726–1732. [CrossRef]
19. Kong, X.J.; He, T.; Zhou, J.; Zhao, C.; Li, T.C.; Wu, X.Q.; Wang, K.; Li, J.R. In Situ Porphyrin Substitution in a Zr(IV)-MOF for Stability Enhancement and Photocatalytic CO_2 Reduction. *Small* **2021**, *17*, 2005357. [CrossRef]
20. Gao, Y.; Li, S.; Li, Y.; Yao, L.; Zhang, H. Accelerated photocatalytic degradation of organic pollutant over metal-organic framework MIL-53(Fe) under visible LED light mediated by persulfate. *Appl. Catal. B* **2017**, *202*, 165–174. [CrossRef]
21. Li, M.; Zheng, Z.; Zheng, Y.; Cui, C.; Li, C.; Li, Z. Controlled Growth of Metal-Organic Framework on Upconversion Nanocrystals for NIR-Enhanced Photocatalysis. *ACS Appl. Mater. Interfaces* **2017**, *9*, 2899–2905. [CrossRef] [PubMed]
22. Buru, C.T.; Majewski, M.B.; Howarth, A.J.; Lavroff, R.H.; Kung, C.W.; Peters, A.W.; Goswami, S.; Farha, O.K. Improving the Efficiency of Mustard Gas Simulant Detoxification by Tuning the Singlet Oxygen Quantum Yield in Metal–Organic Frameworks and Their Corresponding Thin Films. *ACS Appl. Mater. Interfaces* **2018**, *10*, 23802–23806. [CrossRef] [PubMed]
23. Gómez-Avilés, A.; Peñas-Garzón, M.; Bedia, J.; Dionysiou, D.D.; Rodríguez, J.J.; Belver, C. Mixed Ti-Zr metal-organic-frameworks for the photodegradation of acetaminophen under solar irradiation. *Appl. Catal. B* **2019**, *253*, 253–262. [CrossRef]
24. Yuan, S.; Liu, T.F.; Feng, D.; Tian, J.; Wang, K.; Qin, J.; Zhang, Q.; Chen, Y.P.; Bosch, M.; Zou, L.; et al. A single crystalline porphyrinic titanium metal-organic framework. *Chem. Sci.* **2015**, *6*, 3926–3930. [CrossRef] [PubMed]
25. Nguyen, H.L.; Vu, T.T.; Le, D.; Doan, T.L.H.; Nguyen, V.Q.; Phan, N.T.S. A Titanium–Organic Framework: Engineering of the Band-Gap Energy for Photocatalytic Property Enhancement. *ACS Catal.* **2016**, *7*, 338–342. [CrossRef]
26. Zeng, L.; Guo, X.; He, C.; Duan, C. Metal–Organic Frameworks: Versatile Materials for Heterogeneous Photocatalysis. *ACS Catal.* **2016**, *6*, 7935–7947. [CrossRef]
27. Zhu, Y.Y.; Lan, G.; Fan, Y.; Veroneau, S.S.; Song, Y.; Micheroni, D.; Lin, W. Merging Photoredox and Organometallic Catalysts in a Metal–Organic Framework Significantly Boosts Photocatalytic Activities. *Angew. Chem. Int. Ed.* **2018**, *57*, 14090–14094. [CrossRef]
28. Chen, Y.; Yang, Y.; Orr, A.A.; Makam, P.; Redko, B.; Haimov, E.; Wang, Y.; Shimon, L.J.W.; Rencus-Lazar, S.; Ju, M.; et al. Self-Assembled Peptide Nano-Superstructure towards Enzyme Mimicking Hydrolysis. *Angew. Chem. Int. Ed.* **2021**, *60*, 17164–17170. [CrossRef]
29. Maji, R.; Mallojjala, S.C.; Wheeler, S.E. Chiral phosphoric acid catalysis: From numbers to insights. *Chem. Soc. Rev.* **2018**, *47*, 1142–1158. [CrossRef]
30. Xia, Z.L.; Xu-Xu, Q.F.; Zheng, C.; You, S.L. Chiral phosphoric acid-catalyzed asymmetric dearomatization reactions. *Chem. Soc. Rev.* **2020**, *49*, 286–300. [CrossRef]

31. Padalkar, V.S.; Seki, S. Excited-state intramolecular proton-transfer (ESIPT)-inspired solid state emitters. *Chem. Soc. Rev.* **2016**, *45*, 169–202. [CrossRef] [PubMed]
32. Wang, W.; Marshall, M.; Collins, E.; Marquez, S.; Mu, C.; Bowen, K.H.; Zhang, X. Intramolecular electron-induced proton transfer and its correlation with excited-state intramolecular proton transfer. *Nat. Commun.* **2019**, *10*, 1170. [CrossRef] [PubMed]
33. Man, Z.; Lv, Z.; Xu, Z.; Liu, M.; He, J.; Liao, Q.; Yao, J.; Peng, Q.; Fu, H. Excitation-Wavelength-Dependent Organic Long-Persistent Luminescence Originating from Excited-State Long-Range Proton Transfer. *J. Am. Chem. Soc.* **2022**, *144*, 12652–12660. [CrossRef] [PubMed]
34. Rueping, M.; Kuenkel, A.; Atodiresei, I. Chiral Bronsted acids in enantioselective carbonyl activations—Activation modes and applications. *Chem. Soc. Rev.* **2011**, *40*, 4539–4549. [CrossRef] [PubMed]
35. Velmurugan, K.; Nandhakumar, R. Binol based "turn on" fluorescent chemosensor for mercury ion. *J. Lumin.* **2015**, *162*, 8–13. [CrossRef]
36. Reid, J.P.; Simon, L.; Goodman, J.M. A Practical Guide for Predicting the Stereochemistry of Bifunctional Phosphoric Acid Catalyzed Reactions of Imines. *Acc. Chem. Res.* **2016**, *49*, 1029–1041. [CrossRef] [PubMed]
37. Zhang, Z.; Wang, Y.; Nakano, T. Photo Racemization and Polymerization of (R)-1,1'-Bi(2-naphthol). *Molecules* **2016**, *21*, 1541. [CrossRef] [PubMed]
38. Posey, V.; Hanson, K. Chirality and Excited State Proton Transfer: From Sensing to Asymmetric Synthesis. *ChemPhotoChem* **2019**, *3*, 580–604. [CrossRef]
39. Kaji, D.; Kitayama, M.; Hara, N.; Yoshida, K.; Wakabayashi, S.; Shizuma, M.; Tsubaki, K.; Imai, Y. Sign control of circularly polarized luminescence by substituent domino effect in binaphthyl-Eu(III) organometallic luminophores. *J. Photoch. Photobio. A* **2020**, *397*, 112490. [CrossRef]
40. Pulis, A.P.; Procter, D.J. C–H Coupling Reactions Directed by Sulfoxides: Teaching an Old Functional Group New Tricks. *Angew. Chem. Int. Ed.* **2016**, *55*, 9842–9860. [CrossRef]
41. Han, J.; Soloshonok, V.A.; Klika, K.D.; Drabowicz, J.; Wzorek, A. Chiral sulfoxides: Advances in asymmetric synthesis and problems with the accurate determination of the stereochemical outcome. *Chem. Soc. Rev.* **2018**, *47*, 1307–1350. [CrossRef] [PubMed]
42. Wojaczyńska, E.; Wojaczyński, J. Modern Stereoselective Synthesis of Chiral Sulfinyl Compounds. *Chem. Rev.* **2020**, *120*, 4578–4611. [CrossRef] [PubMed]
43. Liang, X.; Guo, Z.; Wei, H.; Liu, X.; Lv, H.; Xing, H. Selective photooxidation of sulfides mediated by singlet oxygen using visible-light-responsive coordination polymers. *Chem. Commun.* **2018**, *54*, 13002–13005. [CrossRef]
44. Wei, L.Q.; Ye, B.H. Cyclometalated Ir–Zr Metal–Organic Frameworks as Recyclable Visible-Light Photocatalysts for Sulfide Oxidation into Sulfoxide in Water. *ACS Appl. Mater. Interfaces* **2019**, *11*, 41448–41457. [CrossRef] [PubMed]
45. Meng, Y.; Luo, Y.; Shi, J.L.; Ding, H.; Lang, X.; Chen, W.; Zheng, A.; Sun, J.; Wang, C. 2D and 3D Porphyrinic Covalent Organic Frameworks: The Influence of Dimensionality on Functionality. *Angew. Chem. Int. Ed.* **2020**, *59*, 3624–3629. [CrossRef] [PubMed]
46. Hao, Y.; Papazyan, E.K.; Ba, Y.; Liu, Y. Mechanism-Guided Design of Metal–Organic Framework Composites for Selective Photooxidation of a Mustard Gas Simulant under Solvent-Free Conditions. *ACS Catal.* **2021**, *12*, 363–371. [CrossRef]
47. Liu, M.; Liu, J.; Zhou, K.; Chen, J.; Sun, Q.; Bao, Z.; Yang, Q.; Yang, Y.; Ren, Q.; Zhang, Z. Turn-On Photocatalysis: Creating Lone-Pair Donor-Acceptor Bonds in Organic Photosensitizer to Enhance Intersystem Crossing. *Adv. Sci.* **2021**, *8*, 2100631. [CrossRef]
48. Sadeghfar, F.; Zalipour, Z.; Taghizadeh, M.; Taghizadeh, A.; Ghaedi, M. Photodegradation Processes. In *Interface Science and Technology*; Ghaedi, M., Ed.; Elsevier: Amsterdam, The Netherlands, 2021; Volume 32, pp. 55–124.
49. Li, Y.; Rizvi, S.A.; Hu, D.; Sun, D.; Gao, A.; Zhou, Y.; Li, J.; Jiang, X. Selective Late-Stage Oxygenation of Sulfides with Ground-State Oxygen by Uranyl Photocatalysis. *Angew. Chem. Int. Ed.* **2019**, *58*, 13499–13506. [CrossRef]
50. Furukawa, H.; Gándara, F.; Zhang, Y.B.; Jiang, J.; Queen, W.L.; Hudson, M.R.; Yaghi, O.M. Water Adsorption in Porous Metal–Organic Frameworks and Related Materials. *J. Am. Chem. Soc.* **2014**, *136*, 4369–4381. [CrossRef]
51. Zhang, M.; Chen, Y.P.; Bosch, M.; Gentle, T., III; Wang, K.; Feng, D.; Wang, Z.U.; Zhou, H.C. Symmetry-Guided Synthesis of Highly Porous Metal–Organic Frameworks with Fluorite Topology. *Angew. Chem. Int. Ed.* **2014**, *53*, 815–818. [CrossRef]
52. Wang, S.; Wang, J.; Cheng, W.; Yang, X.; Zhang, Z.; Xu, Y.; Liu, H.; Wu, Y.; Fang, M. A Zr metal–organic framework based on tetrakis(4-carboxyphenyl) silane and factors affecting the hydrothermal stability of Zr-MOFs. *Dalton Trans.* **2015**, *44*, 8049–8061. [CrossRef] [PubMed]
53. Ma, J.; Tran, L.D.; Matzger, A.J. Toward Topology Prediction in Zr-Based Microporous Coordination Polymers: The Role of Linker Geometry and Flexibility. *Cryst. Growth Des.* **2016**, *16*, 4148–4153. [CrossRef]
54. Spek, A.L. Single-crystal structure validation with the program *PLATON*. *J. Appl. Crystallogr.* **2003**, *36*, 7–13. [CrossRef]
55. Zhang, J.; Chen, X.; Takanabe, K.; Maeda, K.; Domen, K.; Epping, J.D.; Fu, X.; Antonietti, M.; Wang, X. Synthesis of a Carbon Nitride Structure for Visible-Light Catalysis by Copolymerization. *Angew. Chem. Int. Ed.* **2010**, *49*, 441–444. [CrossRef]
56. Wang, J.; Yu, Y.; Zhang, L. Highly efficient photocatalytic removal of sodium pentachlorophenate with Bi_3O_4Br under visible light. *Appl. Catal. B* **2013**, *136–137*, 112–121. [CrossRef]
57. Xu, H.Q.; Hu, J.; Wang, D.; Li, Z.; Zhang, Q.; Luo, Y.; Yu, S.H.; Jiang, H.L. Visible-Light Photoreduction of CO_2 in a Metal–Organic Framework: Boosting Electron–Hole Separation via Electron Trap States. *J. Am. Chem. Soc.* **2015**, *137*, 13440–13443. [CrossRef]

58. Colomer, I.; Chamberlain, A.E.R.; Haughey, M.B.; Donohoe, T.J. Hexafluoroisopropanol as a highly versatile solvent. *Nat. Rev. Chem.* **2017**, *1*, 1–12. [CrossRef]
59. Pistritto, V.A.; Schutzbach-Horton, M.E.; Nicewicz, D.A. Nucleophilic Aromatic Substitution of Unactivated Fluoroarenes Enabled by Organic Photoredox Catalysis. *J. Am. Chem. Soc.* **2020**, *142*, 17187–17194. [CrossRef]
60. Dellinger, B.; Lomnicki, S.; Khachatryan, L.; Maskos, Z.; Hall, R.W.; Adounkpe, J.; McFerrin, C.; Truong, H. Formation and stabilization of persistent free radicals. *Proc. Combust. Inst.* **2007**, *31*, 521–528. [CrossRef]

Article

Efficient Construction of Symmetrical Diaryl Sulfides via a Supported Pd Nanocatalyst-Catalyzed C-S Coupling Reaction

Hao Jin [1], Penghao Liu [1], Qiaoqiao Teng [1], Yuxiang Wang [1], Qi Meng [1,*] and Chao Qian [2,3,*]

[1] Jiangsu Key Laboratory of Advanced Catalytic Materials and Technology, School of Petrochemical Engineering, Changzhou University, Changzhou 213164, China
[2] Zhejiang Provincial Key Laboratory of Advanced Chemical Engineering Manufacture Technology, College of Chemical and Biological Engineering, Zhejiang University, Hangzhou 310027, China
[3] Institute of Zhejiang University—Quzhou, Quzhou 324000, China
* Correspondence: cczumengqi@163.com (Q.M.); qianchao@zju.edu.cn (C.Q.)

Abstract: Aryl sulfides play an important role in pharmaceuticals, biologically active molecules and polymeric materials. Herein, a general and efficient protocol for Pd@COF-TB (a kind of Pd nanocatalyst supported by a covalent organic framework)/DIPEA-catalyzed one-pot synthesis of symmetrical diaryl sulfides through a C-S coupling reaction from aryl iodides and $Na_2S_2O_3$ is developed. More importantly, the addition of N,N-diisopropylethylamine (DIPEA) can not only enhance the catalytic activity of a Pd@COF-TB nanocatalyst, but also effectively inhibit the formation of biphenyl byproducts, which are a product of Ullmann reaction. Besides, it has been confirmed that the aryl Bunte salts generated in situ from $Na_2S_2O_3$ and aryl iodides are the sulfur sources involved in this C-S coupling reaction. With the strategy proposed in this work, a variety of symmetrical diaryl sulfides could be obtained in moderate to excellent yields with a high tolerance of various functional groups. Moreover, a possible mechanism of this Pd nanoparticle-catalyzed C-S coupling reaction is proposed based on the results of controlling experiments.

Keywords: C-S coupling; DIPEA; $Na_2S_2O_3$; nanocatalyst; Pd@COF-TB

1. Introduction

Sulfur-containing organic compounds have always received considerable attention due to their various functions in biology, chemistry and materials science [1–3]. As an important class of organic sulfur-containing structures, aryl sulfides are of great significance to the pharmaceutical industry and are a common functionality found in numerous drugs in therapeutic areas [4–6]. As shown in Figure 1, Seroquel is a potential antipsychotic agent, and its propensity to produce tardive dyskinesia with chronic administration to humans is markedly less than that of typical antipsychotic agents [7]. nTZDpa is an effective antibiotic against bacterial persistence [8]. Viracept is often used as an anti-human immunodeficiency virus drug together with other drugs [9]. Cinanserin exhibits good efficacy against SARS viruses and the treatment of schizophrenia [10]. In addition, sulfides are also widely used in organic synthesis. For example, Zhao and co-workers developed a new type of chiral organothioether bifunctional catalyst based on an indene skeleton, which could be effectively applied in the asymmetric trifluoromethylthioesterification of alkenoic acid [11]. Shi's group reported a Pd(OAc)$_2$-catalyzed method for the alkenylation of C-H bonds of asymmetric olefins. Here, sulfide was used as a guiding group, which could not only achieve excellent yield, enantioselectivity and stereoselectivity, but also synthesize a class of thiene ligands with a chiral axis structure [12]. Therefore, various synthetic strategies have been developed to synthesize sulfides. The most classical construction method of sulfides is the substitution reaction of halide and thiol through C-S coupling in the presence of a strong base (Scheme 1a) [13–17]. However, thiols are highly toxic substances with unpleasant odors. Therefore, the use of green sulfur sources to develop novel, practical and efficient methods for the construction of sulfides is demanding. After years of research, a variety of green sulfur sources, including thiourea [18,19], sodium sulfide [20,21], thioacetamide [22], thiocyanate [23,24], carbon disulfide [25] and elemental sulfur [26,27] have been successfully used in the synthesis of symmetrical diaryl sulfides via transition metal-catalyzed C-S coupling reactions.

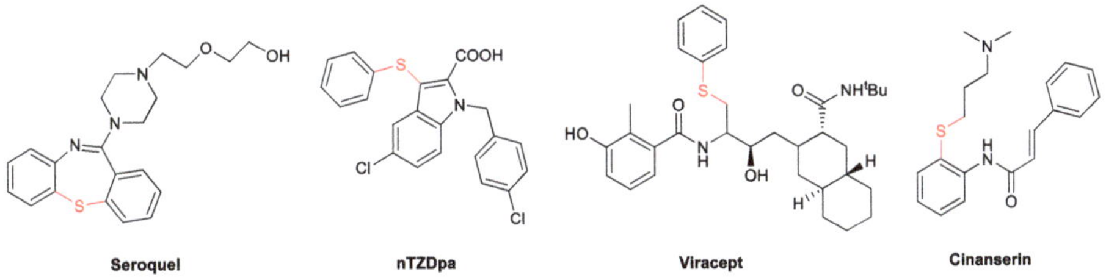

Figure 1. Some pharmaceutical active substances containing an aryl sulfide structure.

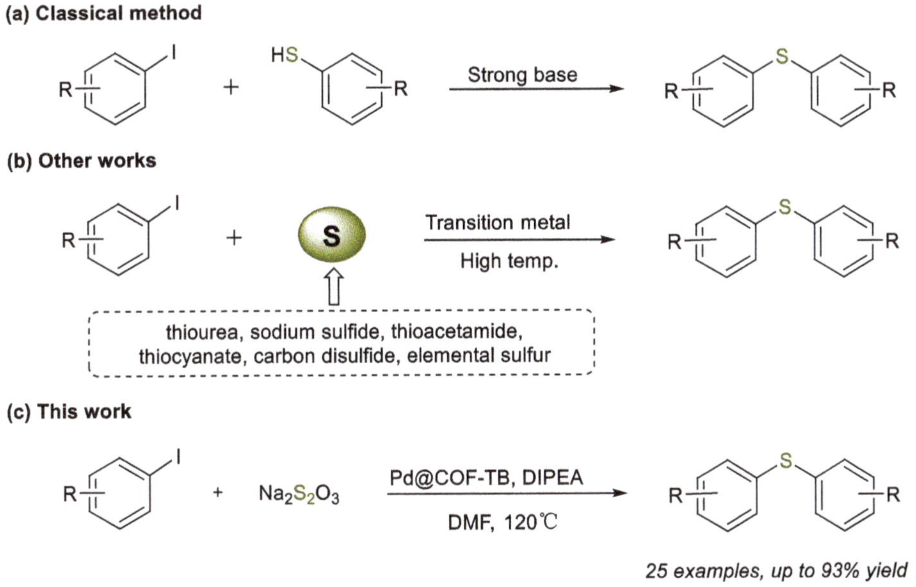

Scheme 1. Approaches for the synthesis of symmetrical diaryl sulfides.

It is well known that sodium thiosulfate ($Na_2S_2O_3$) has been widely used in many C-S coupling reactions due to its low price and odorless and tasteless properties. $Na_2S_2O_3$ mainly transfers the sulfur atom into organic compounds by forming Bunte salts to participate in reactions. Compared with the sulfur atom in thiols, the SO_3Na group attached to the sulfur atom in Bunte salts would change its properties including electronic effects, steric hindrance, and resonance stability [28]. Therefore, Bunte salts are not only used as a substitute for thiols, but also could mediate some reactions that thiols cannot. In recent years, influenced by the rapid development of organosulfur chemistry, Bunte salts, formed in situ by $Na_2S_2O_3$ and halides, are widely used as an efficient sulfur source in various C-S coupling reactions. For example, Yi et al. reported the Pd-catalyzed C-S coupling reaction of aryl halide/aryl triflate and $Na_2S_2O_3$ to synthesize aromatic thiols [29]. In 2013, Jiang's group reported the Pd-catalyzed one-step synthesis of sulfur-containing heterocyclic compounds through the construction of two intramolecular C-S bonds with $Na_2S_2O_3$ as sulfur source [30]. Subsequently, the group extended the system to an intermolecular C-S coupling reaction. Asymmetric sulfides could also be synthesized by the intermolecular C-S coupling reaction of two different halides with $Na_2S_2O_3$ under palladium catalysis [31]. In 2016, Abbasi and co-workers realized the synthesis of symmetric dialkyl disulfides from alkyl halides and $Na_2S_2O_3$ using DMSO as a solvent [32]. In 2019, Li et al. reported a three-component tandem synthesis of isoxazole with $Na_2S_2O_3$ catalyzed by palladium [33]. These examples also suggest that palladium nanoparticles play an important role in catalyzing C-S coupling reactions.

As a continuous work on the application of the Pd@COF-TB nanocatalyst, we herein report a novel, efficient method for the synthesis of symmetrical diaryl sulfides via a Pd@COF-TB/DIPEA-catalyzed C-S coupling reaction from aryl iodides and $Na_2S_2O_3$ (Scheme 1c). A variety of symmetrical diaryl sulfides could be obtained in moderate to excellent yields through this protocol. In this work, DIPEA, apart from providing an alkaline environment, also acts as a palladium ligand, which could not only enhance the catalytic activity of the Pd@COF-TB nanocatalyst, but also effectively inhibit the occurrence of the Ullmann reaction. Furthermore, the reaction mechanisms and the catalytic performance of Pd@COF-TB/DIPEA are discussed.

2. Results and Discussion

2.1. Preparation of Pd@COF-TB Nanocatalyst

The preparation of the Pd@COF-TB nanocatalyst was carried out according to the method established in our previous work [34] and its characterization also has been reported in detail. Therefore, the newly prepared Pd@COF-TB nanocatalyst here was only characterized by Fourier transform infrared spectroscopy (FT-IR) to confirm its correct structure. As shown in Figure 2, the FT-IR absorption spectrum of the newly prepared Pd@COF-TB nanocatalyst was identical to the nanocatalyst prepared in our previous work [34], which means they have the same structure.

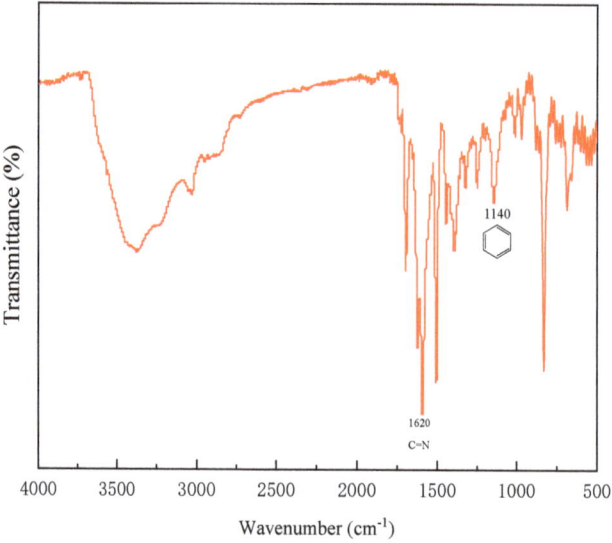

Figure 2. FT-IR spectra of the Pd@COF-TB nanocatalyst.

2.2. Determination of Pd Loading in Pd@COF-TB Nanocatalyst

To further determine the load of Pd in the nanocatalyst, Pd@COF-TB (4 mg) was dissolved in nitric acid (10 mL) and diluted 100 times after complete dissolution. As shown in Table 1, the content of palladium in the solution was obtained by inductively coupled plasma optical emission spectrometry (ICP-OES). The calculation showed that the palladium loading in the nanocatalyst is 4.7% wt.

Table 1. Pd loading determined by ICP-OES.

Entry	Content of Pd Tested by ICP-OES	Load of Pd in the Nanocatalyst (wt%)
1	0.126 ppm	4.72
2	0.123 ppm	4.61
3	0.127 ppm	4.76

2.3. Optimization of Reaction Conditions

It is well known that aryl halides easily undergo Ullmann reactions to form biaryl compounds (**2a**) under transition metal catalysis [35–38]. Therefore, our initial efforts focused on suppressing the occurrence of the Ullmann reaction in the C-S coupling reaction of aryl iodides with $Na_2S_2O_3$. In order to establish standard reaction conditions, a model reaction between iodobenzene and $Na_2S_2O_3$ to produce diphenyl sulfide (**1a**) was chosen and the effects of different reaction parameters like the solvent, base, temperature and the amount of catalyst were studied. The detailed results are summarized in Table 2. The solvent is understood to impact the reaction rate and thereby a number of solvents like DMF, DMSO, NMP, H_2O, EtOH and PEG_{200} were tested in this model reaction. DMF was found to be best for the synthesis of diphenyl sulfide (Table 2, entry 1). Further, we explored the effect of the base on this reaction (Table 2, Entries 1, 7–13). We found that nitrogen-containing organic bases are significantly better than inorganic bases. The most effective of the bases used was observed to be DIPEA (Table 2, entry 13). Besides, the effect of the catalyst loading on this reaction was investigated (Table 2, Entries 13–18). The results showed that no product was detected without Pd@COF-TB (Table 2, entry 14), which means the Pd@COF-TB nanocatalyst could indeed catalyze the C-S coupling reaction of iodobenzene with $Na_2S_2O_3$. Further, 20 mg was found to be the optimized in this reaction condition for the reaction to carry out (Table 2, entry 13). Besides, when $Pd(OAc)_2$ with the same content of Pd replaced Pd@COF-TB, the yield and selectivity of the desired product **1a** both decreased (Table 2, entry 19), which indicated that COF-TB support could optimize the distribution of Pd nanoparticles and improve their catalytic activity. It was noticed that an increasing temperature was beneficial to the desired product **1a**. However, an excessively high reaction temperature would promote the decomposition of more DMF and generate other byproducts, so the optimal reaction temperature is 120 °C. On the basis of the results, the optimized conditions turned out to be: iodobenzene (0.2 g, 1.0 mmol), $Na_2S_2O_3$ (0.32 g, 2.0 mmol), Pd@COF-TB (20 mg) and DIPEA (0.26 g, 2.0 mmol) in DMF (3 mL) at 120 °C.

Table 2. Optimization of conditions for the C-S coupling reaction [1].

Entry	Solvent	Base	Pd@COF-TB (mg)	Temp (°C)	Yield [2]	
					1a	2a
1	DMF	KOH	20	100	36%	50%
2	DMSO	KOH	20	100	21%	67%
3	NMP	KOH	20	100	33%	52%
4	H_2O	KOH	20	100	12%	35%
5	EtOH	KOH	20	100	28%	70%
6	PEG_{200}	KOH	20	100	32%	59%

Table 2. Cont.

Entry	Solvent	Base	Pd@COF-TB (mg)	Temp (°C)	Yield [2]	
					1a	2a
7	DMF	NaOH	20	100	30%	57%
8	DMF	K_2CO_3	20	100	16%	66%
9	DMF	Cs_2CO_3	20	100	24%	58%
10	DMF	NaOMe	20	100	33%	60%
11	DMF	Et_3N	20	100	59%	17%
12	DMF	DBU	20	100	33%	0
13	DMF	DIPEA	20	100	80%	12%
14	DMF	DIPEA	0	100	0	0
15	DMF	DIPEA	10	100	61%	14%
16	DMF	DIPEA	15	100	70%	12%
17	DMF	DIPEA	25	100	80%	14%
18	DMF	DIPEA	30	100	80%	16%
19 [3]	DMF	DIPEA	Pd(OAc)$_2$, 2 mg	100	51%	35%
20	DMF	DIPEA	20	40	43%	18%
21	DMF	DIPEA	20	60	56%	15%
22	DMF	DIPEA	20	80	63%	16%
23	**DMF**	**DIPEA**	**20**	**120**	**94%**	**3%**
24	DMF	DIPEA	20	140	86%	4%

[1] Typical conditions: iodobenzene (1.0 mmol), Na$_2$S$_2$O$_3$ (2.0 mmol) and base (2.0 mmol) in 3 mL of solvent, 10 h, in N$_2$. [2] The content of Pd in 2 mg Pd(OAc)$_2$ is consistent with that in 20 mg Pd@COF-TB. [3] Determined by GC-MS.

2.4. Substrate Expansion under Optimal Reaction Conditions

To assess the substrate scope of this C-S coupling reaction catalyzed by Pd@COF-TB/DIPEA, we screened a range of commercially available aryl iodides with Na$_2$S$_2$O$_3$ under the optimized conditions. The results are listed in Table 3. Unsubstituted iodobenzene exhibited an excellent reaction result, giving a 93% yield of the desired product (**1a**) under the optimized reaction conditions. Substituted iodobenzenes with an electron-rich group, such as methoxy, methyl, *tert*-butyl, hydroxyl and amino, could provide the corresponding coupling products in good to excellent yields (**1b–k**). Due to the influence of the steric hindrance of the substituents, the reaction result of *para*-substituted iodobenzene is the best, followed by *meta*-substitution, and *ortho*-substitution is the worst (**1b–g**). Besides, electron-deficient iodobenzenes bearing fluoro, chloro, bromo, cyano, trifluoromethyl, nitro group could give the desired products in moderate to good yields. It is worth mentioning that the reactions of these substituted iodobenzenes with an electron-poor group would produce more biaryl byproducts. Furthermore, multi-substituted iodobenzene could also react smoothly and give the corresponding products in good yields through this protocol (**1r, 1s**). Additionally, this Pd@COF-TB/DIPEA-catalyzed strategy could also be applied to heteroaromatic iodides, such as naphthalene (**1t**), thiophene (**1u, 1v**) and pyridine (**1w–1t**). These heterocyclic substrates could generate the desired coupling products in good to excellent yields under the optimized conditions without being affected by the position of iodine. These results show the excellent substrate compatibility and functional group tolerance of this protocol.

Table 3. Substrate scope of symmetrical diaryl sulfides [1].

Entry	Product	Yield
1a	Ph-S-Ph	93%
1b	bis(2-methoxyphenyl) sulfide	73%
1c	bis(3-methoxyphenyl) sulfide	81%
1d	bis(4-methoxyphenyl) sulfide	88%
1e	bis(2-methylphenyl) sulfide	76%
1f	bis(3-methylphenyl) sulfide	86%
1g	bis(4-methylphenyl) sulfide	91%
1h	bis(4-tert-butylphenyl) sulfide	85%
1i	bis(4-hydroxyphenyl) sulfide	73%
1j	bis(4-aminophenyl) sulfide	82%
1k	bis(2-aminophenyl) sulfide	71%
1l	bis(2-fluorophenyl) sulfide	70%
1m	bis(4-chlorophenyl) sulfide	78%
1n	bis(4-bromophenyl) sulfide	74%
1o	bis(4-cyanophenyl) sulfide	68%
1p	bis(4-trifluoromethylphenyl) sulfide	71%
1q	bis(4-nitrophenyl) sulfide	63%
1r	bis(3,5-dimethylphenyl) sulfide	80%
1s	bis(5-chloro-2-hydroxyphenyl) sulfide	64%
1t	bis(1-naphthyl) sulfide	75%
1u	bis(2-thienyl) sulfide	70%
1v	bis(3-thienyl) sulfide	76%
1w	bis(2-pyridyl) sulfide	85%
1x	bis(3-pyridyl) sulfide	89%
1y	bis(4-pyridyl) sulfide	92%

Reagents and conditions: aryl iodide (2.0 mmol), $Na_2S_2O_3$ (0.63 g, 4.0 mmol), Pd@COF-TB (40 mg), and DIPEA (0.52 g, 4.0 mmol) in 6 mL DMF, 120 °C, in N_2. [1] Isolated yield.

2.5. Gram-Scale Synthesis Reaction of Iodobenzene with $Na_2S_2O_3$

We investigated the conversion of iodobenzene as a substrate to diphenyl sulfide (**1a**) by the above method in a scale-up reaction (Scheme 2). It was shown that when iodobenzene was 50 mmol, the desired product (**1a**) could be still obtained with a high yield of 91% and good selectivity of 94%.

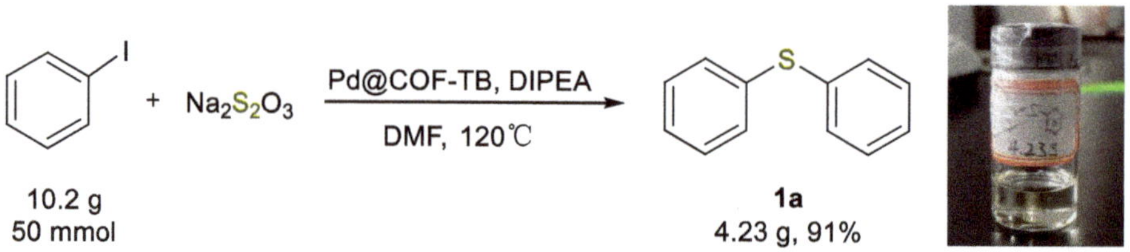

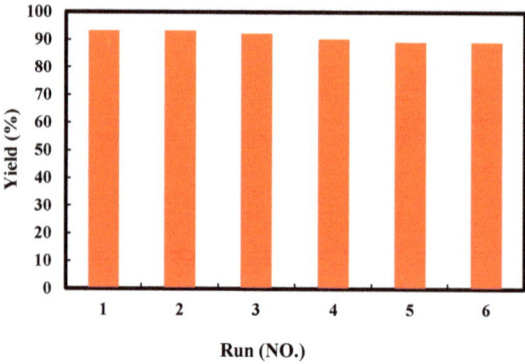

Scheme 2. Gram-scale synthesis reaction of iodobenzene with $Na_2S_2O_3$.

2.6. Catalyst Reuse

In order to test the reusability of the Pd@COF-TB nanocatalyst, cycling reuse tests were performed for the reaction of iodobenzene with $Na_2S_2O_3$ under the optimized reaction conditions in this study. After each run, the nanocatalyst was separated by filtration, washed, dried under vacuum, and then carried out for the next consecutive cycles. As shown in Figure 3, the Pd@COF-TB nanocatalyst could be efficiently recycled and reused for six cycles without a significant decrease in the desired product (**1a**) yield, which means the catalyst has receptable reusability.

Figure 3. Recovery and reuse of the Pd@COF-TB nanocatalyst.

2.7. Mechanism Studies

In order to gain mechanistic insight into this protocol, some controlling experiments were carried out (Scheme 3). As shown in Equation (1), no product was detected in the absence of a base, suggesting that the existence of a base is crucial. After the addition of KOH, the product of the Ullmann reaction (Equation (2), **2a**) was significantly more than the desired C-S coupling product (Equation (2), **1a**). However, the yield and selectivity of diphenyl sulfide were significantly improved when KOH was replaced by DIPEA (Equation (3)). These results show that DIPEA, apart from providing an alkaline environment, is also an excellent ligand for transition metals [39–41], which could enhance the catalytic activity of a Pd@COF-TB nanocatalyst. At the same time, due to the steric hindrance effect, DIPEA could significantly inhibit the occurrence of the Ullmann reaction. Furthermore, we monitored the progress of the C-S coupling reaction of iodobenzene with $Na_2S_2O_3$ (Equation (4)) using LC-MS. The results showed that sodium S-phenyl sulfurothioate (**3a**) was formed in situ from iodobenzene and $Na_2S_2O_3$ in this reaction. The yield of this phenyl Bunte salt showed a trend of increasing first and then decreasing, which means that the Bunte salt was mainly produced in the early stage and then converted into the desired product. In order to verify whether the Bunte salt is the intermediate, sodium S-phenyl sulfurothioate reacted alone under standard conditions. As a result, a trace amount of diphenyl disulfide was found without diphenyl sulfide generated. The yield of diphenyl disulfide was greatly improved when the reaction was exposed to O_2 (Equation (5)). These outcomes suggest that the Bunte salt is a component involved in the C-S coupling reaction for the synthesis of diphenyl sulfide. The reaction must be carried out under anaerobic conditions in order to inhibit the formation of diphenyl disulfide. Finally, treating iodobenzene with sodium S-phenyl sulfurothioate under standard conditions predominately afforded diphenyl sulfide in a 92% yield, along with biphenyl in a 4% yield (Equation (6)). Furtherly, 4-methyldiphenyl sulfide (**4a**) could be obtained through the reaction of 4-iodotoluene and sodium S-phenyl sulfurothioate under standard conditions (Equation (7)). These results indicated that the Bunte salts formed by aryl iodides and $Na_2S_2O_3$ were the real sulfur sources participating in this C-S coupling reaction.

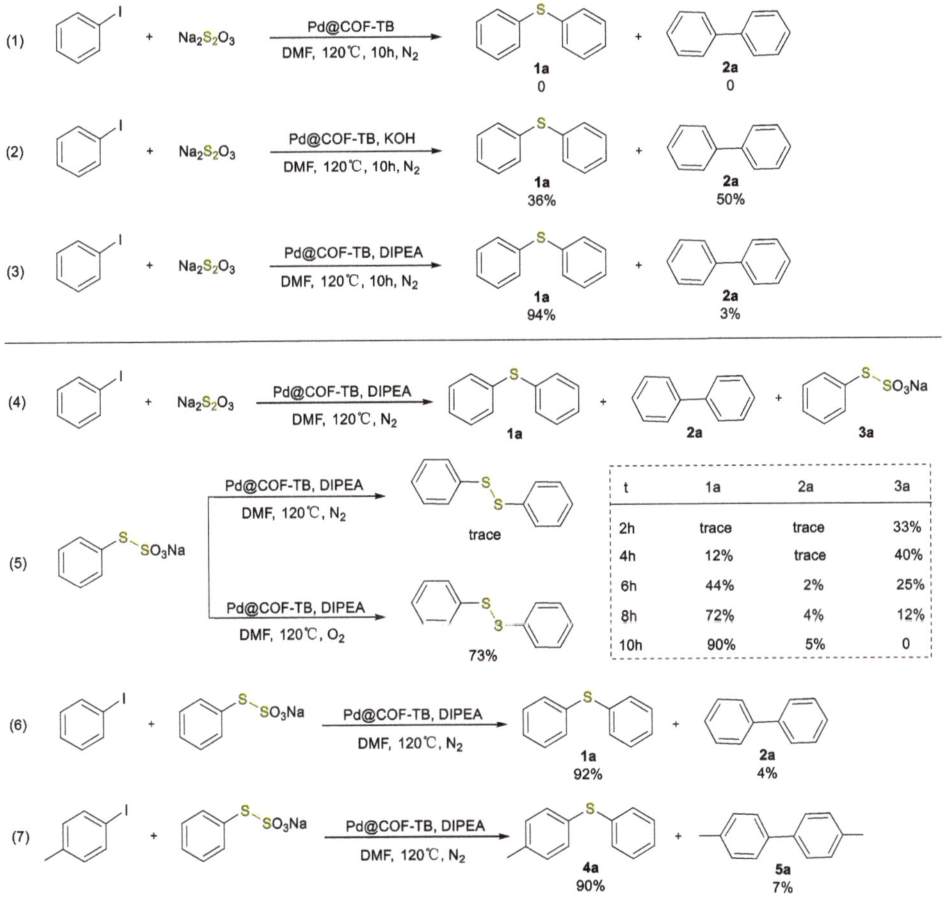

Scheme 3. Controlling experiments. Typical conditions: iodobenzene/4-iodotoluene/sodium S-phenyl sulfurothioate (1.0 mmol, 0.5 mmol in Equations (6) and (7)), $Na_2S_2O_3$ (2.0 mmol), Pd@COF-TB (20 mg), and DIPEA (2 mmol) in 3 mL DMF, 120 °C, in N_2. Yields in Equations (1)–(3) were determined by GC-MS and yields in Equations (4)–(7) were determined by LC-MS.

Based on the above-mentioned experimental results and combined with the previously similar reports [29,31,42,43], a proposed reaction mechanism for the catalytic process of Pd@COF-TB/DIPEA is proposed in Scheme 4. Firstly, a catalytically active Pd$^{(0)}$ species generated in situ from Pd@COF-TB nanoparticles is coordinated with DIPEA to form LPd@COF-TB. Then, organopalladium intermediates (**A**) are generated via the oxidative addition of LPd@COF-TB with iodobenzene. A part of **A** reacts with Na$_2$S$_2$O$_3$ to obtain organopalladium intermediate **B** by eliminating NaI. Through reductive elimination step, sodium S-phenyl sulfurothioate **3a** is obtained from **B**. Subsequently, another part of intermediate **A** reacts with Bunte salt **3a**, producing aryl palladium thiosulfate intermediate **C**, which could be transformed to intermediate **D** via the release of SO$_3$. Finally, through a second reductive elimination, intermediate **D** gave the desired product diphenyl sulfide **1a** and regenerated LPd@COF-TB. It should be noted that the LPd@COF-TB generated in the two reductive elimination processes would continue to undergo oxidative addition reaction with unreacted iodobenzene again. In addition, we speculated that due to the steric hindrance effect of DIPEA, the phenyl group could not directly replace the -I of intermediate **A** in the process of **A** + **C** → **D**, thus inhibiting the occurrence of the Ullmann reaction.

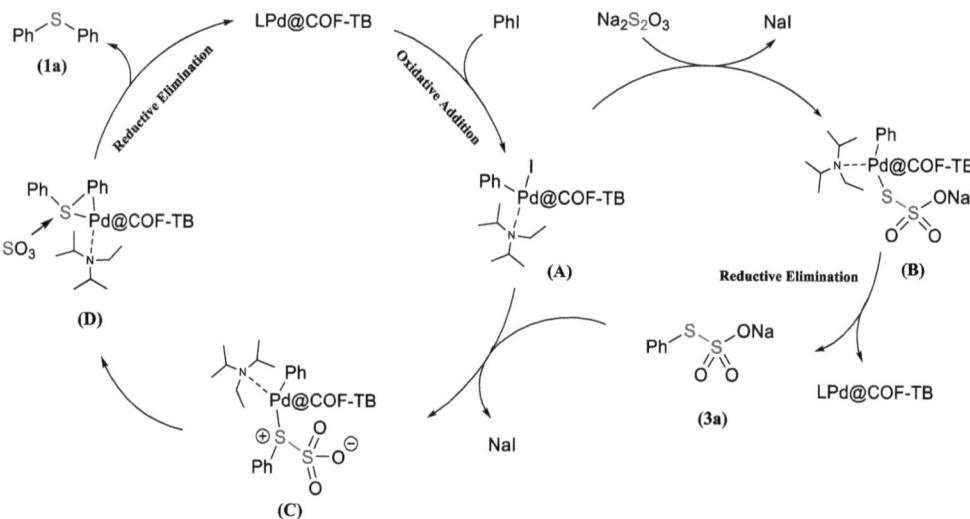

Scheme 4. Proposed mechanism.

3. Materials and Methods

3.1. Materials

All solvents and reagents were purchased at the highest commercial quality grade and were used without further purification, unless otherwise stated. All reactions were carried out under an atmosphere of nitrogen, unless otherwise stated. Purification by column chromatography was performed using E. Merck silica (60, particle size 0.040–0.045 mm). The FT-IR spectra were obtained via Nicolet IS50 FT-IR spectrometers. The Pd loading in the nanocatalyst was analyzed by an Agilent 720ES type inductively coupled plasma optical emission spectroscopy (ICP-OES) instrument. NMR spectra were recorded at room temperature on Bruker AVANCE III spectrometers. GC-MS analysis was recorded on an Agilent 5977B MSD Series spectrometer. HRMS (high-resolution mass spectra) were recorded on a Shimadzu LCMS-IT-TOF mass spectrometer.

3.2. Preparation of the Pd@COF-TB Nanocatalyst

A Schlenk tube was used for the addition of COF-TB (200 mg), Pd(OAc)$_2$ (20 mg) and Diox (60 mL), followed by agitation at 70 °C for 15 h. After that, a yellow–green solid was obtained by high-speed centrifugation. Subsequently, the solid was washed and filtered with acetonitrile, and dried overnight in a fume hood to obtain a yellow–green powder of the Pd@COF-TB nanocatalyst (208.6 mg) [34].

*3.3. Synthesis of Sodium S-phenyl Sulfurothioate (**3a**)*

A Schlenk tube was used for the addition of iodobenzene (2.04 g, 10 mmol), Na$_2$S$_2$O$_3$ (3.15 g, 20 mmol), Pd@COF-TB (0.20 g), DIPEA (2.60 g, 20 mmol) and DMF (15 mL) in a N$_2$ atmosphere. The mixture was then stirred at 120 °C for 4 h. When the reaction was finished, the mixture was cooled to room temperature, quenched with saturated aqueous NaCl (15 mL) and then vigorously stirred at room temperature for another 5 h. Then, the precipitated solid in this system was filtered and washed with saturated aqueous NaCl and n-hexane to give sodium S-phenyl sulfurothioate [44] (0.95 g, 45%) as a white solid. ^{1}H NMR (400 MHz, Methanol-d4) δ 7.81–7.63 (m, 2H), 7.61–7.32 (m, 3H). HRMS (ESI-TOF) m/z calcd. for: C$_6$H$_5$O$_3$S$_2$ [M-Na]$^-$: 188.9680, found: 188.9680.

*3.4. Synthesis of 4-Methyldiphenyl Sulfide (**4a**)*

A Schlenk tube was used for the addition of iodobenzene (0.10 g, 0.5 mmol), sodium S-phenyl Sulfurothioate (0.09 g, 0.5 mmol), Pd@COF-TB (20 mg), DIPEA (0.26 g, 2.0 mmol) and DMF (3 mL) in a N$_2$ atmosphere. The mixture was then stirred at 120 °C for 10 h. When the reaction was finished, the mixture was cooled to room temperature, quenched by H$_2$O (3 mL) and extracted with ethyl acetate (3 mL × 3). Then the combined extract was washed with saturated aqueous NaCl (3 mL × 3), dried over anhydrous sodium sulfate and concentrated under vacuum. Purification by column chromatography on silica gel afforded 4-methyldiphenyl sulfide [45] (0.09 g, 89%) as a colorless liquid. ^{1}H NMR (400 MHz, CDCl$_3$) δ 7.29 (d, J = 8.2 Hz, 2H), 7.24 (q, J = 7.8 Hz, 4H), 7.18–7.13 (m, 1H), 7.11 (d, J = 8.2 Hz, 2H), 2.32 (s, 3H). ^{13}C NMR (101 MHz, CDCl$_3$) δ 137.73, 137.31, 132.56, 131.47, 130.34, 130.09, 129.20, 126.39. 21.34. GC-MS (EI) m/z calcd. for: C$_{13}$H$_{12}$S [M]$^+$: 200.07, found: 200.14.

3.5. General Procedure for the Synthesis of Symmetrical Diaryl Sulfide

A Schlenk tube was used for the addition of aryl iodide (2.0 mmol), Na$_2$S$_2$O$_3$ (0.63 g, 4.0 mmol), Pd@COF-TB (40 mg), DIPEA (0.52 g, 4.0 mmol) and DMF (6.0 mL) in N$_2$ atmosphere. The mixture was then stirred at 120 °C and monitored by TLC and HPLC. When the reaction was finished, the mixture was cooled to room temperature, quenched by H$_2$O (6 mL) and extracted with ethyl acetate (5 mL × 3). Then, the combined extract was washed with saturated aqueous NaCl (5 mL × 3), dried over anhydrous sodium sulfate and concentrated under vacuum. Purification by column chromatography on silica gel afforded the desired products and their detailed characterization data are reported in the Supporting Information.

4. Conclusions

In summary, we have developed a general and efficient Pd@COF-TB/DIPEA-catalyzed one-pot synthesis of symmetrical diaryl sulfides through a C-S coupling reaction with aryl iodides as the starting materials and Na$_2$S$_2$O$_3$ as the sulfur source. An array of symmetrical diaryl sulfides have been synthesized in moderate to excellent yields through this protocol. As a good ligand, DIPEA, apart from providing an alkaline environment, could not only enhance the catalytic activity of Pd@COF-TB nanocatalyst, but also effectively inhibit the formation of biphenyl byproducts. Additionally, it has been confirmed that the aryl Bunte salts generated in situ from Na$_2$S$_2$O$_3$ and aryl iodides are the real sulfur sources involved in this C-S coupling reaction. Lastly, a proposed mechanism of this Pd@COF-TB/DIPEA-catalyzed C-S coupling reaction was revealed in detail through many controlling experiments. This work has expanded the application of Pd@COF-TB nanocatalyst in a C-S coupling reaction, and further studies on other applications of Pd@COF-TB nanocatalyst are ongoing in our laboratory.

Supplementary Materials: The following supporting information can be downloaded at: https://www.mdpi.com/article/10.3390/ijms232315360/s1. References are cited in [25,45–52].

Author Contributions: Investigation, methodology and writing—original draft preparation, H.J.; investigation, P.L.; review and editing, Q.T.; data curation, Y.W.; conceptualization and supervision, Q.M. and C.Q. All authors have read and agreed to the published version of the manuscript.

Funding: This research received no external funding.

Institutional Review Board Statement: Not applicable.

Informed Consent Statement: Not applicable.

Data Availability Statement: Additional figures are available in the Supplementary Materials.

Acknowledgments: The authors are grateful for the support from the Analysis and Testing Center of Changzhou University.

Conflicts of Interest: The authors declare no conflict of interest.

References

1. Poon, S.Y.; Wong, W.Y.; Cheah, K.W.; Shi, J.X. Spatial extent of the singlet and triplet excitons in luminescent angular-shaped transition-metal diynes and polyynes comprising non-π-conjugated group 16 main group elements. *Chem. Eur. J.* **2006**, *12*, 2550–2563. [CrossRef] [PubMed]
2. Correa, A.; Mancheño, O.G.; Bolm, C. Iron-catalysed carbon–heteroatom and heteroatom–heteroatom bond forming processes. *Chem. Soc. Rev.* **2008**, *37*, 1108–1117. [CrossRef] [PubMed]
3. Liu, H.; Jiang, X. Transfer of sulfur: From simple to diverse. *Chem. Asian J.* **2013**, *8*, 2546–2563. [CrossRef] [PubMed]
4. Liu, G.; Link, J.T.; Pei, Z.; Reilly, E.B.; Leitza, S.; Nguyen, B.; Marsh, K.C.; Okasinski, G.F.; von Geldern, T.W.; Ormes, M. Discovery of novel p-arylthio cinnamides as antagonists of leukocyte function-associated antigen-1/intracellular adhesion molecule-1 interaction. 1. Identification of an additional binding pocket based on an anilino diaryl sulfide lead. *J. Med. Chem.* **2000**, *43*, 4025–4040. [CrossRef] [PubMed]
5. Amorati, R.; Fumo, M.G.; Menichetti, S.; Mugnaini, V.; Pedulli, G.F. Electronic and hydrogen bonding effects on the chain-breaking activity of sulfur-containing phenolic antioxidants. *J. Org. Chem.* **2006**, *71*, 6325–6332. [CrossRef]
6. Denes, F.; Pichowicz, M.; Povie, G.; Renaud, P. Thiyl radicals in organic synthesis. *Chem. Rev.* **2014**, *114*, 2587–2693. [CrossRef]
7. Migler, B.M.; Warawa, E.J.; Malick, J.B. Seroquel: Behavioral effects in conventional and novel tests for atypical antipsychotic drug. *Psychopharmacology* **1993**, *112*, 299–307. [CrossRef]
8. Kim, W.; Steele, A.D.; Zhu, W.; Csatary, E.E.; Fricke, N.; Dekarske, M.M.; Jayamani, E.; Pan, W.; Kwon, B.; Sinitsa, I.F. Discovery and optimization of nTZDpa as an antibiotic effective against bacterial persisters. *ACS Infect. Dis.* **2018**, *4*, 1540–1545. [CrossRef] [PubMed]
9. Kaldor, S.W.; Kalish, V.J.; Davies, J.F.; Shetty, B.V.; Fritz, J.E.; Appelt, K.; Burgess, J.A.; Campanale, K.M.; Chirgadze, N.Y.; Clawson, D.K. Viracept (nelfinavir mesylate, AG1343): A potent, orally bioavailable inhibitor of HIV-1 protease. *J. Med. Chem.* **1997**, *40*, 3979–3985. [CrossRef]
10. Holden, J.M.; Itil, T.; Keskiner, A.; Gannon, P. A clinical trial of an antiserotonin compound, cinanserin, in chronic schizophrenia. *J. Clin. Pharmacol.* **1971**, *11*, 220–226. [CrossRef]
11. Liu, X.; An, R.; Zhang, X.; Luo, J.; Zhao, X. Enantioselective trifluoromethylthiolating lactonization catalyzed by an indane-based chiral sulfide. *Angew. Chem. Int. Ed.* **2016**, *55*, 5846–5850. [CrossRef] [PubMed]
12. Jin, L.; Zhang, P.; Li, Y.; Yu, X.; Shi, B.-F. Atroposelective synthesis of conjugated diene-based axially chiral styrenes via Pd (II)-catalyzed thioether-directed alkenyl C–H olefination. *J. Am. Chem. Soc.* **2021**, *143*, 12335–12344. [CrossRef]
13. Ranu, B.C.; Jana, R. Ionic liquid as catalyst and reaction medium: A simple, convenient and green procedure for the synthesis of thioethers, thioesters and dithianes using an inexpensive ionic liquid, [pmIm]Br. *Adv. Synth. Catal.* **2005**, *347*, 1811–1818. [CrossRef]
14. Baig, R.B.N.; Varma, R.S. A highly active and magnetically retrievable nanoferrite–DOPA–copper catalyst for the coupling of thiophenols with aryl halides. *Chem. Commun.* **2012**, *48*, 2582–2584. [CrossRef] [PubMed]
15. Fernández-Rodríguez, M.A.; Shen, Q.; Hartwig, J.F. A general and long-lived catalyst for the palladium-catalyzed coupling of aryl halides with thiols. *J. Am. Chem. Soc.* **2006**, *128*, 2180–2181. [CrossRef]
16. Zong, C.; Liu, J.; Chen, S.; Zeng, R.; Zou, J. Efficient C-S cross-coupling of thiols with aryl iodides catalyzed by $Cu(OAc)_2 \cdot H_2O$ and 2,2′-Biimidazole. *Chin. J. Chem.* **2014**, *32*, 212–218. [CrossRef]

17. Kwong, F.Y.; Buchwald, S.L. A general, efficient, and inexpensive catalyst system for the coupling of aryl iodides and thiols. *Org. Lett.* **2002**, *4*, 3517–3520. [CrossRef]
18. Hajipour, A.R.; Karimzadeh, M.; Azizi, G. Highly efficient and magnetically separable nano-$CuFe_2O_4$ catalyzed S-arylation of thiourea by aryl/heteroaryl halides. *Chin. Chem. Lett.* **2014**, *25*, 1382–1386. [CrossRef]
19. Ashraf, M.A.; Liu, Z.; Peng, W.-X. Trisaminomethane-cobalt complex supported on Fe_3O_4 magnetic nanoparticles as an efficient recoverable nanocatalyst for oxidation of sulfides and C–S coupling reactions. *Appl. Organomet. Chem.* **2020**, *34*, e5260. [CrossRef]
20. Chen, J.; Zhang, Y.; Liu, L.; Yuan, T.; Yi, F. Efficient copper-catalyzed double S-arylation of aryl halides with sodium sulfide in PEG-400. *Phosphorus Sulfur* **2012**, *187*, 1284–1290. [CrossRef]
21. Li, Y.; Nie, C.; Wang, H.; Li, X.; Verpoort, F.; Duan, C. A highly efficient method for the copper-catalyzed selective synthesis of diaryl chalcogenides from easily available chalcogen sources. *Eur. J. Org. Chem.* **2011**, *2011*, 7331–7338. [CrossRef]
22. Ghorbani-Choghamarani, A.; Seydyosefi, Z.; Tahmasbi, B. Zirconium oxide complex anchored on boehmite nanoparticles as highly reusable organometallic catalyst for C–S and C–O coupling reactions. *Appl. Organomet. Chem.* **2018**, *32*, e4396. [CrossRef]
23. Ke, F.; Qu, Y.; Jiang, Z.; Li, Z.; Wu, D.; Zhou, X. An efficient copper-catalyzed carbon−sulfur bond formation protocol in water. *Org. Lett.* **2011**, *13*, 454–457. [CrossRef] [PubMed]
24. Cai, M.; Xiao, R.; Yan, T.; Zhao, H. A simple and green synthesis of diaryl sulfides catalyzed by an MCM-41-immobilized copper (I) complex in neat water. *J. Organomet. Chem.* **2014**, *749*, 55–60. [CrossRef]
25. Zhao, P.; Yin, H.; Gao, H.; Xi, C. Cu-catalyzed synthesis of diaryl thioethers and S-cycles by reaction of aryl iodides with carbon disulfide in the presence of DBU. *J. Org. Chem.* **2013**, *78*, 5001–5006. [CrossRef] [PubMed]
26. Azadi, G.; Taherinia, Z.; Naghipour, A.; Ghorbani-Choghamarani, A. Synthesis of sulfides via reaction of aryl/alkyl halides with S_8 as a sulfur-transfer reagent catalyzed by Fe_3O_4-magnetic-nanoparticles-supported L-Histidine-Ni (II). *J. Sulfur Chem.* **2017**, *38*, 303–313. [CrossRef]
27. Ghorbani-Choghamarani, A.; Taherinia, Z. The first report on the preparation of peptide nanofibers decorated with zirconium oxide nanoparticles applied as versatile catalyst for the amination of aryl halides and synthesis of biaryl and symmetrical sulfides. *New J. Chem.* **2017**, *41*, 9414–9423. [CrossRef]
28. Liu, F.; Jiang, L.; Qiu, H.; Yi, W. Bunte salt CH_2FSSO_3Na: An efficient and odorless reagent for monofluoromethylthiolation. *Org. Lett.* **2018**, *20*, 6270–6273. [CrossRef]
29. Yi, J.; Fu, Y.; Xiao, B.; Cui, W.-C.; Guo, Q.-X. Palladium catalyzed synthesis of aryl thiols: Sodium thiosulfate as a cheap and nontoxic mercapto surrogate. *Tetrahedron Lett.* **2011**, *52*, 205–208. [CrossRef]
30. Qiao, Z.; Liu, H.; Xiao, X.; Fu, Y.; Wei, J.; Li, Y.; Jiang, X. Efficient access to 1,4-benzothiazine: Palladium-catalyzed double C–S bond formation using $Na_2S_2O_3$ as sulfurating reagent. *Org. Lett.* **2013**, *15*, 2594–2597. [CrossRef]
31. Qiao, Z.; Wei, J.; Jiang, X. Direct cross-coupling access to diverse aromatic sulfide: Palladium-catalyzed double C–S bond construction using $Na_2S_2O_3$ as a sulfurating reagent. *Org. Lett.* **2014**, *16*, 1212–1215. [CrossRef] [PubMed]
32. Abbasi, M.; Mohammadizadeh, M.R.; Saeedi, N. The synthesis of symmetrical disulfides by reacting organic halides with $Na_2S_2O_3 \cdot 5H_2O$ in DMSO. *New J. Chem.* **2016**, *40*, 89–92. [CrossRef]
33. Li, J.; Wu, Y.; Hu, M.; Li, C.; Li, M.; He, D.; Jiang, H. A palladium-catalyzed three-component cascade S-transfer reaction in ionic liquids. *Green Chem.* **2019**, *21*, 4084–4089. [CrossRef]
34. Jin, H.; Zhang, C.; Liu, P.; Ge, X.; Zhou, S. Covalent organic framework-supported Pd nanoparticles: An efficient and reusable heterogeneous catalyst for Suzuki–Miyaura coupling reactions. *Appl. Organomet. Chem.* **2022**, *36*, e6642. [CrossRef]
35. Wang, Z.-J.; Wang, X.; Lv, J.-J.; Feng, J.-J.; Xu, X.; Wang, A.-J.; Liang, Z. Bimetallic Au-Pd nanochain networks: Facile synthesis and promising application in biaryl synthesis. *New J. Chem.* **2017**, *41*, 3894–3899. [CrossRef]
36. Gong, X.; Wu, J.; Meng, Y.; Zhang, Y.; Ye, L.-W.; Zhu, C. Ligand-free palladium catalyzed Ullmann biaryl synthesis: "household" reagents and mild reaction conditions. *Green Chem.* **2019**, *21*, 995–999. [CrossRef]
37. Vasconcelos, S.N.S.; Reis, J.S.; de Oliveira, I.M.; Balfour, M.N.; Stefani, H.A. Synthesis of symmetrical biaryl compounds by homocoupling reaction. *Tetrahedron* **2019**, *75*, 1865–1959. [CrossRef]
38. Jiang, H.; Xu, J.; Zhang, S.; Cheng, H.; Zang, C.; Bian, F. Efficient photocatalytic chemoselective and stereoselective C-C bond formation over AuPd@N-rich carbon nitride. *Catal. Sci. Technol.* **2021**, *11*, 219–229. [CrossRef]
39. Inkster, J.A.H.; Guerin, B.; Ruth, T.J.; Adam, M.J. Radiosynthesis and bioconjugation of [^{18}F]FPy5yne, a prosthetic group for the ^{18}F labeling of bioactive peptides. *J. Labelled Compd. Rad.* **2008**, *51*, 444–452. [CrossRef]
40. Huo, J.-P.; Xiong, J.-F.; Mo, G.-Z.; Peng, P.; Wang, Z.-Y. Synthesis of chiral 2(5H)-furanone derivatives with 1,3-butadiyne structure. *Res. Chem. Intermed.* **2013**, *39*, 4321–4335. [CrossRef]
41. Jia, L.; Li, Q.; Bayaguud, A.; Huang, Y.; She, S.; Chen, K.; Wei, Y. Diversified polyoxovanadate derivatives obtained by copper(I)-catalysed azide-alkyne cycloaddition reaction: Their synthesis and structural characterization. *Dalton Trans.* **2018**, *47*, 577–584. [CrossRef] [PubMed]
42. Li, Y.; Pu, J.; Jiang, X. A highly efficient Cu-Catalyzed S-transfer reaction: From amine to sulfide. *Org. Lett.* **2014**, *16*, 2692–2695. [CrossRef] [PubMed]
43. Ma, X.; Yu, J.; Yan, R.; Yan, M.; Xu, Q. Promoting effect of crystal water leading to catalyst-free synthesis of heteroaryl thioether from heteroaryl chloride, sodium thiosulfate pentahydrate, and alcohol. *J. Org. Chem.* **2019**, *84*, 11294–11300. [CrossRef] [PubMed]
44. Tan, W.; Jänsch, N.; Öhlmann, T.; Meyer-Almes, F.-J.; Jiang, X. Thiocarbonyl surrogate via combination of potassium sulfide and chloroform for dithiocarbamate construction. *Org. Lett.* **2019**, *21*, 7484–7488. [CrossRef] [PubMed]

45. Xu, J.; Lu, F.; Sun, L.; Huang, M.; Jiang, J.; Wang, K.; Ouyang, D.; Lu, L.; Lei, A. Electrochemical reductive cross-coupling of acyl chlorides and sulfinic acids towards the synthesis of thioesters. *Green Chem.* **2022**, *24*, 7350–7354. [CrossRef]
46. Ghodsinia, S.S.E.; Akhlaghinia, B. Cu I anchored onto mesoporous SBA-16 functionalized by aminated 3-glycidyloxypropyltrimethoxysilane with thiosemicarbazide (SBA-16/GPTMS-TSC-Cu I): A heterogeneous mesostructured catalyst for S-arylation reaction under solvent-free conditions. *Green Chem.* **2019**, *21*, 3029–3049.
47. Li, X.; Du, J.; Zhang, Y.; Chang, H.; Gao, W.; Wei, W. Synthesis and nano-Pd catalyzed chemoselective oxidation of symmetrical and unsymmetrical sulfides. *Org. Biomol. Chem.* **2019**, *17*, 3048–3055.
48. Kollár, L.; Rao, Y.V.R.; Zugó, A.; Pongrácz, P. Palladium-catalysed thioetherification of aryl and alkenyl iodides using 1, 3, 5-trithiane as sulfur source. *Tetrahedron* **2022**, *104*, 132602.
49. Liu, Y.; Kim, J.; Seo, H.; Park, S.; Chae, J. Copper (II)-Catalyzed Single-Step Synthesis of Aryl Thiols from Aryl Halides and 1, 2-Ethanedithiol. *Adv. Synth. Catal.* **2015**, *357*, 2205–2212.
50. Zhang, Y.; Liu, L.; Chen, J. Efficient synthesis of diaryl sulfides by copper-catalysed coupling of aryl halides and thioacetate in water. *J. Chem. Res.* **2013**, *37*, 19–21.
51. Csokai, V.; Gruen, A.; Balázs, B.; Tóth, G.; Horváth, G.; Bitter, I. Unprecedented cyclizations of calix [4] arenes with glycols under the mitsunobu protocol, part 2.1 O, O-and O, S-bridged calixarenes. *Org. Lett.* **2004**, *6*, 477–480.
52. May, L.; Müller, T.J.J. Electron-Rich Phenothiazine Congeners and Beyond: Synthesis and Electronic Properties of Isomeric Dithieno [1, 4] thiazines. *Chem. Eur. J.* **2020**, *26*, 12111–12118.

Article

Palladium Decorated, Amine Functionalized Ni-, Cd- and Co-Ferrite Nanospheres as Novel and Effective Catalysts for 2,4-Dinitrotoluene Hydrogenation

Viktória Hajdu [1], Emőke Sikora [1], Ferenc Kristály [2], Gábor Muránszky [1], Béla Fiser [1], Béla Viskolcz [1], Miklós Nagy [1,*] and László Vanyorek [1,*]

1. Institute of Chemistry, University of Miskolc, Miskolc-Egyetemváros, 3515 Miskolc, Hungary
2. Institute of Mineralogy and Geology, University of Miskolc, Miskolc-Egyetemváros, 3515 Miskolc, Hungary
* Correspondence: nagy.miklos@uni-miskolc.hu (M.N.); kemvanyi@uni-miskolc.hu (L.V.)

Abstract: 2,4-diaminotoluene (TDA) is one of the most important polyurethane precursors produced in large quantities by the hydrogenation of 2,4-dinitrotoluene using catalysts. Any improvement during the catalysis reaction is therefore of significant importance. Separation of the catalysts by filtration is cumbersome and causes catalyst loss. To solve this problem, we have developed magnetizable, amine functionalized ferrite supported palladium catalysts. Cobalt ferrite ($CoFe_2O_4$-NH_2), nickel ferrite ($NiFe_2O_4$-NH_2), and cadmium ferrite ($CdFe_2O_4$-NH_2) magnetic catalyst supports were produced by a simple coprecipitation/sonochemical method. The nanospheres formed contain only magnetic (spinel) phases and show catalytic activity even without noble metals (palladium, platinum, rhodium, etc.) during the hydrogenation of 2,4-dinitrotoluene, 63% (n/n) conversion is also possible. By decorating the supports with palladium, almost 100% TDA selectivity and yield were ensured by using Pd/$CoFe_2O_4$-NH_2 and Pd/$NiFe_2O_4$-NH_2 catalysts. These catalysts possess highly favorable properties for industrial applications, such as easy separation from the reaction medium without loss by means of a magnetic field, enhanced reusability, and good dispersibility in aqueous medium. Contrary to non-functionalized supports, no significant leaching of precious metals could be detected even after four cycles.

Keywords: ferrite; magnetic catalyst; hydrogenation; TDA; nanostructure

1. Introduction

The catalytic hydrogenation of 2,4-dinitrotoluene (2,4-DNT) is one of the most widely used industrial processes for the synthesis of 2,4-toluenediamine (2,4-TDA). TDA is an important intermediate in the formation of toluene diisocyanate (TDI, 2.49 Mtons produced in 2021) used in the production of polyurethanes, a compound used primarily in flexible polyurethane foams, elastomers, coatings, and adhesives. Transition metals (Pd, Pt, Ni, etc.) or transition metal oxides on carbon support (carbon nanotubes, graphene, carbon black) are most commonly used as catalysts in the catalytic hydrogenation of DNT [1–7]. Their advantage is that they adsorb the various organic substances well and bind the catalytically active metal particles (platinum, palladium). Furthermore, they are well dispersed in a liquid medium, which makes it easy for the reactant molecules to have access to the catalytically active metals on the surface of the support particles. Specific surface area is a key factor in heterogeneous catalysis, which increases dramatically with decreasing particle size according to the square-cubic law. However, catalysts of small particle size form a stable dispersion in the reaction medium, therefore their economical recovery cannot be achieved without loss by conventional filtration and centrifugation operations. This issue can be almost completely eliminated by using magnetizable catalysts, efficient catalyst recovery can be achieved by a simple magnetic separation operation [8]. Possible supports with magnetic properties are: ferrites, chromium dioxide, magnetite, maghemite. They are also used in many separation operations, such as DNA purification [9,10], heavy metal ions binding [11,12], removal of organic contaminants from water [13,14], and even in the field of catalysis [15–18]. Weng and co-workers [19] converted nickel-laden electroplating slurry into $NiFe_2O_4$ nanomaterial using sodium carbonate by a hydrothermal washing strategy. The prepared nanoparticles showed stable electrochemical Li storage performance. This new strategy can provide a sustainable approach for the conversion of heavy metals in industrial waste into high-value functional materials and for the selective recycling of heavy metals. Ebrahimi and co-workers [20] prepared superparamagnetic $CoFe_2O_4$NPs@Mn-Organic Framework core-shell nanocomposites by a layer-by-layer method. The structures exhibit high temperature stability and good magnetization. This magnetic nanometal-organic framework is an excellent candidate in targeted drug-delivery systems.

It is important to ensure that the metal oxide with magnetic properties as a catalyst support is well dispersed in the reaction medium. Since magnetic nanoparticles have a strong agglomeration tendency, in order to solve this problem, it is necessary to modify the surface of the magnetic nanoparticles with different functional groups (NH_2, OH, SiH, SH groups) [21–26]. In some cases, without surface modification, the products of hydrogenolysis can deactivate the supported precious metal catalysts due to their poisoning effects, therefore a higher amount of catalyst is necessary to complete the reaction [27,28]. Since heterogeneous catalysis involves the adsorption of reactants (intermediates and products as well), surface functionality may be crucial for reaching good conversion and selectivity. Amino-groups on the surface may promote the binding of nitro-compounds (reactant), via H-bonding interactions. In addition, the functional groups can substantially enhance the binding of catalytically active noble metals on the surface during the preparation of the catalyst [29]. Sharma and co-workers have successfully prepared a supported ruthenium nanoparticle on amino-functionalized Fe_3O_4 ($Fe_3O_4/NH_2/Ru$) that exhibits excellent catalytic activity for the transfer hydrogenation of nitro compounds using $NaBH_4$ as hydrogen donor. In addition, the catalyst can be easily recovered by an external magnetic field, recycled five times and reused without loss of activity [30].

In the preparation of conventional (non-magnetic) oxide- (SiO_2, Al_2O_3 etc.) or C-based catalysts, after the catalytically active metal salt or its oxide has been deposited on the support surface, heat treatment in an inert atmosphere (nitrogen, argon) is required, followed by an activation step (most often reduction with hydrogen gas) to form an elemental catalytically active form of the precious metal. As this last step takes place at high temperatures (300–400 °C), this process is not feasible in the case of metal oxides with magnetic properties (magnetite, maghemite, chromium dioxide, ferrites) because they are not thermally stable at such high temperatures. In this case, an alternative catalyst preparation method should be used. Precious metal deposition by sonochemistry on the surface of ferrite particles may be an ideal solution. Thus, ultrasonic cavitation in the alcoholic solution of noble metal precursors reduces the noble metal nanoparticles on the surface of the ferrite particles [31]. Intense ultrasonic irradiation yields vapor bubbles, or rather cavities in the liquid as the boiling point decreases. These vapor bubbles collapse in the fluid in the high pressure ranges, releasing large amounts of energy in these microvolumes, which can cover the energy requirements of chemical reactions [32]. This way, catalytically active magnetic nanoparticles containing noble metals can be efficiently produced.

The aim of this work Is the development of stable and selective catalysts of high activity, which are easily separable from the reaction-medium, without loss. Hereby, we report the preparation and possible application of amine-functionalized magnetic, Co-, Ni-, or Cd-ferrite based catalysts decorated with Pd. The composition, morphology, and surface of the nanoparticles have been examined in detail. In addition, their applicability in the industrially important hydrogenation of 2,4-dinitrotoluene has been studied.

2. Results and Discussion

The amine-functionalized cobalt-, cadmium-, and nickel ferrite nanoparticles were investigated by HRTEM (Figure 1a–c). The HRTEM pictures clearly show the spherical shape of the ferrite nanoparticles, which are composed of smaller, individual nanoparticles of size 4–10 nm (Figures S3–S5 in the Supporting Information). The average crystallite size of the ferrite nanoparticles (the building blocks of the spherical aggregates) was also calculated based on the XRD patterns by the mean column length calibrated method using the full width at half maximum (FWHM) and the width of the Lorentzian component of the fitted profiles. The sizes of the individual nanoparticles, which can be measured in the HRTEM pictures are coherent with the XRD measurements. The average size of the nanoparticles, which build up the NiFe$_2$O$_4$-NH$_2$ is 6 ± 2 nm. In the case of the CoFe$_2$O$_4$-NH$_2$ and CdFe$_2$O$_4$-NH$_2$ nanospheres (building blocks), the average sizes are 4 ± 2 nm and 8 ± 3 nm, respectively.

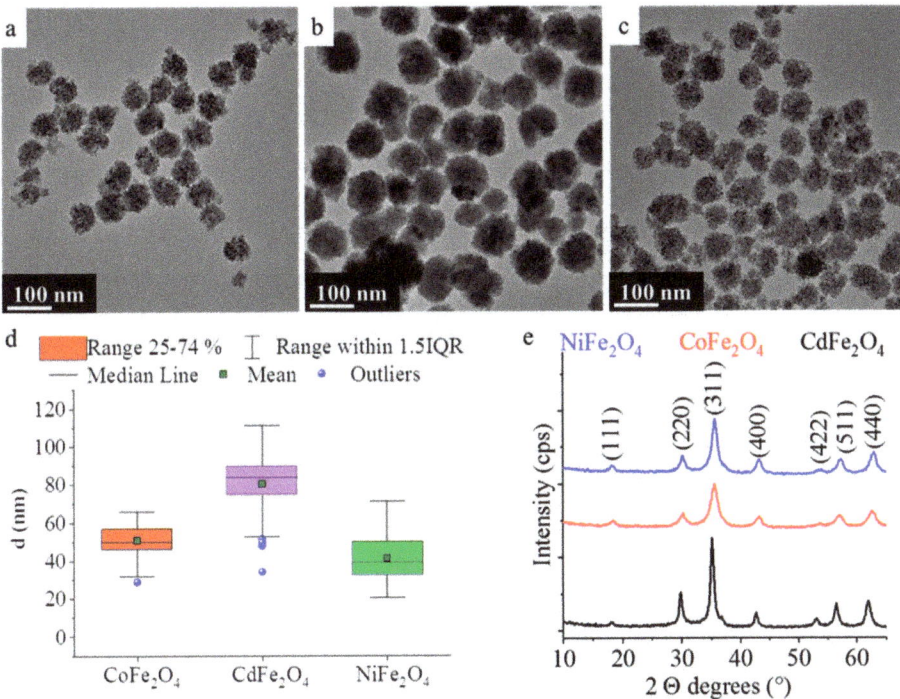

Figure 1. HRTEM pictures of the CoFe$_2$O$_4$-NH$_2$ (**a**), CdFe$_2$O$_4$-NH$_2$ (**b**) and NiFe$_2$O$_4$-NH$_2$ (**c**). Box plot diagrams for the particle size analysis (**d**) and the XRD pattern of the ferrites (**e**).

In contrast, the average particle sizes of the aggregate ferrite spheres, namely CoFe$_2$O$_4$-NH$_2$, NiFe$_2$O$_4$-NH$_2$, and CdFe$_2$O$_4$-NH$_2$ are 51 ± 8 nm, 42 ± 13 nm, and 80 ± 14 nm (Figure 1d and Table 1). The particle size distribution is broad in all three cases, based on the interquartile range width in the box plot diagram in Figure 1d. The cadmium-ferrite sample contains the biggest aggregates, however, the mean and median particle sizes are very close to each other, these are 80 ± 14 nm and 84 nm (Table 1).

Table 1. Results of the size analysis (in nm) of the ferrite nanospheres (based on HRTEM pictures).

(nm)	Mean	SD	Min.	Max.	1st Quartile	3rd Quartile	Median	P90	P95
CoFe$_2$O$_4$-NH$_2$	50.9	8.2	28.8	66.0	46.3	56.9	50.0	61.2	62.6
CdFe$_2$O$_4$-NH$_2$	80.2	14.2	34.0	111.4	74.5	89.5	83.7	92.1	100.7
NiFe$_2$O$_4$-NH$_2$	41.5	12.9	20.5	71.2	32.4	50.0	39.3	61.6	66.4

In heterogeneous catalysis the specific surface area of the nanoparticles (supports) is a key feature, since it affects the sorption of the different chemical species during the reaction. The specific surface areas of the ferrites were determined by carbon dioxide adsorption measurements using the Dubinin–Astakhov model. The highest surface area was measured in case of the $CoFe_2O_4$-NH_2 (279 m^2/g), much higher than those of the other two samples: 72 m^2/g ($CdFe_2O_4$-NH_2) and 93 m^2/g ($NiFe_2O_4$-NH_2) The specific surface areas are below that of the activated carbon supports (up to 1000 m^2/g).

XRD measurements revealed the spinel structure of the prepared magnetic particles (Figure 1e). The following reflexions can be identified on the diffractograms: (111), (220), (311), (400), (422), (511), and (440) at 18.1°, 30.1°, 35.5°, 43.2°, 53.8°, 57.2°, and 62.8° two Theta degrees which belong to $NiFe_2O_4$ spinel (PDF 10-0325). In the case of the $CoFe_2O_4$ phase the above listed reflexions are located at 18.3°, 30.3°, 35.4°, 43.1°, 53.6°, 57.0°, and 62.5° two Theta degrees (PDF 22-1086). The reflexions of the $CdFe_2O_4$ phase are found at 18.2°, 29.8°, 35.0°, 42.6°, 53.0°, 56.4°, and 61.9° two Theta degrees (PDF 89-2810). Based on the XRD analysis we can say that the synthetized ferrite samples contained only the appropriate spinel phase, other non-magnetic oxides are not found in addition to the ferrites.

Electron diffraction measurements on the individual ferrite nanospheres further supported the exclusive presence of spinel phases, since the measured d spacing were perfectly correlated with the d-values in X-ray databases (Figure 2).

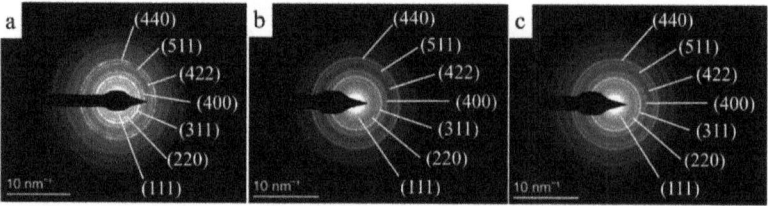

Figure 2. Electron diffraction patterns of the $CoFe_2O_4$-NH_2 (**a**), $CdFe_2O_4$-NH_2 (**b**) and $NiFe_2O_4$-NH_2 (**c**) with the Miller indexes (PDF 89-2810; PDF 22-1086 and PDF 10-0325).

To map the surface functionality and structure of the amine functionalized ferrite nanoparticles Fourier-transform infrared (FTIR) spectra were recorded. In the spectra presented in Figure 3a the following bands can be identified: two bands between 500 cm^{-1} and 600 cm^{-1} and between 400 cm^{-1} and 450 cm^{-1} which belong to the tetrahedral and octahedral complexes of the spinel structures, respectively [33,34]. The band at 592 cm^{-1} belongs to the vibration of Fe^{3+}–O^{2-} in the sublattice A-site. The presence of the absorption band at 416 cm^{-1} can be assigned to the trivalent metal-oxygen vibration at the octahedral B-sites [33,34]. The band at 1052 cm^{-1} belongs to C-N stretching in the case of the three ferrites [35,36]. In the case of all ferrite samples, a band at 1615 cm^{-1} can be identified as the N–H stretching vibration of the free amino functional groups. Additional absorption bands are found at 871 cm^{-1}, 1048 cm^{-1}, 2874 cm^{-1}, and 2929 cm^{-1} which belong to the C–O vibrations, the C–N, and symmetric and asymmetric C–H stretching vibrations, respectively [35,36]. The presence of νC-O and νC-H bands suggest that adsorbed organic molecules (ethylene glycol and ethanol amine) were anchored on the surface of the ferrite particles. The stretching vibration band of the N–H bonds overlaps with the vibration band of -OH groups. The bending vibration mode of the -OH groups resulted a band at 1393 cm^{-1}. As a result of the polar functional groups on the surface, the magnetic ferrite nanoparticles are easily dispersed in aqueous medium. Moreover, they can be easily removed upon the action of an external magnetic field (Figure 3b). Additional advantage of the amine functional groups is that they improve the stabilization and distribution of the Pd nanoparticles, simultaneously as electron donor, the amino groups may enhance the surface electron density on the palladium particles.

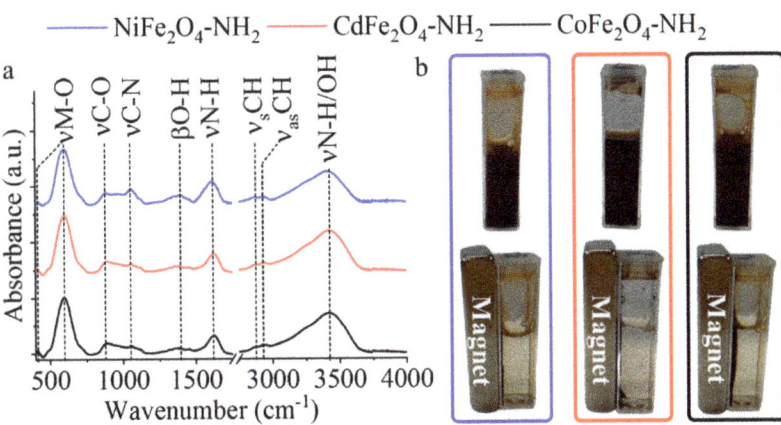

Figure 3. (a) FTIR spectra of the NiFe$_2$O$_4$-NH$_2$, CdFe$_2$O$_4$-NH$_2$ and CoFe$_2$O$_4$-NH$_2$ supports. (b) Demonstration of the good dispersibility and separability of the nanoparticles by magnetic field.

Bonds containing carbon can be identified in the FTIR spectra, thus the total carbon content of the palladium-decorated ferrite samples was examined by CHNS elemental analysis. The carbon content was the highest in the case of the Pd/NiFe$_2$O$_4$-NH$_2$ (6.3 wt%), the other two catalysts contain lower amounts of carbon: 2.5 wt% (Pd/CoFe$_2$O$_4$-NH$_2$) and 2.9 wt% (Pd/CdFe$_2$O$_4$-NH$_2$) (Table 2). The presence of carbon can be explained by the anchoring of the ethylene glycol and ethanol amine molecules on the surface of the ferrite nanoparticles. The total nitrogen content, which originates from the amino functional groups, was also measured. The highest N content was found in the case of the nickel-ferrite supported catalyst: 1.4 wt%, while 0.4 wt% and 0.5 wt% was found in the case of the cobalt and cadmium ferrites, respectively (Table 2). The palladium content of the freshly prepared catalysts was measured by ICP-OES. The Pd contents were found to be 5.05 wt% (Pd/CoFe$_2$O$_4$-NH$_2$), 5.25 wt% (Pd/CdFe$_2$O$_4$-NH$_2$) and 3.96 wt% (Pd/NiFe$_2$O$_4$-NH$_2$).

Table 2. Carbon and nitrogen content (based on CHNS elemental analysis) of the catalysts, as well as their palladium content before use and after four catalytic cycles.

	C wt%	N wt%	Pd wt% (Before Use)	Pd wt% (After Use)
Pd/CoFe$_2$O$_4$-NH$_2$	2.5	0.4	5.05	4.68
Pd/CdFe$_2$O$_4$-NH$_2$	2.9	0.5	5.25	5.10
Pd/NiFe$_2$O$_4$-NH$_2$	6.3	1.4	3.96	3.78

The palladium containing ferrite catalysts were investigated by HRTEM (Figure 4a–c). In the TEM pictures, the same spherical aggregates as the original catalyst-support can be identified. It should be noted here that the aggregates did not disintegrate despite the high-energy sonication during the palladium anchoring on ferrites. The phase identification of the Pd/NiFe$_2$O$_4$-NH$_2$, Pd/CoFe$_2$O$_4$-NH$_2$, and Cd/NiFe$_2$O$_4$-NH$_2$ catalysts was carried out based on XRD measurements (Figure 4d–f). In addition to the nickel-ferrite, cobalt-ferrite, and the cadmium-ferrite magnetic catalyst-supports elemental palladium could also be identified on the diffractograms. The (111) and (200) reflexions of the elemental palladium are located at 40.2° and 46.7° two Theta degrees in the case of all catalysts (PDF 46-1043). The ultrasound-assisted decomposition of the palladium, did not change the chemical composition of the ferrite phases, indicating that the ferrites have sufficient chemical stability.

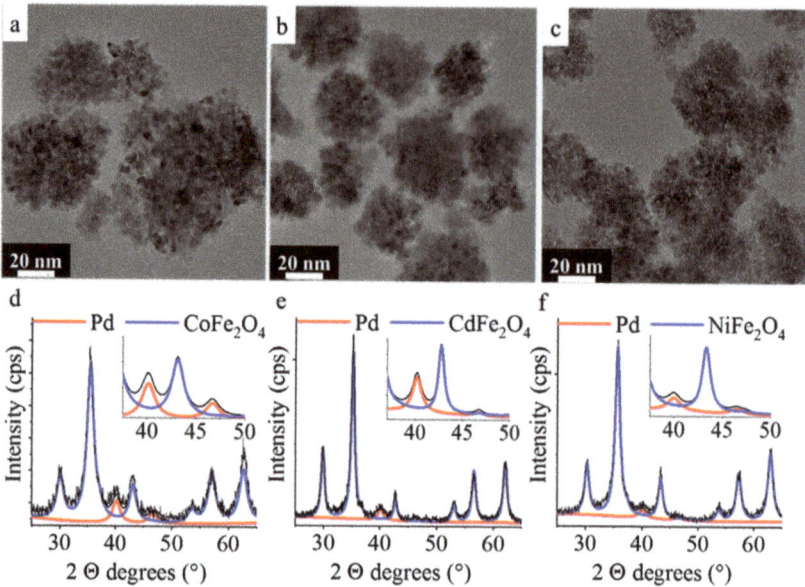

Figure 4. HRTEM pictures and XRD patterns of the Pd/CoFe$_2$O$_4$-NH$_2$ (**a,d**), Pd/CdFe$_2$O$_4$-NH$_2$ (**b,e**), and Pd/NiFe$_2$O$_4$-NH$_2$ catalysts (**c,f**).

The visual identification of palladium nanoparticles is difficult and uncertain next to the ferrite nanoparticles due to their small particle size, thus HAADF pictures and element mapping were made to confirm their presence (Figure 5 and Figure S6a,b in the SI). In the HAADF pictures, the Pd nanoparticles are slightly brighter compared to the ferrite nanoparticles. Moreover, the element mapping also confirmed the position of the palladium particles next to the ferrites. Palladium nanoparticles (highlighted in yellow) are located on the surface of the spherical CoFe$_2$O$_4$ aggregates (Figure 5) as was also confirmed in the case of Pd/CdFe$_2$O$_4$-NH$_2$ and Pd/NiFe$_2$O$_4$-NH$_2$ (Figure S6a,b). The presence of cobalt and iron are also detectable and are marked with green and red color.

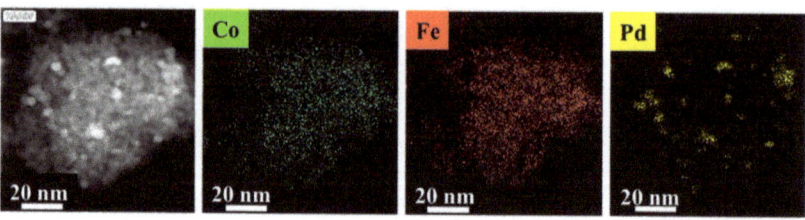

Figure 5. Element maps of the Pd/CoFe$_2$O$_4$-NH$_2$ catalyst.

The particle size of the Pd nanoparticles was measured in the HAADF pictures using ImageJ software (version 1.53t) and the scalebars (Figures S7–S9). The measured diameters of Pd particles were 5.1 ± 0.6 nm (Pd/CoFe$_2$O$_4$-NH$_2$), 3.8 ± 0.5 nm (Pd/CdFe$_2$O$_4$-NH$_2$), and 4.0 ± 0.8 nm (Pd/NiFe$_2$O$_4$-NH$_2$). Very similar particle sizes were resulted based on the XRD measurements, 4 ± 2 nm, 6 ± 2 nm, and 4 ± 2 nm in case of the cobalt-ferrite, cadmium-ferrite, and nickel-ferrite-supported palladium catalysts.

Catalytic Tests of the Magnetic Catalysts for Hydrogenation of 2,4-DNT

Before the catalytic measurements of the palladium decorated, amine-functionalized ferrites, the palladium-free magnetic supports were tested in the hydrogenation of 2,4-DNT at 333 K temperature for 240 min. The tests revealed that even without the noble metal the amine-functionalized cobalt ferrite, cadmium ferrite, and nickel ferrite catalyst supports showed activity in TDA synthesis. The highest DNT conversion (63.01% n/n) was achieved in the case of the $CoFe_2O_4$-NH_2 sample. The other two spinels were less active, 19.4% n/n ($CdFe_2O_4$-NH_2) and 26.7% n/n ($NiFe_2O_4$-NH_2) DNT conversions were measured. In order to increase the DNT conversion and TDA yield, the use of palladium is necessary.

The catalytic activity of the palladium decorated ferrite supported catalysts was compared in the hydrogenation of 2,4-dinitrotoluene to 2,4-diaminotoluene at three different reaction temperatures (at 303 K, 313 K, and 333 K). Interestingly, total DNT conversion was achieved in a short time (40 min) by using the $Pd/CoFe_2O_4$-NH_2 and $Pd/NiFe_2O_4$-NH_2 catalysts at 333 K hydrogenation temperature (Figure 6a,c). In addition, close to or above 99 $n/n\%$ TDA yields were reached after 120 min at 333 K in the case of the $Pd/CoFe_2O_4$-NH_2 (98.9%) and $Pd/NiFe_2O_4$-NH_2 (99.9%) catalysts (Figure 6d,f). The catalytic activity of the $Pd/CdFe_2O_4$-NH_2 sample was found to be far below the two above mentioned catalysts (Figure 6b,e). More than 99 $n/n\%$ DNT conversion was reached after four hours, and the maximum TDA yield was only 65.5 $n/n\%$ after 240 min at 333 K.

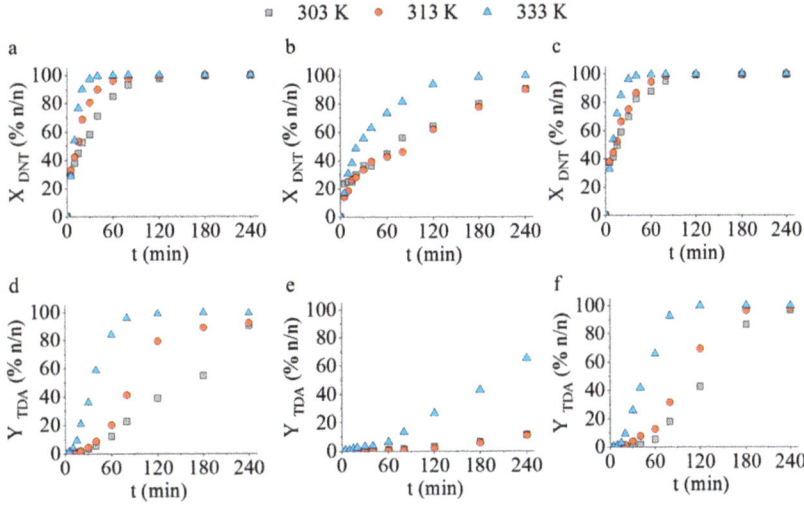

Figure 6. DNT conversions and TDA yields vs. time of hydrogenation at 303 K, 313 K, and 333 K using $Pd/CoFe_2O_4$-NH_2 (**a,d**), $Pd/CdFe_2O_4$-NH_2 (**b,e**), and $Pd/NiFe_2O_4$-NH_2 (**c,f**) catalysts.

GC-MS measurements revealed the presence of two semi-hydrogenated intermediates, namely 4-amino-2-nitrotoluene (4A2NT) and 2-amino-4-nitrotoluene (2A4NT), which were not transformed to 2,4-TDA in the case of the $Pd/CdFe_2O_4$ catalyst. By using the cobalt- or nickel-based catalysts ($Pd/CoFe_2O_4$-NH_2 and $Pd/NiFe_2O_4$-NH_2) the two semi-hydrogenated compounds were completely transformed to 2,4-TDA after 120 min hydrogenation at 333 K, resulting a >99 $n/n\%$ selectivity (2,4-TDA) for both catalysts (Figure 7). In contrast, the selectivity was only 65% for the cadmium ferrite-supported catalyst. Byproducts were not detectable after the tests.

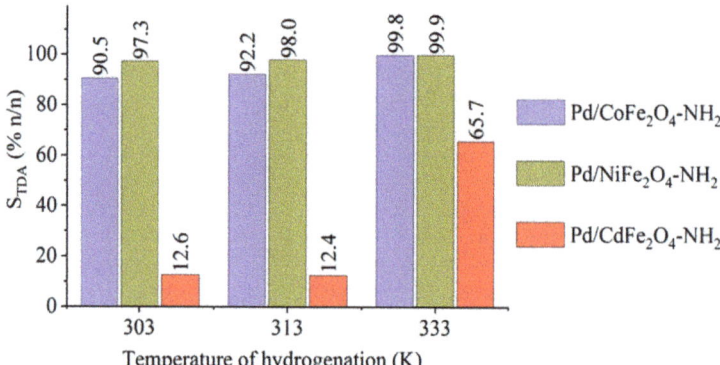

Figure 7. The maximum selectivity of TDA at three reaction temperature for the magnetic catalysts.

Since the Pd/CoFe$_2$O$_4$-NH$_2$ and Pd/NiFe$_2$O$_4$-NH$_2$ catalysts had the highest DNT conversion, TDA yield, and selectivity, these two catalysts were selected for reuse-tests. The catalysts were tested in four cycles at 333 K temperature at 20 bar hydrogen pressure. As can be seen in Figure 8, the results are almost identical for each reuse cycle (4), i.e., there is no visible difference during the whole reaction time between the DNT conversions and TDA yields. The catalytic activity remained constant and excellent even after repeated use. The results suggest that this type of catalyst is stable, namely no Pd loss occurred, possibly due to a strong interaction between the ferrite catalyst support and the palladium particles. The results of ICP-OES measurements are in line with those of the reuse experiments, since the palladium content of the reused catalysts was almost the same, as that of the fresh (non-used) Pd/NiFe$_2$O$_4$-NH$_2$ and Pd/CoFe$_2$O$_4$-NH$_2$ samples (Table 3). This catalyst stability and the no apparent loss of the noble metal may be explained by the presence of the amine functional groups anchoring the Pd nanoparticles on the ferrite nanospheres. This theory is further supported by our previous results, where significant loss in catalytic activity and noble metal was observed in the case of non-functionalized ferrites [36].

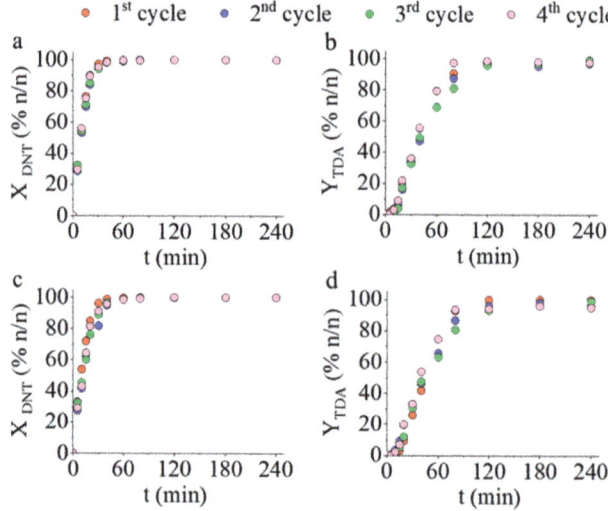

Figure 8. Reuse tests of the Pd/CoFe$_2$O$_4$-NH$_2$ (**a**,**b**) and Pd/NiFe$_2$O$_4$-NH$_2$ (**c**,**d**) catalysts. Change of the 2,4-DNT conversion and 2,4-TDA yield vs. time of hydrogenation.

Table 3. Amounts of the respective transition metal nitrates dissolved in the reaction media for the synthesis of the NH$_2$-functionalized ferrite nanoparticles.

	Ni(NO$_3$)$_2$·6H$_2$O	Cd(NO$_3$)$_2$·4H$_2$O	Co(NO$_3$)$_2$·6H$_2$O	Fe(NO$_3$)$_3$·9H$_2$O
NiFe$_2$O$_4$-NH$_2$	2.91 g (1 mmol)	-	-	8.08 g (2 mmol)
CdFe$_2$O$_4$-NH$_2$	-	3.08 g (1 mmol)	-	8.08 g (2 mmol)
CoFe$_2$O$_4$-NH$_2$	-	-	2.91 g (1 mmol)	8.08 g (2 mmol)

The activity, yield, and selectivity of our catalysts are as high as of those found in the literature. Malyala and co-workers [37] investigated a powdered Y zeolite (10% Ni/HY) catalyst containing 10% Ni for the hydrogenation of 2,4 DNT with 85% TDA yield and selectivity. Ren et al. [38] investigated several Pd/C and Pt/ZrO$_2$ catalysts. The Pd/C catalyst's yield and selectivity were 98% and 99.2%, respectively. The Pt/ZrO$_2$ catalysts yield varied between 97.1% and 98.9% depending upon Pt-content and reduction temperature. The selectivity values are the same as yields since conversion was 100% in each case. In addition, our catalysts are magnetic, therefore they can be easily separated from the reaction medium and even the support itself shows catalytic activity.

3. Materials and Methods

The amine-functionalized ferrite nanoparticles were made from nickel(II) nitrate hexahydrate, Ni(NO$_3$)$_2$·6H$_2$O, MW: 290.79 g/mol (Thermo Fisher GmbH, D-76870 Kandel, Germany), cadmium(II) nitrate tetrahydrate, Cd(NO$_3$)$_2$·4H$_2$O, MW: 308.46 g/mol (Acros Organics Ltd., B-2440 Geel, Belgium), cobalt(II) nitrate hexahydrate, Co(NO$_3$)$_2$·6H$_2$O, MW: 291.03 g/mol and iron(III) nitrate nonahydrate, Fe(NO$_3$)$_3$·9H$_2$O, MW: 404.00 g/mol (VWR Int. Ltd., B-3001 Leuven, Belgium), respectively. As dispersion medium ethylene glycol, HOCH$_2$CH$_2$OH, (VWR Int. Ltd., F-94126 Fontenay-sous-Bois, France) was applied. For coprecipitation and functionalization of the ferrites, ethanolamine, NH$_2$CH$_2$CH$_2$OH (Merck KGaA, D-64271 Darmstadt, Germany) and sodium acetate, CH$_3$COONa (ThermoFisher GmbH, D-76870 Kandel, Germany) were used. Palladium(II) nitrate dihydrate, Pd(NO$_3$)$_2$·2H$_2$O, MW: 266.46 g/mol (Thermo Fisher Scientific Ltd., D-76870 Kandel, Germany) as Pd precursor and Patosolv®, a mixture of 90 vol% ethanol and 10 vol% isopropanol (Molar Chem. Ltd., H-2314 Halásztelek, Hungary) were used during the preparation of the magnetic catalysts. The starting material, main product, and the standards used for the catalytic tests and calibration of the GC measurements were as follows: 2,4-dinitrotoluene (DNT), C$_7$H$_6$N$_2$O$_4$, MW:182.13 g/mol, 2,4-diaminotoluene (TDA), C$_7$H$_{10}$N$_2$, MW: 122.17 g/mol, 4-methyl-3-nitroaniline, 2-methyl-3-nitroaniline and 2-methyl-5-nitroaniline, C$_7$H$_8$N$_2$O$_2$, MW: 152.15 g/mol, (Sigma Aldrich Chemie Gmbh, D-89555 Steinheim, Germany). As an internal standard, nitrobenzene, C$_6$H$_5$NO$_2$, MW: 123.11 g/mol (Merck KGaA, D-64293 Darmstadt, Germany) was used.

3.1. Preparation of the Amine-Functionalized Magnetic Spinel Nanoparticles and the Pd-Catalyst

Amine-functionalized ferrite magnetic catalyst supports were synthesized according to a modified coprecipitation method. In 50 mL ethylene glycol iron(III) nitrate nonahydrate and the nitrate salt of the respective transition metal (Co, Ni or Cd) were dissolved (Table 3). Sodium acetate 12.30 g (15 mmol) was dissolved in another 100 mL ethylene-glycol and it was heated to 100 °C in three necked flask under reflux and continuous stirring. The solution of the metal precursors was added to the glycol-based sodium acetate solution, followed by the addition of 35 mL ethanol amine. After 12 h continuous agitation and reflux, the cooled solution was separated by centrifugation (4200 rpm, 10 min). The solid phase was washed by distilled water several times and the magnetic ferrite was easily separated by a magnet from the aqueous phase. Finally, the ferrite was also rinsed with anhydrous ethanol, and was dried by lyophilization. These ferrite samples were used as magnetic catalyst support for the preparation of palladium decorated spinel catalysts.

For the deposition of the palladium nanoparticles onto the surface of the ferrite crystals, a Hielscher UIP100 Hdt homogenizer was used (120 W, 17 kHz). First, palladium(II) nitrate (0.25 g) was dissolved in patosolv (50 mL) containing 2.00 g dispersed ferrite. Then the solution was treated by ultrasonic cavitation for two minutes. During the process, elemental palladium nanoparticles formed from the Pd(II)-ions as a result of the reducing action of the alcohol (patosolv).

3.2. Catalytic Test: Hydrogenation of 2,4-Dintitro Toluene

2,4-dinitrotoluene, DNT (c_n:50 mmol/dm^3 in methanolic solution, V_{tot} = 150 mL) was hydrogenated by using 0.10 g magnetic Pd catalysts in a Büchi Uster Picoclave reactor of 200 mL volume under constant agitation at 1000 rpm (Figures S1 and S2 in the SI).

The pressure of the hydrogenation was kept at 20 bar in all experiments and the reaction temperature was set to 303 K, 313 K, and 333 K, respectively. The sampling was carried out after 0, 5, 10, 15, 20, 30, 40, 60, 80, 120, 180, and 240-min hydrogenation. As internal standard, 5.0 µL nitrobenzene was added to 1.00 mL sample.

3.3. Characterization Techniques

High-resolution transmission electron microscopy (HRTEM, Talos F200X G2 electron microscope with field emission electron gun, X-FEG, accelerating voltage: 20–200 kV) was applied for examination of particle size and morphology. For the imaging and electron diffraction a SmartCam digital search camera (Ceta 16 Mpixel, 4 k × 4 k CMOS camera) and high-angle annular dark-field (HAADF) detector were used. For the TEM measurement, the samples were dispersed in distilled water, and this aqueous dispersion was dropped on 300 mesh copper grids (Ted Pella Inc., 4595 Redding, CA 96003, USA). The qualitative and quantitative analysis of the ferrite and the palladium catalysts were carried out by X-ray diffraction (XRD) measurements by applying the Rietveld method. Bruker D8 diffractometer (Cu-Kα source) in parallel beam geometry (Göbel mirror) with Vantec detector was used. The average crystallite size of the oxide domains was calculated from the mean column length calibrated method by using full width at half maximum (FWHM) and the width of the Lorentzian component of the fitted profiles. The carbon content originating from the residual ethanolamine and ethylene glycol on the surface of the ferrites, was determined by a Vario Macro CHNS element analyzer by using the phenanthrene standard (C: 93.538%, H: 5.629%, N: 0.179%, S: 0.453%) from Carlo Erba Inc. Helium (99.9990%) was the carrier gas, and oxygen (99.995%) was used as the oxidative atmosphere. The functional groups on the surface of the amine-functionalized ferrite nanoparticles were identified by Fourier transform infrared spectroscopy (FTIR), with a Bruker Vertex 70 spectroscope. Spectra were recorded in transmission mode, in KBr pellet (10 mg ferrite sample was pelletized with 250 mg potassium bromide). The palladium contents of the magnetic ferrite catalysts were measured by a Varian 720 ES inductively coupled optical emission spectrometer (ICP-OES). For the ICP-OES measurements, the catalysts were dissolved in aqua regia. The specific surface area of the catalysts was examined based on the Dubinin–Astakhov (DA) method, by CO_2 adsorption–desorption experiments at 273 K using a Micromeritics ASAP 2020 sorptometer. The 2,4-toluenediamine containing samples after the hydrogenation tests were analyzed using an Agilent 7890A gas chromatograph coupled with Agilent 5975C Mass Selective detector. RTX-1MS column (30 m × 0.25 mm × 0.25 µm) was applied and the injected sample volume was 1 µL at 200:1 split ratio, while the inlet temperature was set to 473 K. Helium was used as a carrier gas (2.28 mL/min), and the oven temperature was set to 323 K for 3 min and it was heated up to 523 K with a heating rate of 10 K/min and kept there for another 3 min.

The catalytic activity of the magnetic Pd catalysts was compared based on calculating the conversion (X%) of 2,4-dinitrotoluene as follows (Equation (1)):

$$X\% = \frac{n_{DNT(consumed)}}{n_{DNT(initial)}} \cdot 100 \tag{1}$$

The yield (Y%) of TDA was calculated based on the following equation (Equation (2)):

$$Y\% = \frac{n_{TDA(formed)}}{n_{TDA(theoretical)}} \cdot 100 \qquad (2)$$

where n_{TDA} is the corresponding molar amount of the product (2,4-diaminotoluene).

Aniline selectivity (S%) of the catalysts was also calculated by using the DNT conversion and TDA yield (Equation (3)):

$$S\% = \frac{Y}{X} \cdot 100 \qquad (3)$$

4. Conclusions

Amine-functionalized, cobalt-, nickel-, and cadmium ferrite magnetic catalyst supports were synthesized based on a modified coprecipitation method. HRTEM investigations showed that the support nanoparticles are spherical and composed of smaller individual nanoparticles of size 4–10 nm. In contrast, the average particle sizes of the aggregate ferrite spheres, namely $CoFe_2O_4$-NH_2, $NiFe_2O_4$-NH_2, and $CdFe_2O_4$-NH_2 are 50.9 ± 8.2 nm, 41.5 ± 12.9 nm, and 80.2 ± 14.2 nm. Based on XRD and electron diffraction measurements, the exclusive presence of spinel phases could be detected without any non-magnetic oxides. The surface of the nanoparticles contains NH_2-groups, which promote the dispersibility of the particles in aqueous medium. The good dispersibility is a key factor in catalytic applications, since aggregation reduces the active surface of the particles. Nevertheless, the magnetic nanoparticles are easily recollected from the reaction medium by the action of an external magnetic field, which can greatly reduce operating costs and catalyst loss during separation. The palladium-free magnetic supports were tested in the hydrogenation of 2,4-DNT and moderate catalytic activity in TDA synthesis was found. The highest DNT conversion (63.01% n/n) was achieved in the case of the $CoFe_2O_4$-NH_2 sample, indicating that the use of palladium is necessary. Palladium nanoparticles were deposited onto the ferrite surfaces utilizing a fast and facile sonochemical method, yielding an immediately usable, catalytically active form. The presence of palladium dramatically increased the catalytic performance. Total DNT conversion was achieved in a short time (40 min) by using the Pd/$CoFe_2O_4$-NH_2 and Pd/$NiFe_2O_4$-NH_2 catalysts at 333 K hydrogenation temperature, while TDA yields were (98.9%) and (99.9%) for the cobalt- and nickel-based catalysts after 120 min at 333 K.

No drop in performance was observed during reuse tests, as the conversion, yield, and selectivity values remained unchanged and excellent for four catalytic cycles. ICP-OES measurements revealed no loss in the palladium content of the reused catalysts. This catalyst stability and the no apparent loss of the noble metal may be explained by the presence of the amine functional groups anchoring the Pd nanoparticles on the ferrite nanospheres. In summary, amine functionalization of magnetic ferrite supports can significantly enhance catalytic performance and reusability. Based on our results, the above detailed magnetically separable catalysts may be well used for the hydrogenation of other aromatic nitro compounds.

Supplementary Materials: The following supporting information can be downloaded at: https://www.mdpi.com/article/10.3390/ijms232113197/s1.

Author Contributions: Conceptualization, L.V., B.V. and V.H.; methodology, V.H., E.S. and F.K.; software, M.N. and B.V.; formal analysis, G.M. and F.K.; resources, M.N. and B.V.; data curation, B.F. and M.N.; writing—original draft preparation L.V. and V.H.; writing—review and editing, M.N., B.F., B.V. and L.V.; visualization, V.H. and E.S.; supervision, L.V.; All authors have read and agreed to the published version of the manuscript.

Funding: This research was supported by the European Union and the Hungarian State, co-financed by the European Regional Development Fund in the framework of the GINOP-2.3.4-15-2016-00004 project, aimed to promote the cooperation between the higher education and the industry. Supported by the KDP-2021 Program of the Ministry for Innovation and Technology from the source of the National Research, Development and Innovation Fund.

Institutional Review Board Statement: Not applicable.

Informed Consent Statement: Not applicable.

Data Availability Statement: Data is available upon request from the corresponding authors.

Conflicts of Interest: The authors declare no conflict of interest.

References

1. Neri, G.; Musolino, M.G.; Milone, C.; Visco, A.M.; Di Mario, A. Mechanism of 2,4-dinitrotoluene hydrogenation over Pd/C. *J. Mol. Catal. A Chem.* **1995**, *95*, 235–241. [CrossRef]
2. Neri, G.; Musolino, M.G.; Rotondo, E.; Galvagno, S. Catalytic hydrogenation of 2,4-dinitrotoluene over a Pd/C catalyst: Identification of 2-(hydroxyamino)-4-nitrotoluene (2HA4NT) as reaction intermediate. *J. Mol. Catal. A Chem.* **1996**, *111*, 257–260. [CrossRef]
3. Musolino, M.G.; Milone, C.; Neri, G.; Bonaccorsi, L.; Pietropaolo, R.; Galvagno, S. Selective catalytic hydrogenation of 2,4-dinitrotoluene to nitroarylhydroxylamines on supported metal catalysts. *Stud. Surf. Sci. Catal.* **1997**, *108*, 239–246. [CrossRef]
4. Musolino, X.G.; Neri, G.; Milone, C.; Minicò, S.; Galvagno, S. Liquid chromatographic separation of intermediates of the catalytic hydrogenation of 2,4-dinitrotoluene. *J. Chromatogr. A* **1998**, *818*, 123–126. [CrossRef]
5. Neri, G.; Musolino, M.G.; Milone, C.; Pietropaolo, D.; Galvagno, S. Particle size effect in the catalytic hydrogenation of 2,4-dinitrotoluene over Pd/C catalysts. *Appl. Catal. A Gen.* **2001**, *208*, 307–316. [CrossRef]
6. Neri, G.; Rizzo, G.; Milone, C.; Galvagno, S.; Musolino, M.G.; Capannelli, G. Microstructural characterization of doped-Pd/C catalysts for the selective hydrogenation of 2,4-dinitrotoluene to arylhydroxylamines. *Appl. Catal. A Gen.* **2003**, *249*, 303–311. [CrossRef]
7. Saboktakin, M.R.; Tabatabaie, R.M.; Maharramov, A.; Ramazanov, M.A. Hydrogenation of 2,4-dinitrotoluene to 2,4-diaminotoluene over platinum nanoparticles in a high-pressure slurry reactor. *Synth. Commun.* **2011**, *41*, 1455–1463. [CrossRef]
8. Rossi, L.M.; Costa, N.J.S.; Silva, F.P.; Wojcieszak, R. Magnetic nanomaterials in catalysis: Advanced catalysts for magnetic separation and beyond. *Green Chem.* **2014**, *16*, 2906–2933. [CrossRef]
9. Saiyed, Z.M.; Ramchand, C.N.; Telang, S.D. Isolation of genomic DNA using magnetic nanoparticles as a solid-phase support. *J. Phys. Condens. Matter* **2008**, *20*, 204153. [CrossRef]
10. Vanyorek, L.; Ilosvai, Á.M.; Szőri-Dorogházi, E.; Váradi, C.; Kristály, F.; Prekob, Á.; Fiser, B.; Varga, T.; Kónya, Z.; Viskolcz, B. Synthesis of iron oxide nanoparticles for DNA purification. *J. Dispers. Sci. Technol.* **2019**, *42*, 693–700. [CrossRef]
11. Yantasee, W.; Warner, C.L.; Sangvanich, T.; Addleman, R.S.; Carter, T.G.; Wiacek, R.J.; Fryxell, G.E.; Timchalk, C.; Warner, M.G. Removal of heavy metals from aqueous systems with thiol functionalized superparamagnetic nanoparticles. *Environ. Sci. Technol.* **2007**, *41*, 5114–5119. [CrossRef] [PubMed]
12. Song, J.; Kong, H.; Jang, J. Adsorption of heavy metal ions from aqueous solution by polyrhodanine-encapsulated magnetic nanoparticles. *J. Colloid Interface Sci.* **2011**, *359*, 505–511. [CrossRef] [PubMed]
13. Gutierrez, A.M.; Dziubla, T.D.; Hilt, J.Z. Recent advances on iron oxide magnetic nanoparticles as sorbents of organic pollutants in water and wastewater treatment. *Rev. Environ. Health* **2017**, *32*, 111–117. [CrossRef] [PubMed]
14. Zhu, L.; Li, C.; Wang, J.; Zhang, H.; Zhang, J.; Shen, Y.; Li, C.; Wang, C.; Xie, A. A simple method to synthesize modified Fe3O4 for the removal of organic pollutants on water surface. *Appl. Surf. Sci.* **2012**, *258*, 6326–6330. [CrossRef]
15. Zhang, Q.; Yang, X.; Guan, J. Applications of Magnetic Nanomaterials in Heterogeneous Catalysis. *ACS Appl. Nano Mater.* **2019**, *2*, 4681–4697. [CrossRef]
16. Abu-Dief, A.M.; Abdel-Fatah, S.M. Development and functionalization of magnetic nanoparticles as powerful and green catalysts for organic synthesis. *Beni-Suef Univ. J. Basic Appl. Sci.* **2018**, *7*, 55–67. [CrossRef]
17. Dey, C.; De, D.; Nandi, M.; Goswami, M.M. A high performance recyclable magnetic CuFe2O4 nanocatalyst for facile reduction of 4-nitrophenol. *Mater. Chem. Phys.* **2020**, *242*, 122237. [CrossRef]
18. Pan, Z.; Hua, L.; Qiao, Y.; Yang, H.; Zhao, X.; Feng, B.; Zhu, W.; Hou, Z. Nanostructured Maghemite-Supported Silver Catalysts for Styrene Epoxidation. *Chin. J. Catal.* **2011**, *32*, 428–435. [CrossRef]
19. Weng, C.; Sun, X.; Han, B.; Ye, X.; Zhong, Z.; Li, W.; Liu, W.; Deng, H.; Lin, Z. Targeted conversion of Ni in electroplating sludge to nickel ferrite nanomaterial with stable lithium storage performance. *J. Hazard. Mater.* **2020**, *393*, 122296. [CrossRef]
20. Ebrahimi, A.K.; Barani, M.; Sheikhshoaie, I. Fabrication of a new superparamagnetic metal-organic framework with core-shell nanocomposite structures: Characterization, biocompatibility, and drug release study. *Mater. Sci. Eng. C* **2018**, *92*, 349–355. [CrossRef]

21. Kainz, Q.M.; Reiser, O. Polymer- and dendrimer-coated magnetic nanoparticles as versatile supports for catalysts, scavengers, and reagents. *Acc. Chem. Res.* **2014**, *47*, 667–677. [CrossRef] [PubMed]
22. Aoopngan, C.; Nonkumwong, J.; Phumying, S.; Promjantuek, W.; Maensiri, S.; Noisa, P.; Pinitsoontorn, S.; Ananta, S.; Srisombat, L. Amine-Functionalized and Hydroxyl-Functionalized Magnesium Ferrite Nanoparticles for Congo Red Adsorption. *ACS Appl. Nano Mater.* **2019**, *2*, 5329–5341. [CrossRef]
23. Yavari, S.; Mahmodi, N.M.; Teymouri, P.; Shahmoradi, B.; Maleki, A. Cobalt ferrite nanoparticles: Preparation, characterization and anionic dye removal capability. *J. Taiwan Inst. Chem. Eng.* **2016**, *59*, 320–329. [CrossRef]
24. Bohara, R.A.; Thorat, N.D.; Yadav, H.M.; Pawar, S.H. One-step synthesis of uniform and biocompatible amine functionalized cobalt ferrite nanoparticles: A potential carrier for biomedical applications. *N. J. Chem.* **2014**, *38*, 2979–2986. [CrossRef]
25. Chang, Y.-C.; Chen, D.-H. Catalytic reduction of 4-nitrophenol by magnetically recoverable Au nanocatalyst. *J. Hazard. Mater.* **2009**, *165*, 664–669. [CrossRef]
26. Ren, N.; Dong, A.-G.; Cai, W.-B.; Zhang, Y.-H.; Yang, W.-L.; Huo, S.-J.; Chen, Y.; Xie, S.-H.; Gao, Z.; Tang, Y. Mesoporous microcapsules with noble metal or noble metal oxide shells and their application in electrocatalysis. *J. Mater. Chem.* **2004**, *14*, 3548–3552. [CrossRef]
27. Kovács, E.; Faigl, F.; Mucsi, Z.; Nyerges, M.; Hegedűs, L. Hydrogenolysis of N- and O-protected hydroxyazetidines over palladium: Efficient and selective methods for ring opening and deprotecting reactions. *J. Mol. Catal. A Chem.* **2014**, *395*, 217–224. [CrossRef]
28. Kovács, E.; Thurner, A.; Farkas, F.; Faigl, F.; Hegedűs, L. Hydrogenolysis of N-protected aminooxetanes over palladium: An efficient method for a one-step ring opening and debenzylation reaction. *J. Mol. Catal. A Chem.* **2011**, *339*, 32–36. [CrossRef]
29. Sharma, H.; Bhardwaj, M.; Kour, M.; Paul, S. Highly efficient magnetic Pd(0) nanoparticles stabilized by amine functionalized starch for organic transformations under mild conditions. *Mol. Catal.* **2017**, *435*, 58–68. [CrossRef]
30. Liu, Y.; Yin, D.; Xin, Q.; Lv, M.; Bian, B.; Li, L.; Xie, C.; Yu, S.; Liu, S. Preparation of highly dispersed Ru nanoparticles supported on amine-functionalized magnetic nanoparticles: Efficient catalysts for the reduction of nitro compounds. *Solid State Sci.* **2020**, *101*, 106100. [CrossRef]
31. Vanyorek, L.; Prekob, Á.; Hajdu, V.; Muránszky, G.; Fiser, B.; Sikora, E.; Kristály, F.; Viskolcz, B. Ultrasonic cavitation assisted deposition of catalytically active metals on nitrogen-doped and non-doped carbon nanotubes–A comparative study. *J. Mater. Res. Technol.* **2020**, *9*, 4283–4291. [CrossRef]
32. Suslick, K.S. Sonochemistry. In *Kirk-Othmer Encyclopedia of Chemical Technology*; John Wiley & Sons, Inc.: Hoboken, NJ, USA, 2000.
33. Waldron, R.D. Infrared Spectra of Ferrites. *Phys. Rev.* **1955**, *99*, 1727. [CrossRef]
34. Zhou, B.; Zhang, Y.-W.; Liao, C.-S.; Yan, C.-H.; Chen, L.-Y.; Wang, S.-Y. Rare-earth-mediated magnetism and magneto-optical Kerr effects in nanocrystalline $CoFeMn_{0.9}RE_{0.1}O_4$ thin films. *J. Magn. Magn. Mater.* **2004**, *280*, 327–333. [CrossRef]
35. Bruce, I.J.; Taylor, J.; Todd, M.; Davies, M.J.; Borioni, E.; Sangregorio, C.; Sen, T. Synthesis, characterisation and application of silica-magnetite nanocomposites. *J. Magn. Magn. Mater.* **2004**, *284*, 145–160. [CrossRef]
36. Wang, W.; Su, W.; Tan, M. Endogenous Fluorescence Carbon Dots Derived from Food Items. *Innov.* **2020**, *1*, 100009. [CrossRef]
37. Malyala, R.V.; Chaudhari, R.V. Hydrogenation of 2,4-dinitrotoluene using a supported Ni catalyst: Reaction kinetics and semibatch slurry reactor modeling. *Ind. Eng. Chem. Res.* **1999**, *38*, 906–915. [CrossRef]
38. Ren, X.; Li, J.; Wang, S.; Zhang, D.; Wang, Y. Preparation and catalytic performance of ZrO_2-supported Pt single-atom and cluster catalyst for hydrogenation of 2,4-dinitrotoluene to 2,4-toluenediamine. *J. Chem. Technol. Biotechnol.* **2020**, *95*, 1675–1682. [CrossRef]

Article

General Construction of Amine via Reduction of N=X (X = C, O, H) Bonds Mediated by Supported Nickel Boride Nanoclusters

Da Ke [1,2] and Shaodong Zhou [1,2,*]

[1] Zhejiang Provincial Key Laboratory of Advanced Chemical Engineering Manufacture Technology, College of Chemical and Biological Engineering, Zhejiang University, Hangzhou 310027, China
[2] Institute of Zhejiang University—Quzhou, Zheda Rd. #99, Quzhou 324000, China
* Correspondence: szhou@zju.edu.cn

Abstract: Amines play an important role in synthesizing drugs, pesticides, dyes, etc. Herein, we report on an efficient catalyst for the general construction of amine mediated by nickel boride nanoclusters supported by a TS-1 molecular sieve. Efficient production of amines was achieved via catalytic hydrogenation of N=X (X = C, O, H) bonds. In addition, the catalyst maintains excellent performance upon recycling. Compared with the previous reports, the high activity, simple preparation and reusability of the Ni-B catalyst in this work make it promising for industrial application in the production of amines.

Keywords: nickel boride; primary amine; hydrogenation; reductive amination

1. Introduction

Amine constitutes a vital class of chemicals abundantly existing in nature, and are widely used in industry to produce pharmaceutical drugs, agrochemicals, fine chemicals, polymers, dyes, perfumes, pigments, etc. [1–6]. In recent years, great efforts have been conducted on the synthesis of primary amines. At present, primary amines can be prepared via direct amination of alcohols [7,8], reductive amination of aldehydes or ketone compounds [9–11], amination of carboxylic acids [12,13], and reduction of nitriles [14–18], nitro compounds [19–21], or amides [22]. Among these methods, the reduction of N=X (X = C, O, H) bonds plays a key role. Generally, nitriles, nitro compounds and amides can be reduced to primary amines using borane [21,23,24], silane [25], hydrides [26], formats [20,27], alcohols [28], or molecular hydrogen [29]. Since Raney Ni was first prepared in 1905, it has become one of the most important catalysts for reduction. Though Raney Ni is indeed active, it suffers from high inflammability [30]. To improve this, researchers have developed a variety of homogeneous or heterogeneous catalysts. For example, non-precious metals, such as iron [31–36], cobalt [37–47], copper [48,49], nickel [10,11,21,24,50–53], manganese [6,54,55], and noble metals, such as palladium [19,56–58], platinum [59], ruthenium [8,60–62], rhodium [28,63–65], samarium [66], and iridium [67], have been employed to construct hydrogenation catalysts.

Efficient, stable, and economical hydrogenation catalysts to synthesize primary amines continue to be demanding in both academia and industry. Amorphous nickel boride is well known for its short-range ordered and long-range disordered structures, as well as their activity in liquid phase hydrogenation [68]. Li et al. [69] used Ni-B/SiO$_2$ as a catalyst to reduce adiponitrile with good selectivity and a low TOF of 1.2 (Scheme 1). At present, there exists only a few reports on the reduction of unsaturated bonds mediated by nickel boride [70,71]. Additionally, the unique pore structure, large specific surface area and excellent hydrothermal stability of titanium silicalite molecular sieves make them widely used in the chemical industry, environmental protection and energy conversion [72–74].

The diffusion path length and the aforementioned characteristics enable titanium silicalite (TS-1) molecular sieves to perform strongly as catalysts. Herein, we report on a nickel boride catalyst with TS-1 as support, for the reduction of N=X (X = C, O, H) bonds to amines with high efficiency and universality (Scheme 1).

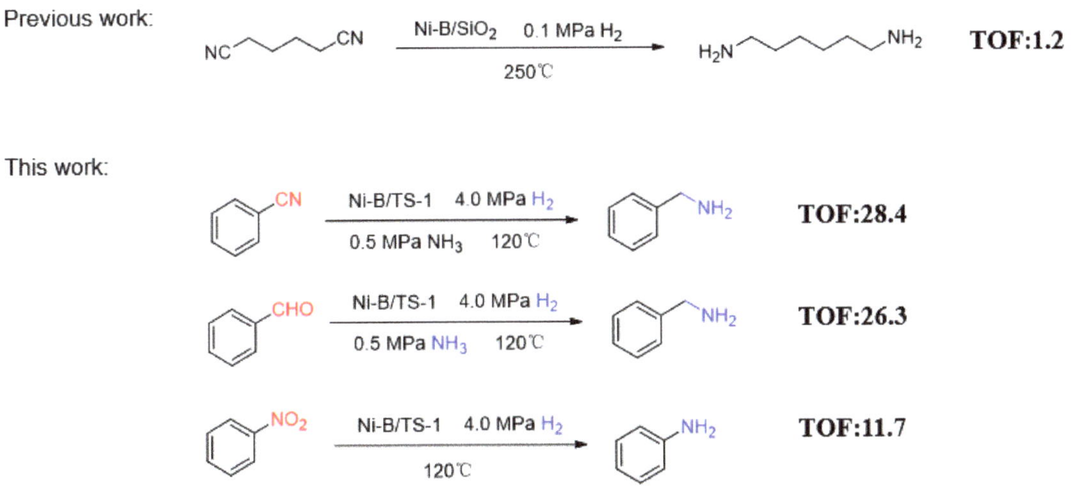

Previous work:

NC~~~CN $\xrightarrow[250\,°C]{\text{Ni-B/SiO}_2 \quad 0.1\text{ MPa H}_2}$ H$_2$N~~~NH$_2$ **TOF:1.2**

This work:

Ph-CN $\xrightarrow[0.5\text{ MPa NH}_3 \quad 120\,°C]{\text{Ni-B/TS-1} \quad 4.0\text{ MPa H}_2}$ Ph-CH$_2$NH$_2$ **TOF:28.4**

Ph-CHO $\xrightarrow[0.5\text{ MPa NH}_3 \quad 120\,°C]{\text{Ni-B/TS-1} \quad 4.0\text{ MPa H}_2}$ Ph-CH$_2$NH$_2$ **TOF:26.3**

Ph-NO$_2$ $\xrightarrow[120\,°C]{\text{Ni-B/TS-1} \quad 4.0\text{ MPa H}_2}$ Ph-NH$_2$ **TOF:11.7**

Scheme 1. The formation of primary amines.

2. Results and Discussion

2.1. Catalyst Evaluation

We first examined the performances of the catalysts prepared under different conditions, including temperature, pressure, additive and solvent. More details are listed in Table S1. Three reactions were selected to evaluate the catalysts, i.e., the hydrogenation of benzonitrile, nitrobenzene, and the reductive amination of benzaldehyde. The detailed results are shown in Tables 1–3.

Table 1. Catalyst screening for benzonitrile [a].

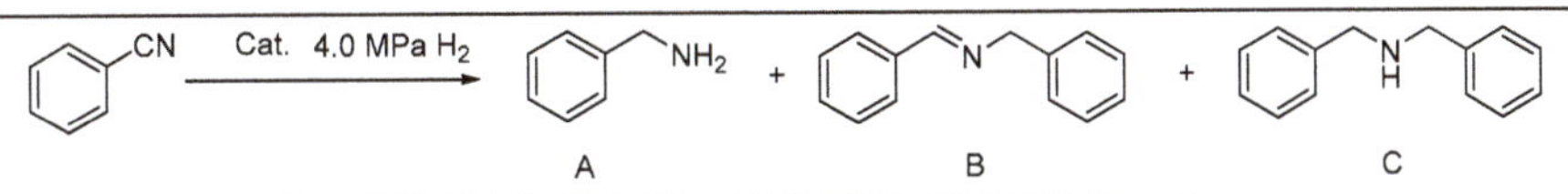

Ph-CN $\xrightarrow{\text{Cat.} \quad 4.0\text{ MPa H}_2}$ Ph-CH$_2$NH$_2$ (A) + Ph-CH=N-CH$_2$-Ph (B) + Ph-CH$_2$-NH-CH$_2$-Ph (C)

	Catalyst (mg)	Conversion (%) [b]	Yield (%) [c]	TOF (h^{-1}) [d]
1	Ni$_{6.2}$-30 (100)	99	74	23.4
2	Ni$_{6.2}$-50 (100)	100	71	21.9
3	Ni$_{6.2}$-70 (100)	99	66	18.8
4	Ni$_{6.2}$-100 (100)	100	60	17.2
5	Ni$_{2.5}$-30 (250)	100	68	19.0
6	Ni$_{12.4}$-30 (50)	100	77	28.4
7	Ni$_{18.6}$-30 (33)	77	51	22.2
8	Ni$_{24.8}$-30 (25)	71	47	12.7

[a] Reaction condition: 5.0 mmol benzonitrile, 4.0 MPa H$_2$, 20 mL of isopropanol, 120 °C. [b] Conversion was calculated by GC. [c] Isolated yield. [d] TOF was the amount of benzonitrile converted by per mol Ni in an hour.

Table 2. Reduction of nitrobenzene by different catalysts [a].

$$\text{C}_6\text{H}_5\text{-NO}_2 \xrightarrow{\text{Cat. 4.0 MPa H}_2} \text{C}_6\text{H}_5\text{-NH}_2$$

	Catalyst (mg)	Conversion (%) [b]	Yield (%) [c]	TOF (h^{-1}) [d]
1	Ni$_{6.2}$-30 (100)	100	96	5.9
2	Ni$_{6.2}$-50 (100)	100	95	5.6
3	Ni$_{6.2}$-70 (100)	100	95	5.1
4	Ni$_{6.2}$-100 (100)	100	94	4.3
5	Ni$_{2.5}$-30 (250)	85	81	2.9
6	Ni$_{12.4}$-30 (50)	100	97	9.5
7	Ni$_{18.6}$-30 (33)	100	95	11.7
8	Ni$_{24.8}$-30 (25)	100	94	10.5

[a] Reaction condition: 5.0 mmol nitrobenzene, 4.0 MPa H$_2$, 20 mL of isopropanol, 120 °C. [b] Conversion was calculated by GC. [c] Isolated yield. [d] TOF was the amount of nitrobenzene converted by per mol Ni in an hour.

Table 3. Reduction of aldehyde-ammonia by different catalysts [a].

$$\text{C}_6\text{H}_5\text{-CHO} \xrightarrow[\text{4.0 MPa H}_2]{\text{Cat. 0.5 MPa NH}_3} \text{C}_6\text{H}_5\text{-CH}_2\text{NH}_2$$

	Catalyst (mg)	Conversion (%) [b]	Yield (%) [c]	TOF (h^{-1}) [d]
1	Ni$_{6.2}$-30 (100)	100	92	21.5
2	Ni$_{6.2}$-50 (100)	100	97	19.0
3	Ni$_{6.2}$-70 (100)	100	96	15.8
4	Ni$_{6.2}$-100 (100)	100	95	14.8
5	Ni$_{2.5}$-30 (250)	100	93	19.0
6	Ni$_{12.4}$-30 (50)	100	96	23.7
7	Ni$_{18.6}$-30 (33)	100	96	26.3
8	Ni$_{24.8}$-30 (25)	100	95	22.6

[a] Reaction condition: 5.0 mmol benzaldehyde, 0.5 MPa NH$_3$ and 4.0 MPa H$_2$, 20 mL of isopropanol, 120 °C. [b] Conversion was calculated by GC. [c] Isolated yield. [d] TOF was the amount of benzaldehyde converted by per mol Ni in an hour.

During the reduction of benzonitrile, the imine intermediate would react with primary amine to generate N-benzylidenebenzylamine (B) and further hydrogenated to dibenzylamine (C). Generally, excessive ammonia can inhibit the side reaction with the primary amine [75]. Moreover, acetylation reactions, using highly acidic or basic additives, can also promote the selectivity of primary amines [76–78]. In the model reaction, ammonia was not added to the reaction system in order to evaluate the intrinsic performance of the catalysts. Surprisingly, highly selective generation of primary amines was facilitated. It turned out that both the preparation temperature and the Ni content affect the performance of the catalyst: lower temperature favors a high activity of the catalyst, while a Ni content ~12% is optimal for the catalytic efficiency.

Methanol, ethanol, isopropanol and toluene were tested as the solvent (Figure S1), and isopropanol outperformed the others.

The reaction temperature and hydrogen pressure were simply screened (Figures S2 and S3), and 120 °C and 4.0 MPa were shown to be optimal.

Next, as shown in Table 2, a longer time was required to convert nitrobenzene completely, indicative for a slightly lower activity of the catalyst toward nitro reduction. Though a higher Ni content (18.6%) affords a higher TOF, the yield may not be favored.

Further, reductive amination of benzaldehyde was carried out using the nickel boride catalysts. To promote the selectivity of the target product, the critical point is to avoid further conversion of the product. To this end, it is necessary to use excessive ammonia to suppress the side reaction. As shown in Table 3, when the same amount of nickel was added, the TOF values did not change much.

Considering both the TOF value and the selectivity of the target product, the $Ni_{12.4}$-30 catalyst was selected for further investigation.

2.2. Characterization of $Ni_{12.4}$-30

In order to clarify the actual content of metallic Ni in the catalyst, the accurate mass content of Ni was obtained through the ICP-OES test. The theoretical nickel content in the catalyst was 12.4%, and the experimental data was 12.1%, which was the normal error range (Table S2). Thus, there was no loss of Ni during the preparation process.

The XRD pattern of $Ni_{12.4}$-30, shown in Figure 1, indicates that there was no obvious change on the TS-1 support after loading, implying that the loaded nickel boride component possesses an amorphous structure, in line with previous findings [70]. The rest moiety of the catalyst did not exhibit other diffraction peaks, regardless of the reduction temperature and Ni loading (Figure S4). It is worth noting that nickel boride may react with ethanol at high temperatures to form metallic nickel [79]. The characteristic diffraction peaks for metallic nickel were not found in the used catalyst, therefore, the stability of the nickel boride structure was thus justified, ruling out the possibility that metallic nickel generated in-situ serves as the active species.

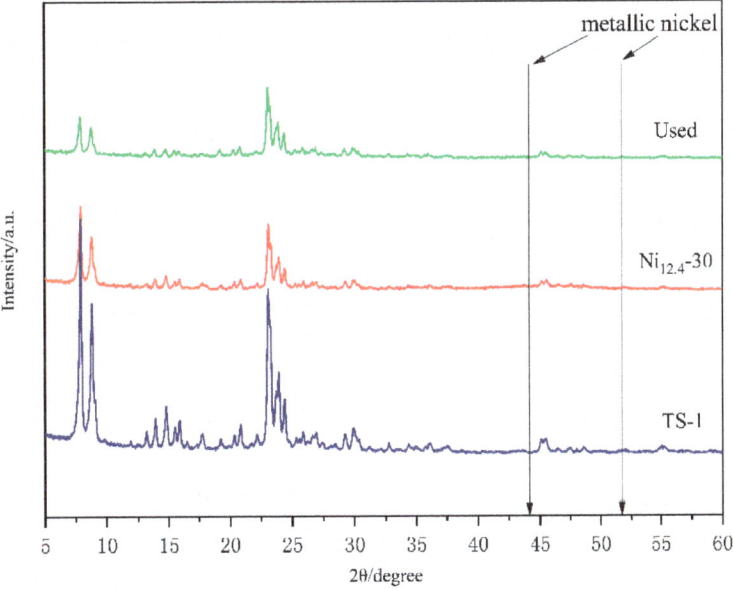

Figure 1. XRD pattern of heterogeneous catalyst $Ni_{12.4}$-30.

In order to further identify the chemical state of the catalyst, $Ni_{12.4}$-30 (both the fresh and the recycled ones) were subjected to XPS analysis, and the results are shown in Figure 2. The signals of high-resolution XPS spectra that emerged at around 860 and 190 eV correspond to Ni and B, respectively [80]. The peaks at 853 and 856 eV in Ni $2p_{3/2}$ are ascribed to the metallic nickel and oxidized nickel. The XPS spectrum of pure nickel boride alloy has only one peak of Ni(0), while the peak of Ni(II) appears when nickel boride is supported, in line with previous reports [69–71,81]. The peaks at 188 and 192 eV in B 1s are assigned to elemental and oxidized boron, respectively. The peaks of pure boron in B 1s at 187 eV (< 188 eV) may result from Ni-B interaction. No significant difference in chemical states of Ni and B appears in the used catalyst, indicative of the catalyst's high stability.

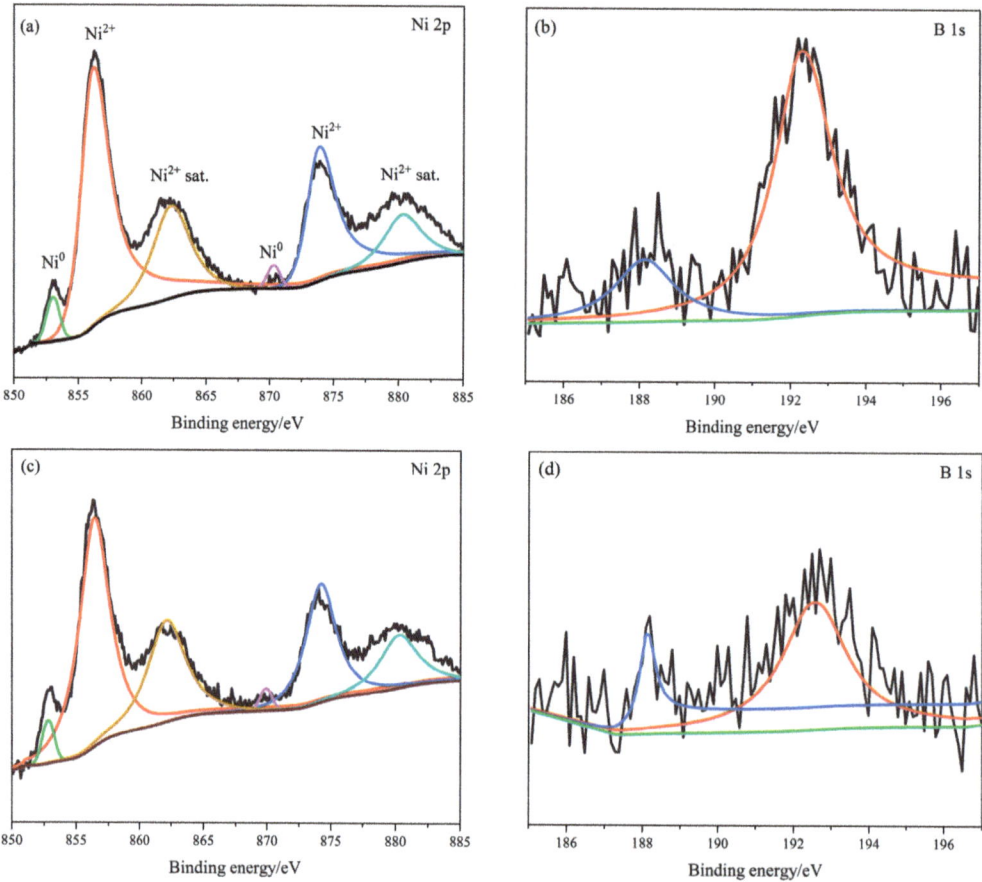

Figure 2. (a) Ni 2p, (b) B 1s XPS spectra of fresh catalyst. (c) Ni 2p, (d) B 1s XPS spectra of used catalyst.

The morphologies of the $Ni_{12.4}$-30 catalyst was investigated using TEM. As shown in Figure 3a, the nickel boride species correspond to nanoparticles ranging 10~40 nm diameter with a mean size of 17 nm. A smaller particle size indicates a higher surface energy, and the diameter 17 nm is much smaller than that of pure nickel boride alloy (60 nm) [82], which benefits from the porous structure of TS-1. Most likely, the high activity of this catalyst generates these results. The SAED was employed to determine the crystal structure of nickel boride. There are halo diffraction rings rather than distinct dots in the SAED image, confirming the amorphous structure of nickel boride, in good agreement with XRD patterns. The EDS revealed that the nickel boride comprised of Ni (60%) and B (40%), similar to Ni_2B.

According to the characteristic results, the high activity of the $Ni_{12.4}$-30 catalyst may result from three aspects. First, the amorphous nickel boride possesses a large number of coordinatively unsaturated active centers on the surface, and a higher surface energy is conducive to the adsorption and conversion of reactants. Second, the electron-transfer from Ni to B causes polarization of the active center and is thus beneficial for Lewis interactions with the reactants. Third, suitable Ni loading dispersed on TS-1 promotes a proper particle size and prevents aggregation, crystallization, and deactivation.

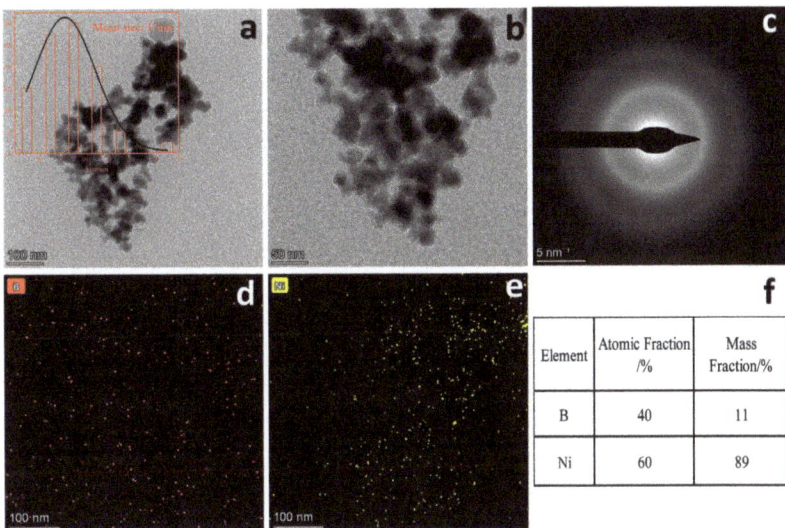

Figure 3. TEM images of Ni$_{12.4}$-30 catalyst: (**a**) 100 nm. (**b**) 50 nm. (**c**) SAED pattern. (**d**) Elemental mapping of B. (**e**) Elemental mapping of Ni. (**f**) Elemental composition.

2.3. The Reduction of Nitrile

In order to test the universality of the selected catalyst (Ni$_{12.4}$-30), the reduction of various nitriles were carried out under optimal conditions. Ammonia was added into the reaction system to avoid side reactions. Consequently, for most aromatic nitriles, ideal conversion (100%) and primary amines yield (>90%) were obtained (Table 4, entries 1–14). However, when picolinonitrile or 2-aminobenzonitrile were the substrate, a much lower rate of conversion occurred. By contrast, when aliphatic nitriles were subjected to the same conditions, the reaction proceeded very inefficiently (Table 4, entries 15–16, 18–20). It was interesting to note that although the performance of adiponitrile, cyclohexanecarbonitrile and butyronitrile were poor, the performance of dodeconitrile was exceptionally good. This abnormal phenomenon might be ascribed the long carbon chain of dodeconitrile.

Table 4. Hydrogenation of nitriles catalyzed by Ni$_{12.4}$-30.

$$R-CN \xrightarrow[0.5\ MPa\ NH_3]{Cat.\ 4.0\ MPa\ H_2} R\frown NH_2$$

	Product	Time (h)	Conversion (%) [a]	Yield (%) [b]
1	Ph-CH$_2$NH$_2$	3.5	100	97
2	4-CH$_3$-Ph-CH$_2$NH$_2$	4.0	100	94
3	4-CH$_3$O-Ph-CH$_2$NH$_2$	2.5	100	95
4	4-Cl-Ph-CH$_2$NH$_2$	3.5	100	95
5	4-NH$_2$-Ph-CH$_2$NH$_2$	4.5	100	97
6	3-CH$_3$-Ph-CH$_2$NH$_2$	4.5	100	96
7	3-CH$_3$O-Ph-CH$_2$NH$_2$	2.5	100	93
8	3-Cl-Ph-CH$_2$NH$_2$	1.5	100	95
9	3-NH$_2$-Ph-CH$_2$NH$_2$	3.0	100	95
10	2-CH$_3$-Ph-CH$_2$NH$_2$	4.0	100	96
11	2-CH$_3$O-Ph-CH$_2$NH$_2$	4.5	100	92
12	2-Cl-Ph-CH$_2$NH$_2$	5.0	100	96
13	2-NH$_2$-Ph-CH$_2$NH$_2$	3.0	100	95

Table 4. Cont.

		R–CN	Cat. 4.0 MPa H_2 / 0.5 MPa NH_3	R⌒NH_2	
		Product	Time (h)	Conversion (%) [a]	Yield (%) [b]
14		H_2N–⌬–CH_2NH_2	5.0	74.9	69
15 [c]		pyridyl-CH_2NH_2	8.0	22.1	17
16 [d]		cyclohexyl-CH_2NH_2	4.0	63.9	55
17		$CH_3(CH_2)_9NH_2$	8.0	100	97
18		H_2N–(CH$_2$)$_n$–NH_2	9.0	68.3	63
19 [e]		$CH_3(CH_2)_3NH_2$·HCl	6.0	40.3	35
20		HO–CH$_2$–N(CH$_3$)–CH$_2$CH$_2$–NH_2	5.0	<5	-

Reaction conditions: 5.0 mmol nitrile, 50 mg $Ni_{12.4}$-30 catalyst (about 2.0 mol% Ni), 0.5 MPa NH_3 and 4.0 MPa H_2, 20 mL of isopropanol, 120 °C. [a] Calculated by GC. [b] Isolated yield. [c] 100 mg catalyst. [d] 110 °C. [e] 20.0 mmol nitrile, 100 mg catalyst, GC yield.

2.4. The Reduction of Nitro Compounds

Further, the catalyst was tested with the hydrogenation of aromatic nitro compounds to primary amines. Under the same conditions to nitrile reduction, more time was needed to convert nitro to amino (see Table 5). In spite of the relatively lower activity toward nitro reduction, the catalyst mediates selective generation of primary amines. The substitution groups on the phenyl ring do not have much effect on the reduction process.

Table 5. Reduction of nitro-aromatic substrates by $Ni_{12.4}$-30.

Ar–NO_2 —Cat. 4.0 MPa H_2→ Ar–NH_2

	Product (R)	Time (h)	Conversion (%) [a]	Yield (%) [b]
1	H	5.0	100	97
2	4-CH_3	6.0	100	95
3	4-F	6.5	100	93
4	3-F	6.0	100	95
5	4-Cl	6.5	100	94
6	4-Br	7.0	100	93
7 [c]	4-OH	5.5	100	94
8	4-NH_2	7.5	100	96

Reaction conditions: 5.0 mmol nitro compound, 50 mg $Ni_{12.4}$-30 catalyst (about 2.0 mol% Ni), 4.0 MPa H_2, 20 mL of isopropanol, 120 °C. [a] Calculated by GC. [b] Isolated yield. [c] 1.0 mmol reactant, 10 mg catalyst.

2.5. The Reduction for Aldehyde and Ammonia

Further, the catalytic performance of $Ni_{12.4}$-30 toward reductive amination of aldehyde was examined. Various aldehydes were employed, and the amination results are shown in Table 6. In general, all selected carbonyl compounds were converted to the corresponding amines with excellent yields upon reductive amination. As compared to aromatic aldehydes, aliphatic substrates are relatively less reactive, thus a slightly longer time is required for them to be completely converted.

Table 6. Reduction of imines generated by aldehyde and ammonia.

	Product	Time (h)	Conversion (%) [a]	Yield (%) [b]
1	Ph-CH$_2$NH$_2$	2.0	100	96
2	4-CH$_3$-Ph-CH$_2$NH$_2$	3.0	100	95
3	4-Cl-Ph-CH$_2$NH$_2$	3.5	100	95
4	4-Br-Ph-CH$_2$NH$_2$	1.5	100	95
5	4-OH-Ph-CH$_2$NH$_2$	2.5	100	92
6	3-CH$_3$-Ph-CH$_2$NH$_2$	2.0	100	98
7	3-Cl-Ph-CH$_2$NH$_2$	3.5	100	92
8	3-Br-Ph-CH$_2$NH$_2$	2.0	100	96
9	3-OH-Ph-CH$_2$NH$_2$	1.5	100	92
10	2-CH$_3$-Ph-CH$_2$NH$_2$	2.5	100	98
11	2-Cl-Ph-CH$_2$NH$_2$	2.0	100	97
12	2-Br-Ph-CH$_2$NH$_2$	1.0	100	97
13	2-OH-Ph-CH$_2$NH$_2$	2.0	100	91
14	~NH$_2$·HCl	4.0	100	>99 [c]

Reaction conditions: 5.0 mmol aldehyde, 50 mg Ni$_{12.4}$-30 catalyst (about 2.0 mol% Ni), 0.5 MPa NH$_3$ and 4.0 MPa H$_2$, 20 mL of isopropanol, 120 °C. [a] Calculated by GC. [b] Isolated yield. [c] GC yield.

In order to test the reusability of the catalyst, the Ni$_{12.4}$-30 species was used fifteen times consecutively with 10 mmol scale. The conversion of benzonitrile, nitrobenzene, and benzaldehyde to amines were all tested. Surprisingly, no obvious loss of activity was observed (Figure 4). Furthermore, the catalytic performance of Ni$_{12.4}$-30 was compared with the commercial Raney Ni (Table 7, see more details in Table S3). For the reduction of benzonitrile, Ni$_{12.4}$-30 catalyst exhibits higher selectivity of benzylamine under ammonia-free conditions. For the other two reactions, there was no obvious difference between Ni$_{12.4}$-30 and Raney Ni.

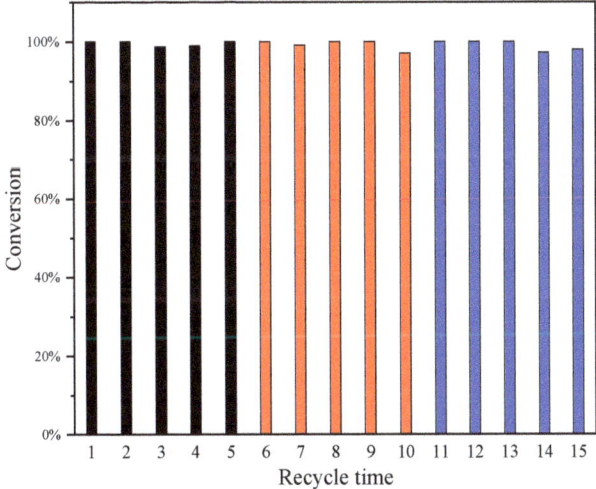

Figure 4. Black: benzonitrile. Red: nitrobenzene. Blue: benzaldehyde. The reaction condition: 10 mmol reagent, 100 mg catalyst, 120 °C 4.0 MPa H$_2$. When the loss of catalyst reached 20%, the catalyst was replenished, and replenishment was carried out at the 7th and 12th recycles.

Table 7. The comparison between two catalysts.

	Catalyst	Time (h)	Conversion (%)	Yield (%)
Benzonitrile	$Ni_{12.4}$-30	3.5	100	77
	Raney Ni	4	100	26
Nitrobenzene	$Ni_{12.4}$-30	5	100	97
	Raney Ni	2	100	97
Benzaldehyde	$Ni_{12.4}$-30	2	100	96
	Raney Ni	3.5	100	92

3. Experimental

3.1. Catalyst Preparation

A typical process for catalyst preparation was followed. $NiCl \cdot 6H_2O$ was dissolve in 50 mL deionized water, the TS-1 molecular sieve was added, and this was stirred for 0.5 h at 30 °C. After that, 1 M $NaBH_4$ solution was added to the suspension while stirring, and stirring continued for 2 h. Finally, the suspension was filtered and washed to obtain a solid catalyst, which was then subjected to vacuum drying at 50 °C for 2 h. The catalyst is named Ni_w-T, in which "w" and "T" represent the mass content of nickel (compared to TS-1) and the temperature for catalyst preparation, respectively (see more details in the Support Information).

3.2. Catalyst Characterization

ICP data was obtained from Agilent-ICPOES730 (Santa Clara, CA, USA). The X-ray diffraction (XRD) patterns were measured at room temperature using D/max-rA with Cu-Kα radiation generated at 10 mA and 40 kV. The X-ray photoelectron spectroscopy (XPS) analysis was carried out by using Thermo Scientific K-Alpha (Waltham, MA, USA) with Al-Kα radiation. The morphological information was measured by a transmission electron microscope (TEM) conducted using a Thermo Scientific Talos F200S coupled with X-ray spectroscopy (EDS).

3.3. Catalyst Activity Measurement

Nitrile, catalyst and solvent were mixed in a 100 mL volume autoclave equipped with PTFE and magnetic pellet. The kettle was filled with 0.5 MPa ammonia gas and heated to 120 °C; at this temperature, 4.0 MPa H_2 was pressed in, and then reaction was started. During the process, the system pressure was controlled between 4.0 ± 0.1 MPa. After reaction (under constant pressure), the autoclave was cooled and degassed, the reaction solution was filtered to recover the catalyst, and the filtrate was concentrated to determine the conversion by GC. ^1H NMR and ^{13}C NMR spectrum data were recorded by a Bruker DRX-400 spectrometer (Billerica, MA, USA) using $CDCl_3$ or DMSO-d_6 as solvent at 298 K. Gas chromatography (GC) was performed on Agilent chromatography with a SE54 column. More details in support information (for spectra, see Figures S4–S72).

4. Conclusions

In conclusion, we have presented a nanostructured nickel boride catalyst than can be used for the efficient reduction of nitrile, nitro compounds, and imine groups. This catalyst was prepared via chemical reduction at room temperature with an average particle size of 17 nm and homogeneous distribution. XRD and SAED justified the amorphous structure of nickel boride. The $Ni_{12.4}$-30 catalyst has been proven to be highly active towards all three reactions, with high TOF values. Furthermore, recycling tests proved that the catalysts are robust for consecutive use. In addition, the performance of the $Ni_{12.4}$-30 catalyst is comparable to commercial Raney Ni, but it is safer for storage. The promising prospect of the nickel boride catalyst for industrial application has thus been proven.

Supplementary Materials: The following supporting information can be downloaded at: https://www.mdpi.com/article/10.3390/ijms23169337/s1.

Author Contributions: Conceptualization, D.K. and S.Z.; methodology, D.K. and S.Z.; writing—original draft preparation, D.K.; writing—review and editing, S.Z. All authors have read and agreed to the published version of the manuscript.

Funding: Generous financial support by the National Natural Science Foundation of China (21878265).

Institutional Review Board Statement: Not applicable.

Informed Consent Statement: Not applicable.

Data Availability Statement: Additional figures are available in the Supplementary Materials.

Conflicts of Interest: The authors declare no conflict of interest.

References

1. Ezelarab, H.A.A.; Abbas, S.H.; Hassan, H.A.; Abuo-Rahma, G.E.A. Recent updates of fluoroquinolones as antibacterial agents. *Arch. Pharm.* **2018**, *351*, e1800141. [CrossRef] [PubMed]
2. Ahmadi, T.; Mohammadi Ziarani, G.; Bahar, S.; Badiei, A. Domino synthesis of quinoxaline derivatives using SBA-Pr-NH2 as a nanoreactor and their spectrophotometric complexation studies with some metals ions. *J. Iran. Chem. Soc.* **2018**, *15*, 1153–1161. [CrossRef]
3. Vo, N.B.; Nguyen, L.A.; Pham, T.L.; Doan, D.T.; Nguyen, T.B.; Ngo, Q.A. Straightforward access to new vinca-alkaloids via selective reduction of a nitrile containing anhydrovinblastine derivative. *Tetrahedron Lett.* **2017**, *58*, 2503–2506. [CrossRef]
4. Wang, P.; Zhao, X.H.; Wang, Z.Y.; Meng, M.; Li, X.; Ning, Q. Generation 4 polyamidoamine dendrimers is a novel candidate of nano-carrier for gene delivery agents in breast cancer treatment. *Cancer Lett.* **2010**, *298*, 34–49. [CrossRef]
5. Yemul, O.; Imae, T. Synthesis and characterization of poly(ethyleneimine) dendrimers. *Colloid Polym. Sci.* **2008**, *286*, 747–752. [CrossRef]
6. Garduño, J.A.; García, J.J. Non-Pincer Mn(I) Organometallics for the Selective Catalytic Hydrogenation of Nitriles to Primary Amines. *ACS Catal.* **2018**, *9*, 392–401. [CrossRef]
7. Mastalir, M.; Stöger, B.; Pittenauer, E.; Puchberger, M.; Allmaier, G.; Kirchner, K. Air Stable Iron(II) PNP Pincer Complexes as Efficient Catalysts for the Selective Alkylation of Amines with Alcohols. *Adv. Synth. Catal.* **2016**, *358*, 3824–3831. [CrossRef]
8. Baumann, W.; Spannenberg, A.; Pfeffer, J.; Haas, T.; Kockritz, A.; Martin, A.; Deutsch, J. Utilization of common ligands for the ruthenium-catalyzed amination of alcohols. *Chem. Eur. J.* **2013**, *19*, 17702–17706. [CrossRef]
9. Zhuang, X.; Liu, J.; Zhong, S.; Ma, L. Selective catalysis for the reductive amination of furfural toward furfurylamine by graphene-co-shelled cobalt nanoparticles. *Green Chem.* **2022**, *24*, 271–284. [CrossRef]
10. Zhang, Y.; Yang, H.; Chi, Q.; Zhang, Z. Nitrogen-Doped Carbon-Supported Nickel Nanoparticles: A Robust Catalyst to Bridge the Hydrogenation of Nitriles and the Reductive Amination of Carbonyl Compounds for the Synthesis of Primary Amines. *ChemSusChem* **2019**, *12*, 1246–1255. [CrossRef]
11. Hahn, G.; Kunnas, P.; de Jonge, N.; Kempe, R. General synthesis of primary amines via reductive amination employing a reusable nickel catalyst. *Nat. Catal.* **2018**, *2*, 71–77. [CrossRef]
12. Coeck, R.; De Vos, D.E. One-pot reductive amination of carboxylic acids: A sustainable method for primary amine synthesis. *Green Chem.* **2020**, *22*, 5105–5114. [CrossRef]
13. Citoler, J.; Derrington, S.R.; Galman, J.L.; Bevinakatti, H.; Turner, N.J. A biocatalytic cascade for the conversion of fatty acids to fatty amines. *Green Chem.* **2019**, *21*, 4932–4935. [CrossRef]
14. Antil, N.; Kumar, A.; Akhtar, N.; Newar, R.; Begum, W.; Dwivedi, A.; Manna, K. Aluminum Metal–Organic Framework-Ligated Single-Site Nickel(I) Hydride for Heterogeneous Chemoselective Catalysis. *ACS Catal.* **2021**, *11*, 3943–3957. [CrossRef]
15. Wang, C.; Jia, Z.; Zhen, B.; Han, M. Supported Ni Catalyst for Liquid Phase Hydrogenation of Adiponitrile to 6-Aminocapronitrile and Hexamethyenediamine. *Molecules* **2018**, *23*, 92. [CrossRef] [PubMed]
16. Konnerth, H.; Prechtl, M.H.G. Nitrile hydrogenation using nickel nanocatalysts in ionic liquids. *New J. Chem.* **2017**, *41*, 9594–9597. [CrossRef]
17. Cheng, H.; Meng, X.; Wu, C.; Shan, X.; Yu, Y.; Zhao, F. Selective hydrogenation of benzonitrile in multiphase reaction systems including compressed carbon dioxide over Ni/Al_2O_3 catalyst. *J. Mol. Catal. A Chem.* **2013**, *379*, 72–79. [CrossRef]
18. Segobia, D.J.; Trasarti, A.F.; Apesteguía, C.R. Hydrogenation of nitriles to primary amines on metal-supported catalysts: Highly selective conversion of butyronitrile to n-butylamine. *Appl. Cat. A Gen.* **2012**, *445–446*, 69–75. [CrossRef]
19. Liu, Y.; He, Y.; Quan, Z.; Cai, H.; Zhao, Y.; Wang, B. Mild palladium-catalysed highly efficient hydrogenation of C–N, C–NO2, and C–O bonds using H_2 of 1 atm in H_2O. *Green Chem.* **2019**, *21*, 830–838. [CrossRef]
20. Martina, K.; Baricco, F.; Tagliapietra, S.; Moran, M.J.; Cravotto, G.; Cintas, P. Highly efficient nitrobenzene and alkyl/aryl azide reduction in stainless steel jars without catalyst addition. *New J. Chem.* **2018**, *42*, 18881–18888. [CrossRef]
21. Göksu, H.; Ho, S.F.; Metin, Ö.; Korkmaz, K.; Mendoza Garcia, A.; Gültekin, M.S.; Sun, S. Tandem Dehydrogenation of Ammonia Borane and Hydrogenation of Nitro/Nitrile Compounds Catalyzed by Graphene-Supported NiPd Alloy Nanoparticles. *ACS Catal.* **2014**, *4*, 1777–1782. [CrossRef]

22. Zhang, T.; Zhang, Y.; Zhang, W.; Luo, M. A Convenient and General Reduction of Amides to Amines with Low-Valent Titanium. *Adv. Synth. Catal.* **2013**, *355*, 2775–2780. [CrossRef]
23. Amberchan, G.; Snelling, R.A.; Moya, E.; Landi, M.; Lutz, K.; Gatihi, R.; Singaram, B. Reaction of Diisobutylaluminum Borohydride, a Binary Hydride, with Selected Organic Compounds Containing Representative Functional Groups. *J. Org. Chem.* **2021**, *86*, 6207–6227. [CrossRef] [PubMed]
24. Zen, Y.-F.; Fu, Z.-C.; Liang, F.; Xu, Y.; Yang, D.-D.; Yang, Z.; Gan, X.; Lin, Z.-S.; Chen, Y.; Fu, W.-F. Robust Hydrogenation of Nitrile and Nitro Groups to Primary Amines Using Ni2P as a Catalyst and Ammonia Borane under Ambient Conditions. *Asian J. Org. Chem.* **2017**, *6*, 1589–1593. [CrossRef]
25. Maddani, M.R.; Moorthy, S.K.; Prabhu, K.R. Chemoselective reduction of azides catalyzed by molybdenum xanthate by using phenylsilane as the hydride source. *Tetrahedron* **2010**, *66*, 329–333. [CrossRef]
26. Zeynizadeh, B.; Mousavi, H.; Mohammad Aminzadeh, F. A hassle-free and cost-effective transfer hydrogenation strategy for the chemoselective reduction of arylnitriles to primary amines through in situ-generated nickelII dihydride intermediate in water. *J. Mol. Struct.* **2022**, *1255*. [CrossRef]
27. Liu, L.; Li, J.; Ai, Y.; Liu, Y.; Xiong, J.; Wang, H.; Qiao, Y.; Liu, W.; Tan, S.; Feng, S.; et al. A ppm level Rh-based composite as an ecofriendly catalyst for transfer hydrogenation of nitriles: Triple guarantee of selectivity for primary amines. *Green Chem.* **2019**, *21*, 1390–1395. [CrossRef]
28. Podyacheva, E.; Afanasyev, O.I.; Vasilyev, D.V.; Chusov, D. Borrowing Hydrogen Amination Reactions: A Complex Analysis of Trends and Correlations of the Various Reaction Parameters. *ACS Catal.* **2022**, *12*, 7142–7198. [CrossRef]
29. Lévay, K.; Hegedűs, L. Recent Achievements in the Hydrogenation of Nitriles Catalyzed by Transitional Metals. *Curr. Org. Chem.* **2019**, *23*, 1881–1900. [CrossRef]
30. Debellefon, C.; Fouilloux, P. Homogeneous and heterogeneous hydrogenation of nitriles in a liquid-phase—Chemical, mechanistic, and catalytic aspects. *Catal. Rev.* **1994**, *36*, 459–506. [CrossRef]
31. Chandrashekhar, V.G.; Senthamarai, T.; Kadam, R.G.; Malina, O.; Kašlík, J.; Zbořil, R.; Gawande, M.B.; Jagadeesh, R.V.; Beller, M. Silica-supported Fe/Fe–O nanoparticles for the catalytic hydrogenation of nitriles to amines in the presence of aluminium additives. *Nat. Catal.* **2021**, *5*, 20–29. [CrossRef]
32. Chakraborty, S.; Milstein, D. Selective Hydrogenation of Nitriles to Secondary Imines Catalyzed by an Iron Pincer Complex. *ACS Catal.* **2017**, *7*, 3968–3972. [CrossRef]
33. Lange, S.; Elangovan, S.; Cordes, C.; Spannenberg, A.; Jiao, H.; Junge, H.; Bachmann, S.; Scalone, M.; Topf, C.; Junge, K.; et al. Selective catalytic hydrogenation of nitriles to primary amines using iron pincer complexes. *Catal. Sci. Technol.* **2016**, *6*, 4768–4772. [CrossRef]
34. Chakraborty, S.; Leitus, G.; Milstein, D. Selective hydrogenation of nitriles to primary amines catalyzed by a novel iron complex. *Chem. Commun.* **2016**, *52*, 1812–1815. [CrossRef]
35. Mérel, D.S.; Do, M.L.T.; Gaillard, S.; Dupau, P.; Renaud, J.-L. Iron-catalyzed reduction of carboxylic and carbonic acid derivatives. *Coordin. Chem. Rev.* **2015**, *288*, 50–68. [CrossRef]
36. Bornschein, C.; Werkmeister, S.; Wendt, B.; Jiao, H.; Alberico, E.; Baumann, W.; Junge, H.; Junge, K.; Beller, M. Mild and selective hydrogenation of aromatic and aliphatic (di)nitriles with a well-defined iron pincer complex. *Nat. Commun.* **2014**, *5*, 4111. [CrossRef] [PubMed]
37. Sheng, M.; Yamaguchi, S.; Nakata, A.; Yamazoe, S.; Nakajima, K.; Yamasaki, J.; Mizugaki, T.; Mitsudome, T. Hydrotalcite-Supported Cobalt Phosphide Nanorods as a Highly Active and Reusable Heterogeneous Catalyst for Ammonia-Free Selective Hydrogenation of Nitriles to Primary Amines. *ACS Sustain. Chem. Eng.* **2021**, *9*, 11238–11246. [CrossRef]
38. Mitsudome, T.; Sheng, M.; Nakata, A.; Yamasaki, J.; Mizugaki, T.; Jitsukawa, K. A cobalt phosphide catalyst for the hydrogenation of nitriles. *Chem. Sci.* **2020**, *11*, 6682–6689. [CrossRef]
39. Formenti, D.; Mocci, R.; Atia, H.; Dastgir, S.; Anwar, M.; Bachmann, S.; Scalone, M.; Junge, K.; Beller, M. A State-of-the-Art Heterogeneous Catalyst for Efficient and General Nitrile Hydrogenation. *Chem. Eur. J.* **2020**, *26*, 15589–15595. [CrossRef]
40. Murugesan, K.; Senthamarai, T.; Sohail, M.; Alshammari, A.S.; Pohl, M.M.; Beller, M.; Jagadeesh, R.V. Cobalt-based nanoparticles prepared from MOF-carbon templates as efficient hydrogenation catalysts. *Chem. Sci.* **2018**, *9*, 8553–8560. [CrossRef]
41. Ferraccioli, R.; Borovika, D.; Surkus, A.-E.; Kreyenschulte, C.; Topf, C.; Beller, M. Synthesis of cobalt nanoparticles by pyrolysis of vitamin B12: A non-noble-metal catalyst for efficient hydrogenation of nitriles. *Catal. Sci. Technol.* **2018**, *8*, 499–507. [CrossRef]
42. Dai, H.; Guan, H. Switching the Selectivity of Cobalt-Catalyzed Hydrogenation of Nitriles. *ACS Catal.* **2018**, *8*, 9125–9130. [CrossRef]
43. Tokmic, K.; Jackson, B.J.; Salazar, A.; Woods, T.J.; Fout, A.R. Cobalt-Catalyzed and Lewis Acid-Assisted Nitrile Hydrogenation to Primary Amines: A Combined Effort. *J. Am. Chem. Soc.* **2017**, *139*, 13554–13561. [CrossRef] [PubMed]
44. Adam, R.; Bheeter, C.B.; Cabrero-Antonino, J.R.; Junge, K.; Jackstell, R.; Beller, M. Selective Hydrogenation of Nitriles to Primary Amines by using a Cobalt Phosphine Catalyst. *ChemSusChem* **2017**, *10*, 842–846. [CrossRef] [PubMed]
45. Shao, Z.; Fu, S.; Wei, M.; Zhou, S.; Liu, Q. Mild and Selective Cobalt-Catalyzed Chemodivergent Transfer Hydrogenation of Nitriles. *Angew. Chem. Int. Ed.* **2016**, *55*, 14653–14657. [CrossRef] [PubMed]
46. Chen, F.; Topf, C.; Radnik, J.; Kreyenschulte, C.; Lund, H.; Schneider, M.; Surkus, A.E.; He, L.; Junge, K.; Beller, M. Stable and Inert Cobalt Catalysts for Highly Selective and Practical Hydrogenation of C≡N and C=O Bonds. *J. Am. Chem. Soc.* **2016**, *138*, 8781–8788. [CrossRef]

47. Mukherjee, A.; Srimani, D.; Chakraborty, S.; Ben-David, Y.; Milstein, D. Selective Hydrogenation of Nitriles to Primary Amines Catalyzed by a Cobalt Pincer Complex. *J. Am. Chem. Soc.* **2015**, *137*, 8888–8891. [CrossRef]
48. Segobia, D.J.; Trasarti, A.F.; Apesteguía, C.R. Chemoselective hydrogenation of unsaturated nitriles to unsaturated primary amines: Conversion of cinnamonitrile on metal-supported catalysts. *Appl. Catal. A-Gen.* **2015**, *494*, 41–47. [CrossRef]
49. van der Waals, D.; Pettman, A.; Williams, J.M.J. Copper-catalysed reductive amination of nitriles and organic-group reductions using dimethylamine borane. *RSC Adv.* **2014**, *4*, 51845–51849. [CrossRef]
50. Lv, Y.; Hao, F.; Liu, P.; Xiong, S.; Luo, H. Liquid phase hydrogenation of adiponitrile over acid-activated sepiolite supported K–La–Ni trimetallic catalysts. *React. Kinet. Mech. Cat.* **2016**, *119*, 555–568. [CrossRef]
51. Konnerth, H.; Prechtl, M.H. Selective partial hydrogenation of alkynes to (Z)-alkenes with ionic liquid-doped nickel nanocatalysts at near ambient conditions. *Chem. Commun.* **2016**, *52*, 9129–9132. [CrossRef] [PubMed]
52. Jia, Z.; Zhen, B.; Han, M.; Wang, C. Liquid phase hydrogenation of adiponitrile over directly reduced Ni/SiO$_2$ catalyst. *Catal. Commun.* **2016**, *73*, 80–83. [CrossRef]
53. Cao, Y.; Niu, L.; Wen, X.; Feng, W.; Huo, L.; Bai, G. Novel layered double hydroxide/oxide-coated nickel-based core–shell nanocomposites for benzonitrile selective hydrogenation: An interesting water switch. *J. Catal.* **2016**, *339*, 9–13. [CrossRef]
54. Weber, S.; Stoger, B.; Kirchner, K. Hydrogenation of Nitriles and Ketones Catalyzed by an Air-Stable Bisphosphine Mn(I) Complex. *Org. Lett.* **2018**, *20*, 7212–7215. [CrossRef] [PubMed]
55. Elangovan, S.; Topf, C.; Fischer, S.; Jiao, H.; Spannenberg, A.; Baumann, W.; Ludwig, R.; Junge, K.; Beller, M. Selective Catalytic Hydrogenations of Nitriles, Ketones, and Aldehydes by Well-Defined Manganese Pincer Complexes. *J. Am. Chem. Soc.* **2016**, *138*, 8809–8814. [CrossRef]
56. Ma, K.; Liao, W.; Shi, W.; Xu, F.; Zhou, Y.; Tang, C.; Lu, J.; Shen, W.; Zhang, Z. Ceria-supported Pd catalysts with different size regimes ranging from single atoms to nanoparticles for the oxidation of CO. *J. Catal.* **2022**, *407*, 104–114. [CrossRef]
57. Yoshimura, M.; Komatsu, A.; Niimura, M.; Takagi, Y.; Takahashi, T.; Ueda, S.; Ichikawa, T.; Kobayashi, Y.; Okami, H.; Hattori, T.; et al. Selective Synthesis of Primary Amines from Nitriles under Hydrogenation Conditions. *Adv. Synth. Catal.* **2018**, *360*, 1726–1732. [CrossRef]
58. Saito, Y.; Ishitani, H.; Ueno, M.; Kobayashi, S. Selective Hydrogenation of Nitriles to Primary Amines Catalyzed by a Polysilane/SiO$_2$-Supported Palladium Catalyst under Continuous-Flow Conditions. *ChemistryOpen* **2017**, *6*, 211–215. [CrossRef]
59. Lu, S.; Wang, J.; Cao, X.; Li, X.; Gu, H. Selective synthesis of secondary amines from nitriles using Pt nanowires as a catalyst. *Chem. Commun.* **2014**, *50*, 3512–3515. [CrossRef]
60. Muratsugu, S.; Kityakarn, S.; Wang, F.; Ishiguro, N.; Kamachi, T.; Yoshizawa, K.; Sekizawa, O.; Uruga, T.; Tada, M. Formation and nitrile hydrogenation performance of Ru nanoparticles on a K-doped Al$_2$O$_3$ surface. *Phys. Chem. Chem. Phys.* **2015**, *17*, 24791–24802. [CrossRef] [PubMed]
61. Segobia, D.J.; Trasarti, A.F.; Apesteguia, C.R. Conversion of butyronitrile to butylamines on noble metals: Effect of the solvent on catalyst activity and selectivity. *Catal. Sci. Technol.* **2014**, *4*, 4075–4083. [CrossRef]
62. Xie, X.F.; Liotta, C.L.; Eckert, C.A. CO2-protected amine formation from nitrile and imine hydrogenation in gas-expanded liquids. *Ind. Eng. Chem. Res.* **2004**, *43*, 7907–7911. [CrossRef]
63. Rajesh, K.; Dudle, B.; Blacque, O.; Berke, H. Homogeneous Hydrogenations of Nitriles Catalyzed by Rhenium Complexes. *Adv. Synth. Catal.* **2011**, *353*, 1479–1484. [CrossRef]
64. Chatterjee, M.; Sato, M.; Kawanami, H.; Yokoyama, T.; Suzuki, T.; Ishizaka, T. An Efficient Hydrogenation of Dinitrile to Aminonitrile in Supercritical Carbon Dioxide. *Adv. Synth. Catal.* **2010**, *352*, 2394–2398. [CrossRef]
65. Monguchi, Y.; Mizuno, M.; Ichikawa, T.; Fujita, Y.; Murakami, E.; Hattori, T.; Maegawa, T.; Sawama, Y.; Sajiki, H. Catalyst-Dependent Selective Hydrogenation of Nitriles: Selective Synthesis of Tertiary and Secondary Amines. *J. Org. Chem.* **2017**, *82*, 10939–10944. [CrossRef] [PubMed]
66. Szostak, M.; Sautier, B.; Spain, M.; Procter, D.J. Electron transfer reduction of nitriles using SmI$_2$-Et$_3$N-H$_2$O: Synthetic utility and mechanism. *Org. Lett.* **2014**, *16*, 1092–1095. [CrossRef] [PubMed]
67. Lopez-De Jesus, Y.M.; Johnson, C.E.; Monnier, J.R.; Williams, C.T. Selective Hydrogenation of Benzonitrile by Alumina-Supported Ir-Pd Catalysts. *Top. Catal.* **2010**, *53*, 1132–1137. [CrossRef]
68. Molnar, A.; Smith, G.V.; Bartok, M. New catalytic materials from amorphous metal-alloys. *Adv. Catal.* **1989**, *36*, 329–383.
69. Li, H.; Xu, Y.; Li, H.; Deng, J.F. Gas-phase hydrogenation of adiponitrile with high selectivity to primary amine over supported Ni-B amorphous catalysts. *Appl. Catal. A-Gen.* **2001**, *216*, 51–58. [CrossRef]
70. Wang, W.-J.; Qiao, M.-H.; Li, H.-X.; Deng, J.-F. Partial Hydrogenation of Cyclopentadiene over Amorphous NiB Alloy on a-Alumina and Titania-Modifiedd a-Alumina. *J. Chem. Technol. Biotechnol.* **1998**, *72*, 280–284. [CrossRef]
71. Chiang, S.-J.; Yang, C.-H.; Chen, Y.-Z.; Liaw, B.-J. High-active nickel catalyst of NiB/SiO$_2$ for citral hydrogenation at low temperature. *Appl. Catal. A-Gen.* **2007**, *326*, 180–188. [CrossRef]
72. Li, Y.; Zhu, G.; Wang, Y.; Chai, Y.; Liu, C. Preparation, mechanism and applications of oriented MFI zeolite membranes: A review. *Microporous Mesoporous Mater.* **2021**, *312*, 110790. [CrossRef]
73. Huybrechts, D.R.C.; Debruycker, L.; Jacobs, P.A. Oxyfunctionalization of alkanes with hydrogen-peroxide on titanium silicalite. *Nature* **1990**, *345*, 240–242. [CrossRef]

74. Lu, J.-Q.; Li, N.; Pan, X.-R.; Zhang, C.; Luo, M.-F. Direct propylene epoxidation with H_2 and O_2 over in modified Au/TS-1 catalysts. *Catal. Commun.* **2012**, *28*, 179–182. [CrossRef]
75. Nishimura, S. *Handbook of Heterogeneous Catalytic Hydrogenation for Organic Synthesis*; J. Wiley: New York, NY, USA, 2001.
76. Liu, Y.; Zhou, K.; Lu, M.; Wang, L.; Wei, Z.; Li, X. Acidic/Basic Oxides-Supported Cobalt Catalysts for One-Pot Synthesis of Isophorone Diamine from Hydroamination of Isophorone Nitrile. *Ind. Eng. Chem. Res.* **2015**, *54*, 9124–9132. [CrossRef]
77. Chojecki, A.; Veprek-Heijman, M.; Müller, T.E.; Schärringer, P.; Veprek, S.; Lercher, J.A. Tailoring Raney-catalysts for the selective hydrogenation of butyronitrile to n-butylamine. *J. Catal.* **2007**, *245*, 237–248. [CrossRef]
78. Gluhoi, A.C.; Mărginean, P.; Stănescu, U. Effect of supports on the activity of nickel catalysts in acetonitrile hydrogenation. *Appl. Catal. A-Gen.* **2005**, *294*, 208–214. [CrossRef]
79. Huang, J.; Han, J.; Wang, R.; Zhang, Y.; Wang, X.; Zhang, X.; Zhang, Z.; Zhang, Y.; Song, B.; Jin, S. Improving Electrocatalysts for Oxygen Evolution Using $Ni_xFe_{3-x}O_4$/Ni Hybrid Nanostructures Formed by Solvothermal Synthesis. *ACS Energy Lett.* **2018**, *3*, 1698–1707. [CrossRef]
80. Li, H.; Li, H.X.; Dai, W.L.; Wang, W.J.; Fang, Z.G.; Deng, J.F. XPS studies on surface electronic characteristics of Ni-B and Ni-P amorphous alloy and its correlation to their catalytic properties. *Appl. Surf. Sci.* **1999**, *152*, 25–34. [CrossRef]
81. Jiang, W.J.; Niu, S.; Tang, T.; Zhang, Q.H.; Liu, X.Z.; Zhang, Y.; Chen, Y.Y.; Li, J.H.; Gu, L.; Wan, L.J.; et al. Crystallinity-Modulated Electrocatalytic Activity of a Nickel(II) Borate Thin Layer on Ni_3B for Efficient Water Oxidation. *Angew. Chem. Int. Ed.* **2017**, *56*, 6572–6577. [CrossRef] [PubMed]
82. Wang, L.; Li, W.; Zhang, M.; Tao, K. The interactions between the NiB amorphous alloy and TiO_2 support in the NiB/TiO_2 amorphous catalysts. *Appl. Catal. A-Gen.* **2004**, *259*, 185–190. [CrossRef]

Article

Copper Foam as Active Catalysts for the Borylation of α, β-Unsaturated Compounds

Kewang Zheng [1,2], Miao Liu [1], Zhifei Meng [1], Zufeng Xiao [1], Fei Zhong [1,*], Wei Wang [1,*] and Caiqin Qin [1]

[1] College of Chemistry and Materials Science, Hubei Engineering University, Xiaogan 432000, China; kewang1104@126.com (K.Z.); liumiao0727@163.com (M.L.); meng16301@163.com (Z.M.); chemhbeu@163.com (Z.X.); qincq@hbeu.edu.cn (C.Q.)

[2] Hubei Key Laboratory of Biological Resources and Environmental Biotechnology, Wuhan University, Wuhan 430079, China

* Correspondence: zhong4536188@hbeu.edu.cn (F.Z.); weiwang@hbeu.edu.cn (W.W.)

Abstract: The use of simple, inexpensive, and efficient methods to construct carbon–boron and carbon–oxygen bonds has been a hot research topic in organic synthesis. We demonstrated that the desired β-boronic acid products can be obtained under mild conditions using copper foam as an efficient heterogeneous catalyst. The structure of copper foam before and after the reaction was investigated by polarized light microscopy (PM), scanning electron microscopy (SEM), and transmission electron microscopy (TEM), and the results have shown that the structure of the catalyst copper foam remained unchanged before and after the reaction. The XPS test results showed that the Cu(0) content increased after the reaction, indicating that copper may be involved in the boron addition reaction. The specific optimization conditions were as follows: CH_3COCH_3 and H_2O were used as mixed solvents, 4-methoxychalcone was used as the raw material, 8 mg of catalyst was used and the reaction was carried out at room temperature and under air for 10 h. The yield of the product obtained was up to 92%, and the catalytic efficiency of the catalytic material remained largely unchanged after five cycles of use.

Citation: Zheng, K.; Liu, M.; Meng, Z.; Xiao, Z.; Zhong, F.; Wang, W.; Qin, C. Copper Foam as Active Catalysts for the Borylation of α, β-Unsaturated Compounds. *Int. J. Mol. Sci.* **2022**, *23*, 8403. https://doi.org/10.3390/ijms23158403

Academic Editor: Shaodong Zhou

Received: 19 July 2022
Accepted: 27 July 2022
Published: 29 July 2022

Publisher's Note: MDPI stays neutral with regard to jurisdictional claims in published maps and institutional affiliations.

Copyright: © 2022 by the authors. Licensee MDPI, Basel, Switzerland. This article is an open access article distributed under the terms and conditions of the Creative Commons Attribution (CC BY) license (https://creativecommons.org/licenses/by/4.0/).

Keywords: porous materials; copper foam; catalytic; aqueous phase reaction; $B_2(pin)_2$; α, β-unsaturated ketone

1. Introduction

The addition reaction of diborons with olefin is an effective method to obtain organoboron compounds [1–3]. Organoboronic acid compounds are important intermediates in drug molecules and organic synthesis, where C-B bonds could be converted into C-C, C-H, and C-O bonds by simple reactions [4–6]. Therefore, the efficient construction of carbon–boron bonds, especially the synthesis of chiral organoboronic acid compounds, has been a hot topic of interest for researchers.

Current methods for the synthesis of C-B bonds include C-X bond boronization [7], C-H bond boronization [8], C-O bond boronization [9], and addition boronization of olefins [10–13]. Of them, addition boronization of olefins has received a lot of attention from organic chemists, due to the easy source of olefin reaction substrates and the variety of reaction types. Earlier studies used simple olefins as substrates and catecholborane as a monofunctional borohydride reagent to construct C-B bonds. The first rhodium-catalyzed olefin addition reaction was reported by Mannig et al. [14] in 1985. Burgess et al. [15–17] subsequently investigated rhodium-based catalytic systems for the asymmetric borohydration of olefins and obtained olefin boron addition products by the addition of chiral ligands with an enantioselectivity of up to 76%. Compared to simple olefins, α, β-unsaturated compounds are more reactive and could be used as substrates to selectively form C-B bonds at the β-position, so methods [18–23] to achieve conjugated boron addition reactions via α, β-unsaturated compounds have been developed successively. Transition metals such as rhodium [24,25], palladium [26,27], platinum [28,29], nickel [30,31] and copper [32,33] are mostly used as catalysts. The use of copper salts or copper complexes as catalysts will greatly reduce the cost of the reaction and therefore have an important place in the conjugation borylation of α, β-unsaturated compounds.

In 2000, Ito et al. [34] has achieved the first conjugated boronization of α, β-unsaturated ketones in DMF using copper sulphonate as a catalyst and bis(pinacolato)diborane polar solvents ($B_2(pin)_2$) as a boron source. In 2008, Lee [35] achieved the asymmetric synthesis of α, β-unsaturated esters and nitriles in a methanol/tetrahydrofuran mixture using cuprous chloride as the catalyst and adding sodium tert-butoxide as the base. In 2016, Ding [36] reported the highly chemoselective catalytic reduction of C=C in α, β-unsaturated ketones with the $CuBr/B_2pin_2$ system. The boron addition reaction in water is more in line with the green requirements [12]. Copper(II) salt has become a new hot spot for the study of boron addition reactions as the copper(II) salt is more stable in water than the copper(I) salt. The groups of Santos [22], Kobayash [37], and Casar et al. [38] have successively studied the boron addition reactions of α, β-unsaturated compounds using various copper(II) salts in water. Among them, the Kobayashi group prepared three ligands that were stable and active in the aqueous phase. In recent years, some Cu(0) nanoparticles have also been used as catalysts [23,39] for boron addition reactions of α, β-unsaturated compounds.

Foam metals are a new type of functional material with a large number of pores within their structure. Foam metals have excellent physical properties, such as porosity, high specific surface area, high mechanical strength, and heat resistance, making them a promising candidate for applications such as carriers for catalytic materials [39–42].

In previous work, we immobilized copper salts or copper nanoparticles on Zeolite [43], chitosan microspheres [44], chitosan films [45], and cellulose [46] and they all had high catalytic activity in the boron addition reaction. As a continuation of the above work, we believe that copper foam can also be used directly as catalysts for boronization processes. Therefore, we report here the performance of copper foam as a highly active and recyclable catalyst for boron addition to α, β-unsaturated acceptors in aqueous media. We demonstrate that the desired β-boronic acid product can be obtained under mild conditions using copper foam as an efficient heterogeneous catalyst.

2. Results and Discussion

2.1. Optimization of Reaction Conditions

Many variables are involved in the yield of the boron addition reaction product. In this experiment, the reaction conditions were optimized in the hope of increasing the yield of the boron addition product. Considering the results of our previous work, we chose the specific reaction conditions set as follows: 4-methoxychalcone (**1a**, 0.2 mmol), copper foam (10 mg), solvent (2 mL), and the reaction was carried out for 12 h in air and at room temperature.

It has been reported that proton-type solvents [37,46], such as methanol, can better promote the boron addition reaction of α, β-unsaturated receptors. Therefore, we first chose the proton solvent as the solvent for the reaction. The different polar solvents are screened in Table 1, Entry 1–5, and the yields of the polar solvents methanol, ethanol i-PrOH, DMF, and water were 67%, 53%, 25%, <5%, and 8%, respectively. When non-protonic solvents ether, toluene, dichloromethane, and tetrahydrofuran were used as solvents for the reaction, none of the reaction yields exceeded 20% (Table 1, Entry 6–9). This also proves that in this reaction, the proton source can accelerate the reaction. When we chose acetone as the reaction solvent, the yield of the product was high (Table 1, Entry 10). Then, we investigated the effect of methanol, acetone, and water solvent mixture on the reaction. The experimental results showed (Entry 11 and Entry 12) that the addition of water can significantly increase the yield. Considering the solvent benefit of water, we then screened the ratio of acetone to water (Entry 13–15), and it can be seen that the yield of the product was 92% at the optimal ratio of acetone to water of 2:1 (v/v). The catalyst dosage has a certain effect on the yield of the reaction. We chose acetone and water (2:1, v/v) as the reaction solvent; the reaction did not proceed without copper foam, and the yield could reach 93% when the catalyst dosage was 8 mg (Entry 16–18). When the reaction time was 10 h, the yield could reach 92%, but a longer reaction time that this yielded basically no change. In summary, the optimum reaction conditions were template substrate 4-methoxychalcone (0.2 mmol), 8 mg catalyst, 1.35 mL CH_3COCH_3 and 0.65 mL H_2O at room temperature for 10 h.

Table 1. Optimization of reaction conditions.

Entry	Copper Foam (mg)	Solvent (2 mL)	Time (h)	Yield (%)
1	10	MeOH	12	67
2	10	EtOH	12	53
3	10	i-PrOH	12	24
4	10	DMF	12	<5%
5	10	H_2O	12	8%
6	10	Et_2O	12	10%
7	10	Toluene	12	8%
8	10	DCM	12	17%
9	10	THF	12	15%
10	10	acetone	12	63%
11	10	MeOH: H_2O (1:1)	12	87%
12	10	acetone:H_2O (1:1)	12	89%
13	10	acetone:H_2O (2:1)	12	92%
14	10	acetone:H_2O (1:2)	12	86%
15	10	acetone:H_2O (3:1)	12	91%
16	-	acetone:H_2O (2:1)	12	-
17	6	acetone:H_2O (2:1)	12	85%
18	8	acetone:H_2O (2:1)	12	92%
19	8	acetone:H_2O (2:1)	6	81%
20	8	acetone:H_2O (2:1)	8	91%
21	8	acetone:H_2O (2:1)	10	92%
22	8	acetone:H_2O (2:1)	14	91%

2.2. Substrate Expansion under Optimal Reaction Conditions

Numerous reports [43–46] have shown that the oxidation reaction of organoboron compounds is an equivalent transformation. To verify the utility of the copper foam catalytic system, we employed a "one-pot" strategy to extend the substrate by direct conversion of α, β-unsaturated acceptors to β-hydroxy compounds under optimal reaction conditions. The substrate expansion is shown in Table 2 below.

First, we investigated the catalytic activity of copper foam for monosubstituted chalcone derivatives (**1a–1g** and **1i–1k**). We found that monosubstituted chalcone derivatives possessing electron-donating groups such as methoxy, methyl, and benzyloxy groups gave high yields of β-hydroxy products (80–94%). Monosubstituted substrates possessing electron-absorbing groups on the benzene ring, such as trifluoromethyl, and fluorine have slightly lower reactivity (80–89%). The substitution group in the interposition has no effect on the activity of the catalyst. We then investigated the catalytic activity of copper foam for the disubstituted chalcone derivatives (**1l–1p**), and excellent yields (82–98%) were obtained for both boron addition reactions. From the above experiments, it can be seen that different positions and types of groups on the benzene ring have little effect on the yields. High yields (46–92%) were obtained for naphthalene ring substrates (**1q**), aliphatic substrates (**1r–1v**), and heterocyclic thiophene substrates (**1x, 1y**). The slightly lower yields for aliphatic and heterocyclic thiophene substrates could be attributed to the reduced activity of the substrates due to the reduced conjugation of the substrates. When α, β, γ, δ-unsaturated ketones substrates (**1w**) were catalyzed with copper foam, we found that 1–4 addition products were generated, which indicates that the copper foam material has good stereoselectivity. In conclusion, the results of the substrate adaptation study showed that the boron addition reactions of the chalcone series of derivatives (**1a–1y**) all yielded the target products in excellent yields under optimal reaction conditions. The ^{1}H NMR and ^{13}C NMR spectra of products **2a, 3aa**, and **3a-x** are shown in Figures S1–S27 in the Supplementary Materials.

Table 2. Substrate expansion of the borylation of α, β-unsaturated compounds reaction.

2.3. Gram-Scale Synthesis Reaction of 1a

We investigated the conversion of 4-methoxychalcone (**1a**) as a substrate to β-hydroxy compound (**3a**) by a one-pot method in a scale-up reaction (Scheme 1). It was shown that when the reactant (**1a**) was 5 mmol, the corresponding product (**3a**) was still obtained in a high yield of 90%.

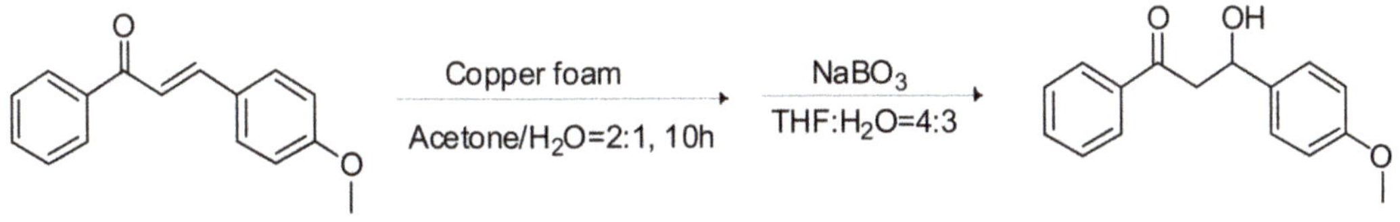

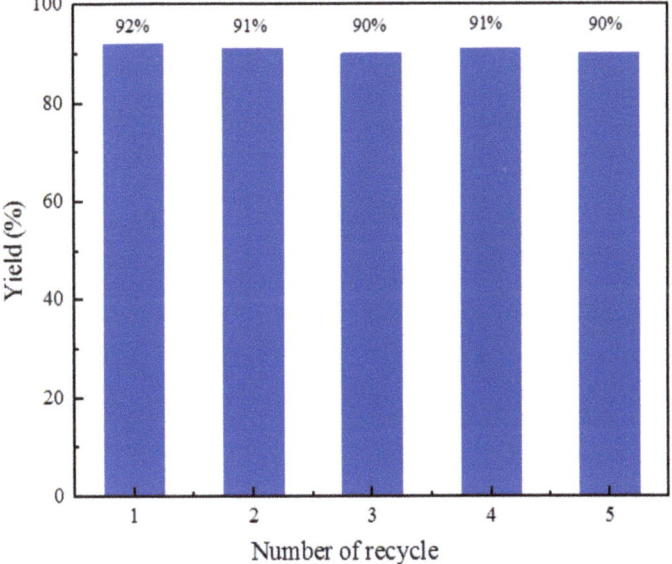

Scheme 1. Gram-scale synthesis reaction of **1a**.

2.4. Catalyst Reuse

To demonstrate the stability of copper foam, cycling reuse tests were performed under optimal conditions in this study. After each cycling test, the copper foam was collected, washed three times alternately with water and acetone, and dried under vacuum conditions of 60 °C for 12 h. As can be seen in Figure 1, the yield of copper foam remained essentially unchanged after five cycles.

Figure 1. Recovery and reuse of copper foam.

2.5. Characterization of Catalytic Materials and Mechanistic Studies

We investigated the state changes in copper foam before and after the reaction by means of polarizing microscope (PM), X-ray diffraction (XRD), scanning electron microscope (SEM), transmission electron microscope (TEM) and X ray photoelectron spectroscopy (XPS) analytical tests.

2.5.1. Polarizing Microscope Analysis

The copper foam was glued on the slide, and the brightness of the light was gradually adjusted from weak to strong to identify the color and glossy nature of the copper foam specimen, and the measurement condition was 100 or 200 times magnification. The results showed that the copper foam had many macroporous structures, the surface color of copper became darker and the luster decreased after the reaction. This indicates that the copper foam was involved in the boron addition reaction. The observation results of polarizing microscopy are shown in Figure 2.

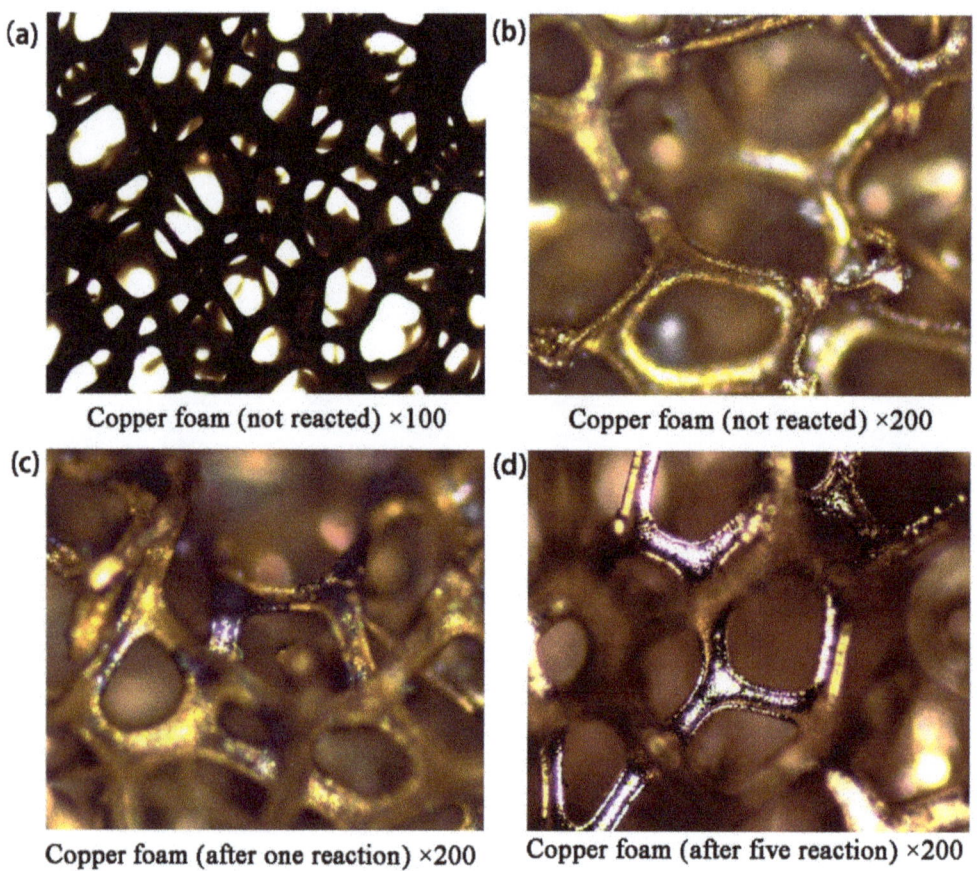

Figure 2. Morphology of copper foam under polarizing microscope.

2.5.2. SEM-EDS Analysis

Figure 3 shows the SEM images of copper foam before and after the reaction at different magnifications. It can be seen from the figure that the pore size of copper foam was approximately 1 mm. Compared with the raw material, the surface of the copper foam skeleton after five reactions was still smooth and the pore structure did not change significantly, indicating that the copper foam has a strong physical stability. The relative contents of copper and oxygen in the copper foam were detected by EDS analysis. Table S1 shows the atomic percentages measured by EDS. Compared with unreacted copper foam, the oxygen content of the reacted copper foam has decreased. The experimental results showed that the ratio of copper to oxygen increased slightly after the reaction, indicating that a small amount of copper oxide may have been turned into copper.

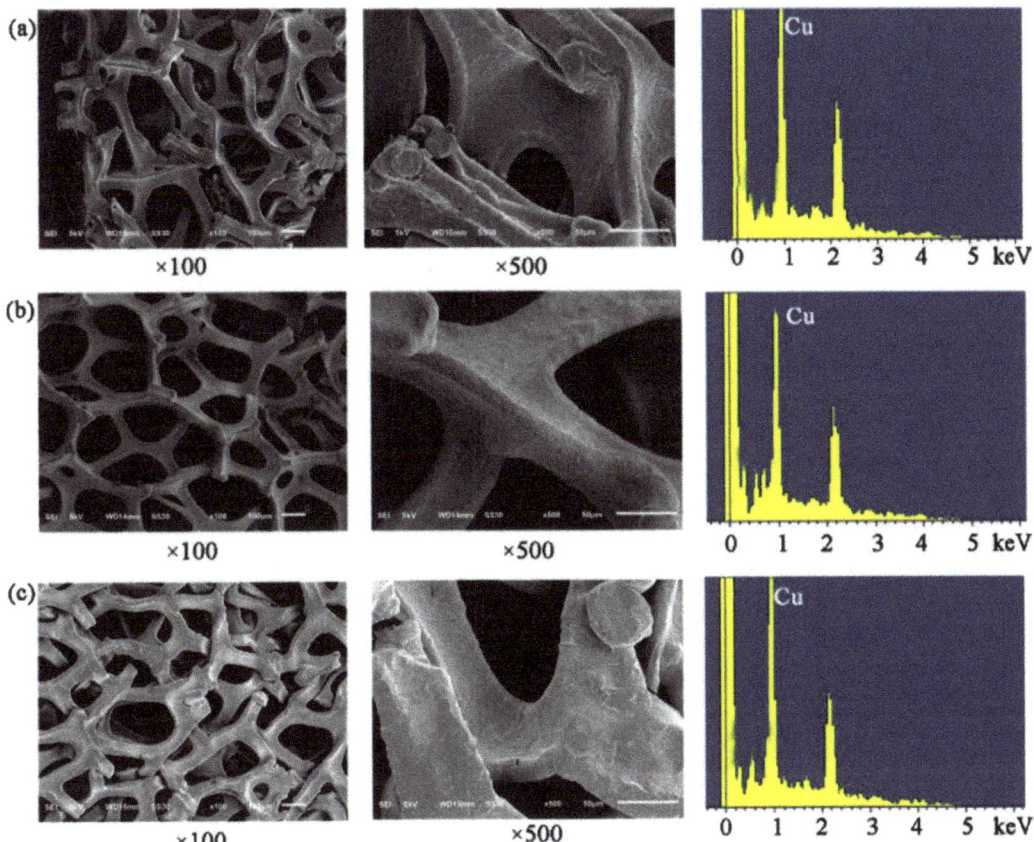

Figure 3. The SEM images of (**a**) copper foam (no reaction), (**b**) copper foam (after one reaction), and (**c**) copper foam (after five reactions).

2.5.3. XRD Analysis

To study the changes in the catalyst before and after the reaction, we characterized the unreacted copper foam, the copper foam after the first reaction, the copper foam after the second reaction, and the copper foam after the fifth reaction by XRD diffraction tests. As can be seen in Figure 4, all copper foams show characteristic diffraction peaks at 43.35°, 50.46°, and 74.17° of 2θ, which correspond to the (111), (200), and (220) crystal planes of the copper standard card, respectively (PDF: 04-0836). Additionally, the (200)/(111) ratios of these copper foams were in the range of 0.43–0.52, which is close to the (200)/(111) ratio of the copper standard card (0.46), indicating that the copper foam surfaces were not oxidized. No shedding of copper was observed in the solution throughout the reaction. Although a deepening of the color of the solution was observed during the reaction, the characteristic peaks of Cu^{2+} could not be observed from XRD because of the small amount of Cu^{2+} formed and dissolved in the solution.

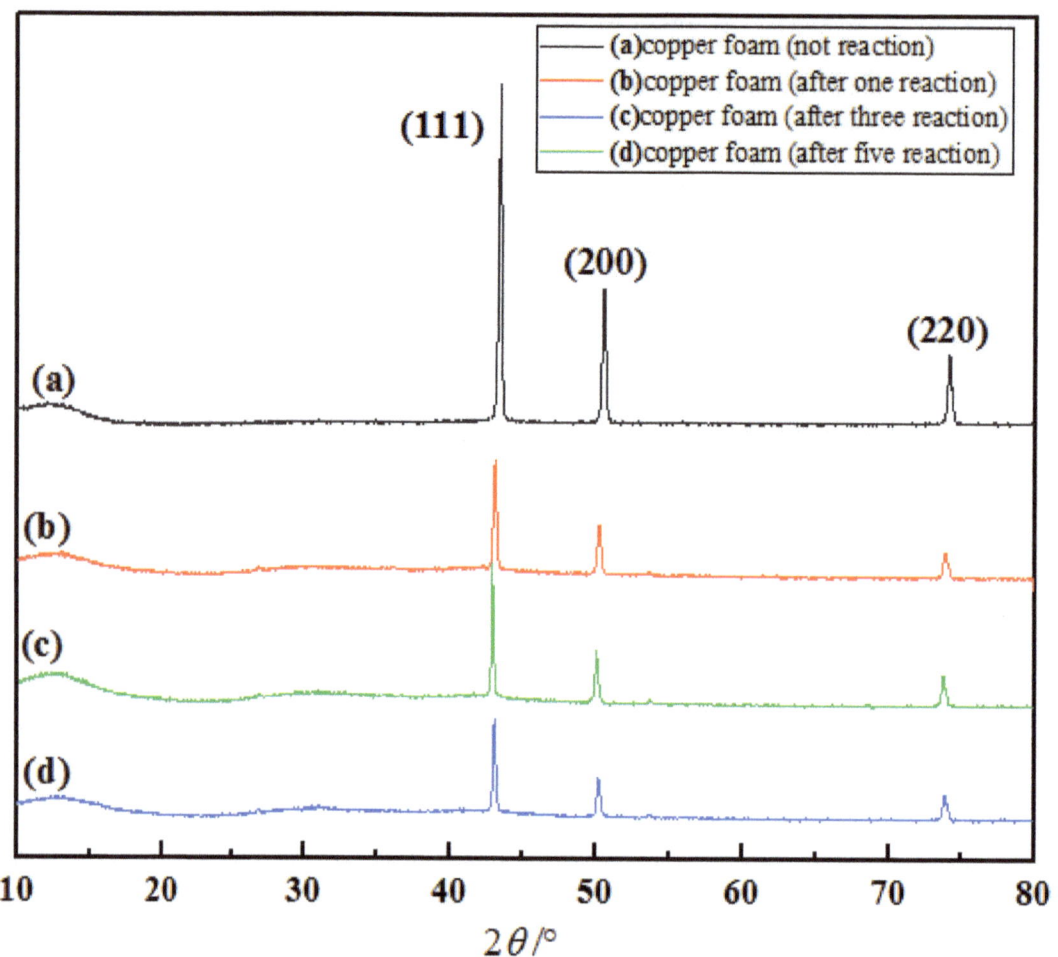

Figure 4. The XRD analysis of (**a**) copper foam (no reaction), (**b**) copper foam (after one reaction), (**c**) copper foam (after three reaction), and (**d**) copper foam (after five reactions).

2.5.4. XPS Analysis

To investigate the valence change and distribution of copper elements, we characterized the unreacted copper foam, the copper foam after the first reaction, and the copper foam after the fifth reaction by XPS tests. As shown in Figure 5a, two main peaks appear in the spectrum at binding energies of 932.6 eV and 952.54 eV, attributed to the Cu 2p3/2 and Cu 2p1/2 peaks of Cu^0, respectively [47], and the binding energy peaks of 934.2 eV and 955.1 eV point to the Cu 2p3/2 and Cu 2p1/2 peaks of Cu^{2+}, respectively. As shown in Figure 5b,c, the Cu 2p3/2 and Cu 2p1/2 peaks attributed to Cu^0 are still evident, while the Cu 2p3/2 and Cu 2p1/2 peaks of divalent Cu gradually decrease. This may be the involvement of the copper foam in the boron addition reaction and the oxide on the copper surface entering the solution.

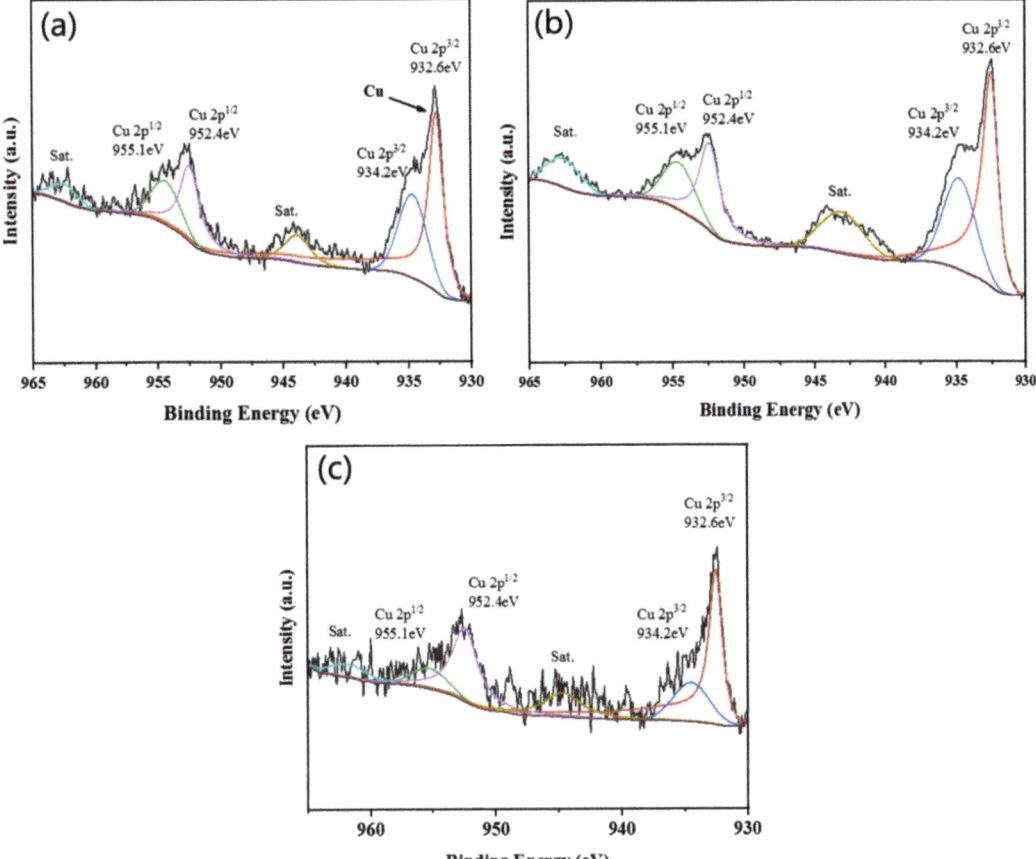

Figure 5. The Xps analysis of (**a**) copper foam (no reaction), (**b**) copper foam (after one reaction), and (**c**) copper foam (after five reactions).

2.5.5. TEM Analysis

The surface structures of copper foam and copper foam after primary reaction were analyzed by transmission electron microscopy. From Figure 6a,b, it can be seen that the crystal structure of the unreacted copper foam is relatively clear with a lattice spacing of 0.22 nm, which should correspond to the (1, 1, 1) crystal plane of face-centered cubic structure copper. As can be seen from Figure 6c,d, the crystal structure of the reacted copper foam is still relatively clear with a lattice spacing of 0.22 nm, which is not significantly different from that of the unreacted copper foam. Combined with the XPS test results, this indicates that the state of the reacted copper is unchanged.

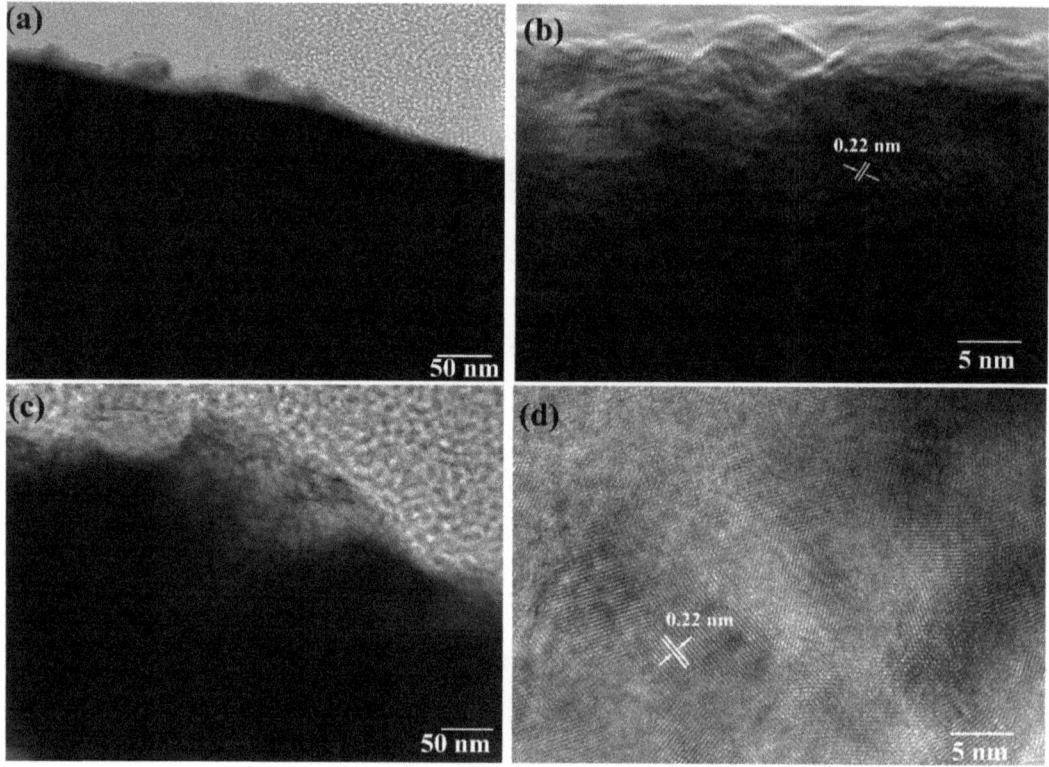

Figure 6. TEM analysis of copper foam. (**a**) TEM image of copper foam (no reaction, 80,000×), (**b**) TEM analysis of lattice spacing of copper foam (no reaction, 800,000×), (**c**) TEM image of copper foam (after one reaction, 80,000×), and (**d**) TEM analysis of lattice spacing of copper foam (after one reaction, 800,000×).

2.5.6. Isotopic Effects of Reactions

To investigate the reaction mechanism of copper foam in catalyzing the boron-addition reaction of α, β-unsaturated acceptors, we performed deuteration experiments using 4-methoxychalcone as raw material using the one-pot method, as shown in Table 3. The results are in general agreement with the previous experiments in Zhou's report [46]. Both MeOH and H_2O were used as solvent and proton sources during the catalytic cycle reaction. The yield of the boron addition product decreased with the addition of deuterium reagent, while at the β-position methylene of the carbonyl group, an H atom was replaced by a D atom. This result suggests that both H and D atoms can participate in the catalytic cycle as proton sources, and the protonation step of the reactants may be the rate-determining step in the overall catalytic process, with the protonation rate of H atoms being greater than that of D atoms.

Table 3. Deuterated experiments of 4-methoxychalcone.

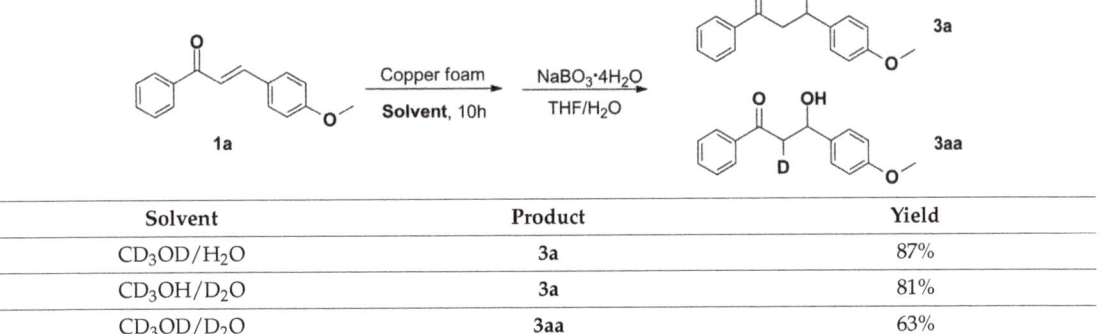

Solvent	Product	Yield
CD$_3$OD/H$_2$O	3a	87%
CD$_3$OH/D$_2$O	3a	81%
CD$_3$OD/D$_2$O	3aa	63%

2.6. Mechanistic Studies

Based on the previous and current experimental results of our group, we propose a possible reaction mechanism as shown in Figure 7. In the first step, B$_2$pin$_2$ breaks the B-B bond catalyzed by copper foam (A) to form the active copper–boron metal complex (B) and the by-product Bpin-OH. In the second step, the active copper–boron metal complex (B) undergoes an addition reaction guided by the carbonyl group in the α, β-unsaturated acceptor to give the intermediate C. Then, the intermediate C undergoes a transition state rearrangement to produce intermediate D. Finally, intermediate D exchanges protons in the solvent to produce the target product E while regenerating the catalytic material.

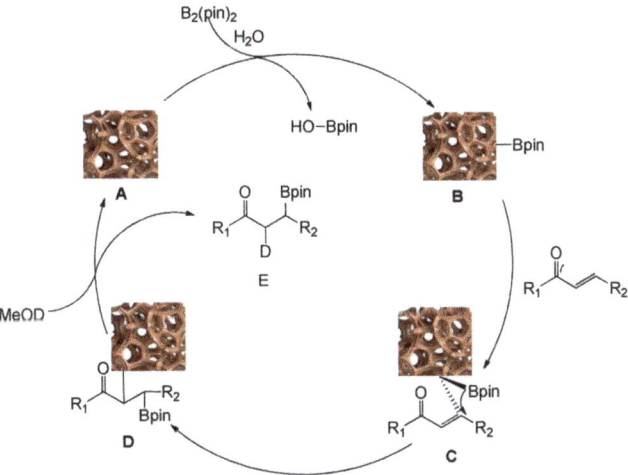

Figure 7. Possible reaction mechanisms.

2.7. Comparison of the Catalytic Performance of Different Copper Catalytic Materials

So far, several copper catalysts with high catalytic performance have been developed and applied to different coupling reactions of diborons with α, β-unsaturated compounds (Table 4). Zhu et al. [48,49] have studied the reaction performance of the catalysts in the presence of chiral ligands L2 or L3 using chitosan@Cu(OH)$_2$ or CuO catalysts and found that Cu(OH)$_2$ and CuO catalyzed the generation of β-addition products almost quantitatively. Zhou et al. [50] have found that the reaction required only 2 h to obtain

95% of the product using carbon black supported Cu^0 material as a catalyst. Other studies from our group [43–46] also showed that both lower-order and higher-order copper are excellent materials for the boron addition reaction. Compared with the above catalysts, the copper foam has a good catalytic effect (92% yield), does not require additional support and ligands, and is more easily separated, making it a more suitable catalyst for a large number of reactions.

Table 4. Comparison of catalytic performance of copper foam catalysts with other copper materials.

Entry	Catalyst	Support	Ligand	Yield (%)
1 [48]	$Cu(OH)_2$	-	L1	93%
2 [48]	$Cu(OH)_2$	Chitosan	L1	90%
3 [48]	$Cu(OH)_2$	Chitosan	L2	100%
4 [48]	$Cu(OH)_2$	Chitosan	-	90%
5 [49]	CuO	-	L3	96%
6 [49]	$Cu(OAc)_2$	Chitosan	-	91%
7 [46]	CuI	Cellulosic	-	96%
8 [50]	Cu_2O	-	-	NR
9 [50]	Cu	Carbon black	-	95%
10 [44]	Cu	Chitosan	-	93%
11 [45]	Cu	Chitosan/PVA	-	87%
12	copper foam	-	-	92%

3. Materials and Methods

3.1. Materials

Copper foam and all of the above reagents were purchased from Energy Chemical. NMR (Bruker Avance III 400 Hz, Berlin, Germany) was used to verify the structure of the products. Scanning Electron Microscopy (JEOL, JSM-6510, Tokyo, Japan) were used to measure the morphology and sizes of the modified chitosan microspheres. The FTIR spectra of the microspheres were obtained using a Nicolet iS5 Spectrophotometer (Thermo, Austin, TX, USA) to investigate possible interactions between the microspheres and the nano-copper. Microscopy (TEM, JEOL-2100F, Tokyo, Japan), X-ray Photoelectron Spectroscopy (XPS, ESCALAB 250xi, Thermo, USA) and X-ray Diffractometry (XRD, S2, RIGAKU, Tokyo, Japan) were used to obtain the elemental valence of the catalyst copper. Purification of the product was carried out using column chromatography (silica gel, 200–300 mesh).

3.2. Arylboronic Acid Self-Coupling Reaction

Copper foam (010 mg), 4-methoxychalcone (**1a**, 0.2 mmol) and bis (Pinacolato) diboron (0.3 mmol) were added to the reaction flask and a solvent mixture of acetone and water (v:v = 1:1) was added to react for 12 h, at room temperature. It was filtered, washed with ethyl acetate and the solvent was evaporated to give a mixture containing β-boronic acid compound **2a**. Then, the mixture was added to a mixed solution of THF and H_2O containing 0.2 mmol of $NaBO_3 \cdot 4H_2O$ and stirred for 4 h, at room temperature. After stopping the reaction, extraction and fractionation, vacuum-drying was performed. The target product **3a** was isolated using a mixture of ethyl acetate and petroleum ether as eluent (v:v = 1:4–1:10). The reaction route is shown in Figure 8. The structure of the product was verified by NMR.

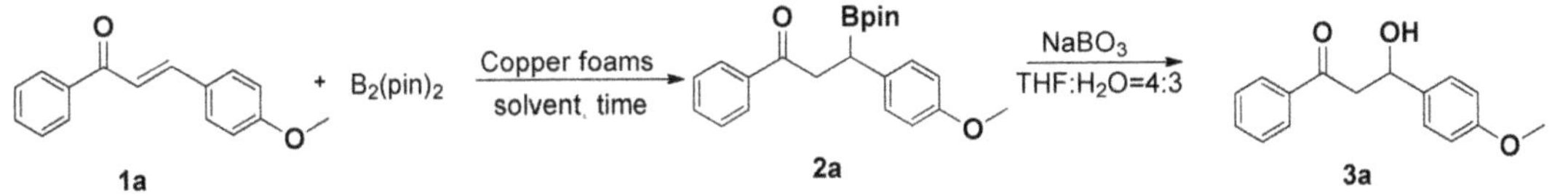

Figure 8. The boron addition reaction general route.

4. Conclusions

In this paper, copper foam was used directly as a catalyst for the borylation addition reaction of α, β-unsaturated compounds under mild conditions. The experimental results show that copper foam is an easy separation, easy recovery and high efficiency catalyst, and the yield of the product is still up to 90% after five cycles. This catalyst is an important reference for the green synthesis and industrial application of C-O compounds.

Supplementary Materials: The supporting information can be downloaded at: https://www.mdpi.com/article/10.3390/ijms23158403/s1.

Author Contributions: Experiment design, W.W. and F.Z.; experiments and writing—draft manuscript, K.Z.; review and editing, Z.M.; data curation, M.L.; writing—review and editing, Z.X., and C.Q. All authors have read and agreed to the published version of the manuscript.

Funding: This research was funded by Hubei Provincial Natural Science Foundation of China (Grant Number 2019CFB363).

Institutional Review Board Statement: Not applicable.

Informed Consent Statement: Not applicable.

Data Availability Statement: Supporting data can be obtained from the corresponding authors.

Acknowledgments: The authors are grateful for the Science and Technology Department of Hubei Province of China (Grant Number 2019CFB363).

Conflicts of Interest: The authors declare no conflict of interest.

References

1. Suzuki, N.; Suzuki, T.; Ota, Y.; Nakano, T.; Kurihara, M.; Okuda, H.; Yamori, T.; Tsumoto, H.; Nakagawa, H.; Miyata, N. Design, synthesis, and biological activity of boronic acid-based histone deacetylase inhibitors. *J. Med. Chem.* **2009**, *52*, 2909–2922. [CrossRef] [PubMed]
2. Adamczyk-Woźniak, A.; Komarovska-Porokhnyavets, O.; Misterkiewicz, B.; Novikov, V.P.; Sporzyński, A. Biological activity of selected boronic acids and their derivatives. *Appl. Organomet. Chem.* **2012**, *26*, 390–393. [CrossRef]
3. Halbus, A.F.; Horozov, T.S.; Paunov, V.N. Strongly enhanced antibacterial action of copper oxide nanoparticles with boronic acid surface functionality. *ACS Appl. Mater. Interfaces* **2019**, *11*, 12232–12243. [CrossRef] [PubMed]
4. Miyaura, N. Organoboron compounds. *Cross-Coupling React.* **2002**, *29*, 11–59.
5. Ishiyama, T.; Miyaura, N. Metal-catalyzed reactions of diborons for synthesis of organoboron compounds. *Chem. Rec.* **2004**, *3*, 271–280. [CrossRef] [PubMed]
6. Tevyashova, A.N.; Chudinov, M.V. Progress in the medical chemistry of organoboron compounds. *Russ. Chem. Rev.* **2021**, *90*, 451. [CrossRef]
7. Manna, S.; Das, K.K.; Nandy, S.; Aich, D.; Paul, S.; Panda, S. A new avenue for the preparation of organoboron compounds via nickel catalysis. *Coord. Chem. Rev.* **2021**, *448*, 214165. [CrossRef]
8. Mkhalid, I.A.; Barnard, J.H.; Marder, T.B.; Murphy, J.M.; Hartwig, J.F. C−H activation for the construction of C−B bonds. *Chem. Rev.* **2010**, *110*, 890–931. [CrossRef]
9. Jin, S.; Dang, H.T.; Haug, G.C.; He, R.; Nguyen, V.D.; Nguyen, V.T.; Arman, H.D.; Schanze, K.S.; Larionov, O.V. Visible light-induced borylation of C–O, C–N, and C–X bonds. *J. Am. Chem. Soc.* **2020**, *142*, 1603–1613. [CrossRef]
10. Lee, K.-S.; Zhugralin, A.; Hoveyda, A. NHC-Catalyzed Boron Conjugate Additions to α,β-Unsaturated Carbonyl Compounds. *Synfacts* **2009**, *2009*, 0898.
11. Gao, M.; Thorpe, S.; Kleeberg, C.; Slebodnick, C.; Marder, T.; Santos, W. Catalytic Regioselective Boration of α, β-Unsaturated Compounds. *Synfacts* **2011**, *2011*, 0995.
12. Shilpa, T.; Neetha, M.; Anilkumar, G. Recent Trends and Prospects in the Copper-Catalysed "on Water" Reactions. *Adv. Synth. Catal.* **2021**, *363*, 1559–1582. [CrossRef]

13. Tsuda, T.; Choi, S.-M.; Shintani, R. Palladium-catalyzed synthesis of dibenzosilepin derivatives via 1, *n*-Palladium migration coupled with anti-Carbopalladation of alkyne. *J. Am. Chem. Soc.* **2021**, *143*, 1641–1650. [CrossRef] [PubMed]
14. Männig, D.; Nöth, H. Catalytic hydroboration with rhodium complexes. *Angew. Chem. Int. Ed. Engl.* **1985**, *24*, 878–879. [CrossRef]
15. Burgess, K.; Ohlmeyer, M.J. Enantioselective hydroboration mediated by homochiral rhodium catalysts. *J. Org. Chem.* **1988**, *53*, 5178–5179. [CrossRef]
16. Burgess, K.; Van der Donk, W.A.; Jarstfer, M.B.; Ohlmeyer, M.J. Further evidence for the role of d. pi.-p. pi. bonding in rhodium-mediated hydroborations. *J. Am. Chem. Soc.* **1991**, *113*, 6139–6144. [CrossRef]
17. Burgess, K.; van der Donk, W.A.; Ohlmeyer, M.J. Enantioselective hydroborations catalyzed by rhodium (+1) complexes. *Tetrahedron Asymmetry* **1991**, *2*, 613–621. [CrossRef]
18. Suginome, M.; Yamamoto, A.; Murakami, M. Palladium-Catalyzed Addition of Cyanoboranes to Alkynes: Regio-and Stereoselective Synthesis of α,β-Unsaturated β-Boryl Nitriles. *Angew. Chem. Int. Ed.* **2005**, *44*, 2380–2382. [CrossRef]
19. Dang, L.; Lin, Z.; Marder, T.B. DFT studies on the borylation of α,β-unsaturated carbonyl compounds catalyzed by phosphine copper (I) boryl complexes and observations on the interconversions between O-and C-bound enolates of Cu, B, and Si. *Organometallics* **2008**, *27*, 4443–4454. [CrossRef]
20. Fleming, W.J.; Müller-Bunz, H.; Lillo, V.; Fernández, E.; Guiry, P.J. Axially chiral PN ligands for the copper catalyzed β-borylation of α,β-unsaturated esters. *Org. Biomol. Chem.* **2009**, *7*, 2520–2524. [CrossRef]
21. Thorpe, S.B.; Guo, X.; Santos, W.L. Regio-and stereoselective copper-catalyzed β-borylation of allenoates by a preactivated diboron. *Chem. Commun.* **2011**, *47*, 424–426. [CrossRef] [PubMed]
22. Thorpe, S.B.; Calderone, J.A.; Santos, W.L. Unexpected copper (II) catalysis: Catalytic amine base promoted β-borylation of α,β-unsaturated carbonyl compounds in water. *Org. Lett.* **2012**, *14*, 1918–1921. [CrossRef] [PubMed]
23. Shegavi, M.L.; Saini, S.; Bhawar, R.; Vishwantha, M.D.; Bose, S.K. Recyclable Copper Nanoparticles-Catalyzed Hydroboration of Alkenes and α,β-Borylation of α,β-Unsaturated Carbonyl Compounds with Bis (Pinacolato) Diboron. *Adv. Synth. Catal.* **2021**, *363*, 2408–2416. [CrossRef]
24. Neely, J.M.; Rovis, T. Rh (III)-Catalyzed regioselective synthesis of pyridines from alkenes and α,β-unsaturated oxime esters. *J. Am. Chem. Soc.* **2013**, *135*, 66–69. [CrossRef] [PubMed]
25. Heravi, M.M.; Dehghani, M.; Zadsirjan, V. Rh-catalyzed asymmetric 1,4-addition reactions to α,β-unsaturated carbonyl and related compounds: An update. *Tetrahedron Asymmetry* **2016**, *27*, 513–588. [CrossRef]
26. Romano, C. Transition Metal-Catalyzed Olefin Isomerization for Remote Functionalization Strategies. Ph.D. Thesis, University of Geneva, Geneva, Switzerland, 2018.
27. Dembitsky, V.M.; Ali, H.A.; Srebnik, M. Recent chemistry of the diboron compounds. *Adv. Organomet. Chem.* **2004**, *51*, 193–250.
28. Lawson, Y.; Norman, N.; Rice, C.; Marder, T. Platinum catalysed 1, 4-diboration of α,β-unsaturated ketones. *Chem. Commun.* **1997**, *21*, 2051–2052. [CrossRef]
29. Cox, A.J. Metal Catalysed Diboration of α,β-Unsaturated Carbonyl Compounds. Master's Thesis, Durham University, Durham, UK, 2002.
30. Kurahashi, T. Ni-Catalyzed C–C Bond Formation with α,β-Unsaturated Carbonyl Compounds and Alkynes. *Bull. Chem. Soc. Jpn.* **2014**, *87*, 1058–1070. [CrossRef]
31. Zaramello, L.; Albuquerque, B.L.; Domingos, J.B.; Philippot, K. Kinetic investigation into the chemoselective hydrogenation of α,β-unsaturated carbonyl compounds catalyzed by Ni (0) nanoparticles. *Dalton Trans.* **2017**, *46*, 5082–5090. [CrossRef]
32. Miwa, Y.; Kamimura, T.; Sato, K.; Shishido, D.; Yoshida, K. Chiral Bicyclic NHC/Cu Complexes for Catalytic Asymmetric Borylation of α,β-Unsaturated Esters. *J. Org. Chem.* **2019**, *84*, 14291–14296. [CrossRef]
33. Che, F.; Wang, M.; Yu, C.; Sun, X.; Xie, D.; Wang, Z.; Zhang, Y. Cu_2O-catalyzed selective 1,2-addition of acetonitrile to α,β-unsaturated aldehydes. *Org. Chem. Front.* **2020**, *7*, 868–872. [CrossRef]
34. Ito, H.; Yamanaka, H.; Tateiwa, J.-I.; Hosomi, A. Boration of an α,β-enone using a diboron promoted by a copper (I)–phosphine mixture catalyst. *Tetrahedron Lett.* **2000**, *41*, 6821–6825. [CrossRef]
35. Lee, J.E.; Yun, J. Catalytic Asymmetric Boration of Acyclic α,β-Unsaturated Esters and Nitriles. *Angew. Chem.* **2008**, *120*, 151–153. [CrossRef]
36. Ding, W.; Song, Q. Chemoselective catalytic reduction of conjugated α,β-unsaturated ketones to saturated ketones via a hydroboration/protodeboronation strategy. *Org. Chem. Front.* **2016**, *3*, 14–18. [CrossRef]
37. Kobayashi, S.; Xu, P.; Endo, T.; Ueno, M.; Kitanosono, T. Chiral Copper (II)-Catalyzed Enantioselective Boron Conjugate Additions to α,β-Unsaturated Carbonyl Compounds in Water. *Angew. Chem. Int. Ed.* **2012**, *51*, 12763–12766. [CrossRef]
38. Stavber, G.; Časar, Z. Basic $CuCO_3$/ligand as a new catalyst for 'on water' borylation of Michael acceptors, alkenes and alkynes: Application to the efficient asymmetric synthesis of β-alcohol type sitagliptin side chain. *Appl. Organomet. Chem.* **2013**, *27*, 159–165. [CrossRef]
39. Deng, F.; Li, S.; Zhou, M.; Zhu, Y.; Qiu, S.; Li, K.; Ma, F.; Jiang, J. A biochar modified nickel-foam cathode with iron-foam catalyst in electro-Fenton for sulfamerazine degradation. *Appl. Catal. B Environ.* **2019**, *256*, 117796. [CrossRef]
40. Liu, Y.; Zhou, W.; Lin, Y.; Chen, L.; Chu, X.; Zheng, T.; Wan, S.; Lin, J. Novel copper foam with ordered hole arrays as catalyst support for methanol steam reforming microreactor. *Appl. Energy* **2019**, *246*, 24–37. [CrossRef]
41. Hu, L.; Zhang, G.; Liu, M.; Wang, Q.; Dong, S.; Wang, P. Application of nickel foam-supported Co_3O_4-Bi_2O_3 as a heterogeneous catalyst for BPA removal by peroxymonosulfate activation. *Sci. Total Environ.* **2019**, *647*, 352–361. [CrossRef]

42. Wang, Y.; Hong, Z.; Mei, D. A thermally autonomous methanol steam reforming microreactor with porous copper foam as catalyst support for hydrogen production. *Int. J. Hydrogen Energy* **2021**, *46*, 6734–6744. [CrossRef]
43. Yan, F.; Zhou, L.; Han, B.; Zhang, Y.; Li, B.; Wang, L.; Zhu, L. Zeolite Immobilized Copper Catalyzed Conjugate Borylation of alpha, beta-Unsaturated Compounds in Aqueous Media. *Chin. J. Org. Chem.* **2021**, *41*, 2074–2081. [CrossRef]
44. Wang, W.; Xiao, Z.; Huang, C.; Zheng, K.; Luo, Y.; Dong, Y.; Shen, Z.; Li, W.; Qin, C. Preparation of Modified Chitosan Microsphere-Supported Copper Catalysts for the Borylation of α,β-Unsaturated Compounds. *Polymers* **2019**, *11*, 1417. [CrossRef] [PubMed]
45. Wen, W.; Han, B.; Yan, F.; Ding, L.; Li, B.; Wang, L.; Zhu, L. Borylation of α,β-Unsaturated Acceptors by Chitosan Composite Film Supported Copper Nanoparticles. *Nanomaterials* **2018**, *8*, 326. [CrossRef] [PubMed]
46. Zhou, L.; Han, B.; Zhang, Y.; Li, B.; Wang, L.; Wang, J.; Wang, X.; Zhu, L. Cellulosic CuI Nanoparticles as a Heterogeneous, Recyclable Catalyst for the Borylation of α,β-Unsaturated Acceptors in Aqueous Media. *Catal. Lett.* **2021**, *151*, 3220–3229. [CrossRef]
47. Yang, Z.; Tuo, Y.; Lu, Q.; Chen, C.; Liu, M.; Liu, B.; Duan, X.; Zhou, Y.; Zhang, J. Hierarchical Cu_3P-based nanoarrays on nickel foam as efficient electrocatalysts for overall water splitting. *Green Energy Environ.* **2020**, *7*, 236–245. [CrossRef]
48. Xu, P.; Li, B.; Wang, L.; Qin, C.; Zhu, L. A green and recyclable chitosan supported catalyst for the borylation of α,β-unsaturated acceptors in water. *Catal. Commun.* **2016**, *86*, 23–26. [CrossRef]
49. Zhu, L.; Kitanosono, T.; Xu, P.; Kobayashi, S. A Cu (II)-based strategy for catalytic enantioselective β-borylation of α,β-unsaturated acceptors. *Chem. Commun.* **2015**, *51*, 11685–11688. [CrossRef]
50. Zhou, X.F.; Sun, Y.Y.; Wu, Y.D.; Dai, J.J.; Xu, J.; Huang, Y.; Xu, H.J. Borylation and selective reduction of α,β-unsaturated ketones under mild conditions catalyzed by Cu nanoparticles. *Tetrahedron* **2016**, *72*, 5691–5698. [CrossRef]

Article

Underlying Mechanisms of Reductive Amination on Pd-Catalysts: The Unique Role of Hydroxyl Group in Generating Sterically Hindered Amine

Zeng Hong [1,2], Xin Ge [3,*] and Shaodong Zhou [1,2,*]

1. Zhejiang Provincial Key Laboratory of Advanced Chemical Engineering Manufacture Technology, College of Chemical and Biological Engineering, Zhejiang University, Hangzhou 310027, China; 12128008@zju.edu.cn
2. Institute of Zhejiang University—Quzhou, 78 Jiuhua Boulevard North, Quzhou 324000, China
3. School of Chemical and Material Engineering, Jiangnan University, Lihu Avenue 1800, Wuxi 214122, China
* Correspondence: gexin@jiangnan.edu.cn (X.G.); szhou@zju.edu.cn (S.Z.)

Abstract: Pd nanospecies supported on porous g-C_3N_4 nanosheets were prepared for efficient reductive amination reactions. The structures of the catalysts were characterized via FTIR, XRD, XPS, SEM, TEM, and TG analysis, and the mechanisms were investigated using in situ ATR–FTIR spectroscopic analysis complemented by theoretical calculation. It transpired that the valence state of the Pd is not the dominating factor; rather, the hydroxyl group of the $Pd(OH)_2$ cluster is crucial. Thus, by passing protons between different molecules, the hydroxyl group facilitates both the generation of the imine intermediate and the reduction of the C=N unit. As a result, the sterically hindered amines can be obtained at high selectivity (>90%) at room temperature.

Keywords: sterically hindered amine; reductive amination; competing mechanisms; in situ ATR–FTIR analysis; Pd-catalyst

Citation: Hong, Z.; Ge, X.; Zhou, S. Underlying Mechanisms of Reductive Amination on Pd-Catalysts: The Unique Role of Hydroxyl Group in Generating Sterically Hindered Amine. *Int. J. Mol. Sci.* **2022**, *23*, 7621. https://doi.org/10.3390/ijms23147621

Academic Editor: Raphaël Schneider

Received: 11 June 2022
Accepted: 8 July 2022
Published: 10 July 2022

Publisher's Note: MDPI stays neutral with regard to jurisdictional claims in published maps and institutional affiliations.

Copyright: © 2022 by the authors. Licensee MDPI, Basel, Switzerland. This article is an open access article distributed under the terms and conditions of the Creative Commons Attribution (CC BY) license (https://creativecommons.org/licenses/by/4.0/).

1. Introduction

Amines constitute an indispensable class of chemicals that are widely used as the raw materials or intermediates in the laboratory and in industry to prepare value-added chemicals, such as pharmaceuticals, agrochemicals, and biomolecules [1,2]. To date, numerous organic methodologies for amine production have been reported, such as the aminolysis of haloalkanes [3], the reaction of N-chloro dialkylamines with alkyl Grignard reagents [4], Buchwald–Hartwig and Ullman-type C–N cross-coupling reactions [5,6], as well as reductive amination. The amination of alcohols via so-called borrowing hydrogen (BH) is another important method which is mediated by rare noble-metal catalysts, based on Ru or Ir mostly [7]. Among all these methods, catalytic reductive aminations using molecular hydrogen as the reductant continue to be in the spotlight of both academic and industrial interests, due to their high atom economy and low pollution [8]. As to the catalysts required, in addition to precious metals such as Pd, Pt, Ru, and Rh [9], continuous efforts are being made on employing earth-abundant metals, such as Ni [10], Co [11], and Fe [12] as the active center. By this means, various amines, such as primary, secondary, and tertiary amines with less steric hindrance, can be prepared under moderate conditions [9].

In spite of the numerous studies reported previously on reductive aminations, the associated mechanisms are still disputable and remain ambiguous. It is generally believed that the transformation starts from the formation of imine upon condensation of the carbonyl group with ammonia or amine, followed by the reduction of imine [8]. However, a previous study pointed out that the imine intermediates do not form in the reductive amination with secondary amines [13]. For tertiary amines, the amine forms either by direct hydrogenolysis of the hemiaminal [13,14] or dehydration to enamine followed by hydrogenation [15]. Most likely, the preference of the reaction pathway depends on both the structure of the reactants and the reaction conditions.

In addition to the ambiguous mechanisms, reductive amination is yet limited for the synthesis of sterically hindered tertiary amines such as non-nucleophilic base and 1,2,2,6,6-Pentamethylpiperidine. In fact, as will be shown later in the Discussion section, we tried a series of commercially available catalysts that are proven to be excellent in mediating the production of most amines via reductive amination; however, most of these catalysts exhibit poor performance in the preparation of N,N-diisopropylbutylamine. Notably, Denis Chusov reported an atom-economical method for the synthesis of sterically hindered tertiary amines based on complementary Rh- and Ru-catalyzed direct reductive amination using carbon monoxide as a deoxygenating agent [16]. Indeed, sterically hindered amine is irreplaceable in organic synthesis and catalysis specifically as non-nucleophilic base [17], light stabilizers (HALS) [18], ligands [19], components of frustrated Lewis pairs [20], etc. As far as we know, the sterically hindered amines have been synthesized at industrial scale by reductive amination of carbonyl compounds with poor atom economy reagents such as borohydride reagents. Further efforts are still highly demanding for developing a more efficient preparation of sterically hindered amine via reductive amination with hydrogen.

On the other hand, Pd nanoparticles (e.g., Pd/C) have been proven to be efficient for reductive aminations with hydrogen, especially for the synthesis of tertiary amines [21]. Despite the fact that the classic Pd/C-H_2 system is relatively less efficient in mediating the production of sterically hindered amine such as 1-cyclohexylpiperidine [22,23], Pd is still highly promising to complete this task benefiting from relatively strong Pd-C, Pd-H, and Pd-N interactions [24]. To tune the activity of the Pd catalysts, the influences of size effect and the support were investigated. The size distribution of the Pd/C catalyst can be controlled by quite a few strategies by tuning the reducing agents [25] and conditions [26], the concentrations of the stabilizing agent, Pd salt and precipitant [27–29], or introducing a second metal [30]. Support modification is another way to promote the catalytic activity of anchored metal nanoclusters [31]. The graphite-like carbon nitride (g-C_3N_4), characterized as an incompletely condensed, N-bridged "poly(tri-s-triazine)" polymer with lamellar structure, possessed unique physicochemical properties due to its appealing electronic band structure. The nitrogen functionalities on the surface might act as strong Lewis base sites and the π-bonded planar-layered configurations are expected to anchor the substrate and metal active species, making it a privileged candidate for hydrogenation reactions [32–34]. Moreover, treatments with various oxidants to introduce acidic sites on the surface of activated carbon may significantly improve the activity and selectivity of catalysts [35–39]. Up to now, however, it remains to be explored how the support affects selectivity for the production of sterically hindered amines via reductive amination.

In this article, we report an efficient Pd(OH)$_2$/g-C_3N_4 catalyst for reductive amination. As compared to the previously reported Pd catalysts, the Pd-based nanoparticles prepared in this work are of higher activity and selectivity. By using a combination of structural characterization, in situ spectroscopic investigation, and theoretical calculation, the elaborate structure of the active center, as well as the root cause for the excellent performance of the Pd(OH)$_2$/g-C_3N_4 catalyst, are revealed.

2. Results and Discussion
2.1. Catalytic Tests of the Prepared Palladium Catalysts

To evaluate the performance of the catalysts studied, the reaction of diisopropylamine with butyraldehyde under hydrogen atmosphere was selected as a model. Various catalysts, including commercially available ones and home-made ones, were employed, and the detailed results are listed in Table 1.

Table 1. Comparisons of the catalytic performance for reductive amination of diisopropylamine with butyraldehyde over various catalysts [a].

Entry	Catalyst	Temp./°C	Time/h	Conv./%	Sel./%	Yield [b]/%
1	Raney Ni	100	7	70	Trace	Trace
2	Raney Co	30	4	55	Trace	Trace
3	Raney Cu	100	7	46	Trace	Trace
4	5.0 wt% Ru/ACs [g]	30	5	Trace	Trace	Trace
5	5.0 wt% Rh/ACs	30	5	Trace	Trace	Trace
6	5.0 wt% Pt/ACs	30	5	51	61	31
7	5.0 wt% Pd/ACs	30	4	94	85	80
8	1.0 wt% Pd/ACs	30	4	90	76	68
9	Pd(AcO)$_2$+Xphos	60	7	28	46	13
10	10.0 wt% Pd(OH)$_2$/ACs	30	4	97	95	92
11	1.2 [c], 1.4 [d], 1.2 [e] wt% Pd(OH)$_2$/ACs	30	4	65/62/33	86 [c]/89 [d]/36 [e]	56/55/12
12	1.1 wt% Pd(OH)$_2$/g-C$_3$N$_4$	30	4	60/77	97/95 [f]	58/73(60)
13	1.2 wt% Pd/g-C$_3$N$_4$	30	4	33	30	10
14	1.2 wt% Pd/ACs	30	4	80	66	53
15	1.3 wt% PdO/ACs	30	4	22	36	8
16	1.1 wt% PdO/g-C$_3$N$_4$	30	4	30	trace	trace

[a] Reaction conditions: diisopropylamine (0.1 mol), butyraldehyde (0.05 mol), catalyst loading (0.2 g, 5.0 wt% refer to butyraldehyde), methanol (2.5 mL), and 1.5 H$_2$ MPa. For Entry 9, Pd(AcO)$_2$ (0.1 mmol), Xphos (0.1 mmol). [b] Conversion of butyraldehyde and yield of product were determined by gas chromatography. All products (i.e., product amine, n-butanol, and aldol condensation products) were determined by GC-MS. [c] Nitric acid treatment. [d] Hydrothermal treatment. [e] No treatment. [f] 4Å molecular sieve (0.6 g, 15%), and the value in the brackets represents isolated yield. [g] the "5.0 wt%" represents the metal content in the catalyst, the same as below in Table 1.

To start with, we took commercially 5.0 wt% Pd/ACs as a catalyst to investigate the effect of various conditions for the reaction. We found that the reaction can be proceeded at room temperature, and lower the ratio of amine to aldehyde (<2) or higher hydrogen pressure (>1.5 MPa) improves the selectivity of C=O reduction. Moreover, further prolongation of the reaction time (>4 h) will not bring about the increase in yield. Last but not least, we used methanol as the solvent with additive amount attributing to its high hydrogenation activity for reduction amination [40]. Thus, the optimal conditions for catalyst screening were determined. As shown in Table 1, classic inexpensive hydrogenation catalysts such as Raney Ni/Co/Cu failed to afford the desired product (Table 1, **Entry 1–3**). Next, for the noble-metal catalysts, only Pt/C and Pd/C successfully brought about the generation of diisopropylbutylamine, and Pd/C performed better (Table 1, **Entry 4–8**); however, the competing reduction of C=O was still unavoidable. In addition, atomically dispersed Pd did not give better performance (Table 1, **Entry 9**). Notably, when commercial 10.0 wt% Pd(OH)$_2$/ACs was employed, the target product was obtained with 92% yield (Table 1, **Entry 10**). This result is consistent with a previous study on Pd(OH)$_2$/ACs-catalyzed synthesis of tertiary amines [41]. However, when reducing the content of Pd loaded on ACs, the activity of the catalyst decreased dramatically: with ~1.0 wt% Pd, although preliminary nitric acid treatment and hydrothermal treatment of the ACs supports kept the Pd(OH)$_2$/ACs catalyst at relatively good selectivity, the reaction rate dropped considerably (Table 1, **Entry 10, 11c,d**). To further improve the performance of the Pd catalysts, we changed the support with g-C$_3$N$_4$, inspired by the finding of Wang et al. that phenol can be selectively reduced to cyclohexanone at Pd@g-C$_3$N$_4$ due to phenol being able to interact with the surface through the hydroxy group to form strong O–H···N or O–H···π interactions [34]. We considered that Pd(OH)$_2$ tends to adsorb on basic sites of g-C$_3$N$_4$ by similar O–H···N or O–H···π interactions. Surprisingly, both the selectivity and the reaction rate maintain simultaneously on 1.1 wt% Pd(OH)$_2$/g-C$_3$N$_4$. By contrast, however, when depositing reduced Pd nanoparticles on g-C$_3$N$_4$, the so-obtained Pd/g-C$_3$N$_4$ exhibits low catalytic activity (Table 1, **Entry 13**). Here, the valence state of Pd seems to be important. Thus, further examination on the performance of PdO species was carried out. In contrast to the Pd/support or the Pd(OH)$_2$/support catalysts, much lower selectivity and yield are given by either PdO/ACs or PdO/g-C$_3$N$_4$ (Table 1, **Entry 15, 16**). Therefore, it is not the high-valence state of Pd that matters; rather, the existence of the hydroxyl group in the Pd-oxide cluster substantially influences the reductive amination processes. Similarly, it has been found previously that the adsorption of hydrogen donors and acceptors on metal catalysts is species-dependent, thus offering opportunities for selectivity control in hydrogen transfer processes [42–44]. Finally, a satisfactory yield was obtained when using 4Å molecular sieve as dehydrating agent (Table 1, **Entry 12f**). Obviously, here, the reactivity of the system is sufficient, while the conversion is dominated by the reaction equilibrium.

2.2. Characterization of the Prepared Pd(OH)$_2$/g-C$_3$N$_4$ and Other Palladium Catalysts

To reveal the root cause for the excellent performance of Pd(OH)$_2$/g-C$_3$N$_4$, structural characterization is prerequisite. The Fourier-transform infrared (FT-IR) was performed. As shown in Figure 1, the absorption peaks at 1245, 1320, and 1408 cm^{-1} correspond to the aromatic C–N stretching vibrations, while 1567 and 1640 cm^{-1} are ascribed to the C=N vibrations. The peak at 808 cm^{-1} is assigned to the breathing mode of the triazine units (Figure 1c) [45,46]. The broad peak at 3000–3400 cm^{-1} is ascribed to the stretching vibration of N–H. When loading Pd to the g-C$_3$N$_4$, the intensity of these peaks does not change obviously but the intensity of the broad peak at 3000–3400 cm^{-1} increases (Figure 1a), which can be attributed to the O–H stretching of the Pd(OH)$_2$. However, intensity of this broad peak decreases with treatment of hydrazine hydrate, indicating that the divalent Pd was reduced partially (Figure 1b). Accordingly, no obvious absorption peak is observed when Pd is loaded on the activated carbons treated by nitric acid or water (Figure 1d–f).

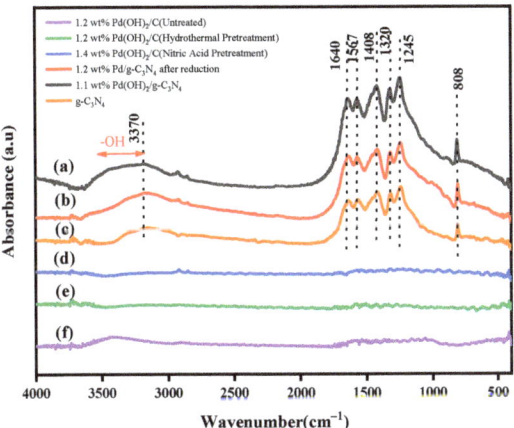

Figure 1. FT-IR spectra of the samples with different of Pd loading determined by ICP-OES.

The phase structure and composition of the selected samples were investigated by X-ray diffraction (XRD). The obtained results are shown in Figure 2. The pattern of g-C$_3$N$_4$ is identified by peaks at 2θ = 27.8° and 12.8°, corresponding to the (002) crystallographic plane and (100) planes of in-planar tris-s-triazine structural packing motifs (Figure 2a,b) [47]. The broad peak at 25.1° indicated the typical amorphous structure of activated carbons (Figure 2c–f). The crystal structure of Pd is identified by the diffraction peaks at 2θ = 40.4°,

47.0°, and 68.6° (JCPDS 87-0645). These peaks can be assigned to the (111), (200), and (220) planes of Pd metal with a face-centered cubic (fcc) structure, respectively. However, no Pd or PdO characteristic peaks were detected for Pd(OH)$_2$/g-C$_3$N$_4$, suggesting that Pd species may exist in the form of subcrystal or nanoparticles. A weak peak at 40.4° can be observed after reduction (Figure 2b). According to Scherrer's formula [48] and the half-width of the Pd (111) peak, the calculated size of Pd NPs in Pd/g-C$_3$N$_4$ is 1.3 nm. When Pd is loaded on ACs, the location of the Pd (111) peak (at 40.4°) remains with a calculated size of Pd 2.4 nm, 3.7 nm, and 2.0 nm. It can be seen that the divalent Pd can be reduced by the reducing groups anchored on the surface of activated carbons during the impregnation process (Figure 2c–e) [49]. Additionally, both PdO/ACs and the commercial 10.0 wt% Pd(OH)$_2$/ACs catalyst exhibited diffraction peaks at two-theta of 33.5°, 33.8°, 60.2°, and 71.4° in their XRD patterns (Figure S1), which can be indexed as the (101), (112), (103), and (211) diffractions of Palladium oxide (JCPDS 06-0515). Furthermore, the PdH$_x$ phase was not observed in any of the XRD patterns, which is considered as the active species for hydrogenation reactions [50].

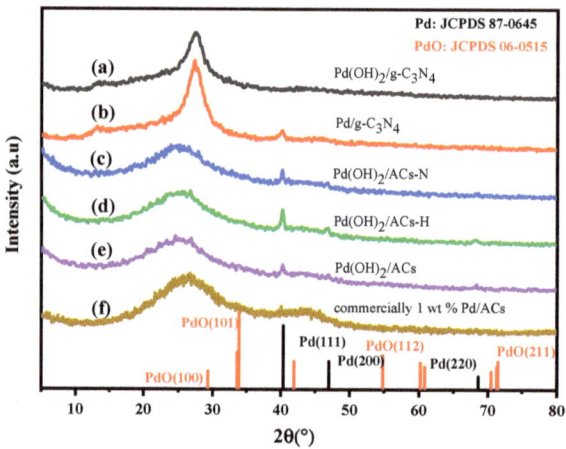

Figure 2. XRD patterns of the selected catalysts. (**a**) 1.1 wt% Pd(OH)$_2$/g-C$_3$N$_4$, (**b**) 1.2 wt% Pd/g-C$_3$N$_4$, (**c**) 1.2 wt% Pd(OH)$_2$/ACs treated by HNO$_3$ (labelled as Pd(OH)$_2$/AC-N), (**d**) 1.4 wt% Pd(OH)$_2$/ACs after hydrothermal treatment (labelled as Pd(OH)$_2$/AC-H), (**e**) 1.2 wt% Pd(OH)$_2$/ACs without treatment, (**f**) commercially 1.0 wt% Pd/ACs.

Next, XPS measurements were performed to explore the chemical properties of the surface and the electronic configurations of the active centers. The survey XPS spectrum in Figure 3a shows that C, O, N, and Pd elements coexist in 1.1 wt% Pd(OH)$_2$/g-C$_3$N$_4$ sample. Figure 3b–d show XPS core level spectra of C 1s, N 1s, and Pd 3d of the 1.1 wt% Pd(OH)$_2$/g-C$_3$N$_4$ sample. All lines in the XPS spectra are corrected with carbon C 1s at 284.8 eV. The peak at 284.8 eV in C 1s is attributed to surface adventitious carbon, whereas the peak at 288.2 eV corresponds to sp^2-bonded carbon in N–C=N of g-C$_3$N$_4$ (Figure 3b). For N 1s, four peaks can be distinguished (Figure 3c). The signal at the binding energy of 398.5 eV is assigned to the pyridinic N (N–C=N), and the peak at 399.9 eV is assigned to the graphitic N. The peaks at 401.0 and 404.6 eV are attributed to the amino group (C–N–H) and some N–O species [33]. The spectra of Pd 3d (Figure 3d) present two doublet peaks, corresponding to the spin–orbital splitting of Pd 3d$_{5/2}$ and Pd 3d$_{3/2}$ for two types of Pd species. The peaks at 335–336 eV can be assigned to reduced Pd and the ones at 336–337 eV to palladium oxide or palladium hydroxides. It has been pointed out previously that Pd(OH)$_2$ on carbon materials is a core–shell structure of C/PdO/OH/H$_2$O [51]. The ratio of Pd^0 to Pd^{2+} was about 31:69 using Gaussian/Lorentzian line shape approximations, suggesting that the Pd atoms in Pd(OH)$_2$ are partially reduced during the preparation (see Figures S3 and S4). After treatment with hydrazine hydrate, the ratio of Pd^0 to Pd^{2+} increases to 70:30 (Figure S2). Thus, the bivalent palladium species was considered as the main active component in the reaction. To our surprise, the content of pyridinic N decreased from 76% to 74% and the content of graphitic N increased from 16% to 19% when the catalyst was treated by hydrazine hydrate (see Figure 2c and Figure S2c). Previous reports revealed that pyridinic coordination sites exhibited the highest metal-loading stability, whereas the graphitic-N coordination sites had the least stable one [33,52]. We infer that the Pd species may aggregate during reduction by hydrazine hydrate. On the other hand, N 1s signals of $Pd/g\text{-}C_3N_4$ shift 0.05 eV towards low binding energy, indicating that the interaction between the $g\text{-}C_3N_4$ and Pd is weakened after reduction. For $Pd(OH)_2/ACs$, the one whose support is hydrothermally treated possesses more phenolic groups on the surface (Figures S3c and S4c). Notably, the Pd $3d_{5/2}$ peak for the Pd species on the hydrothermally treated supports is 0.8 eV higher than that of the nitric acid treatment supports, as depicted in Figures S3d and S4d, indicating the presence of a strong interaction between the support and Pd. According to the literature, Pd species can be stabilized by the deprotonation process between the Pd precursor and phenolic groups (−OH) to produce an oxygen anion (−O−) and facilitate the dispersion of Pd [39,53].

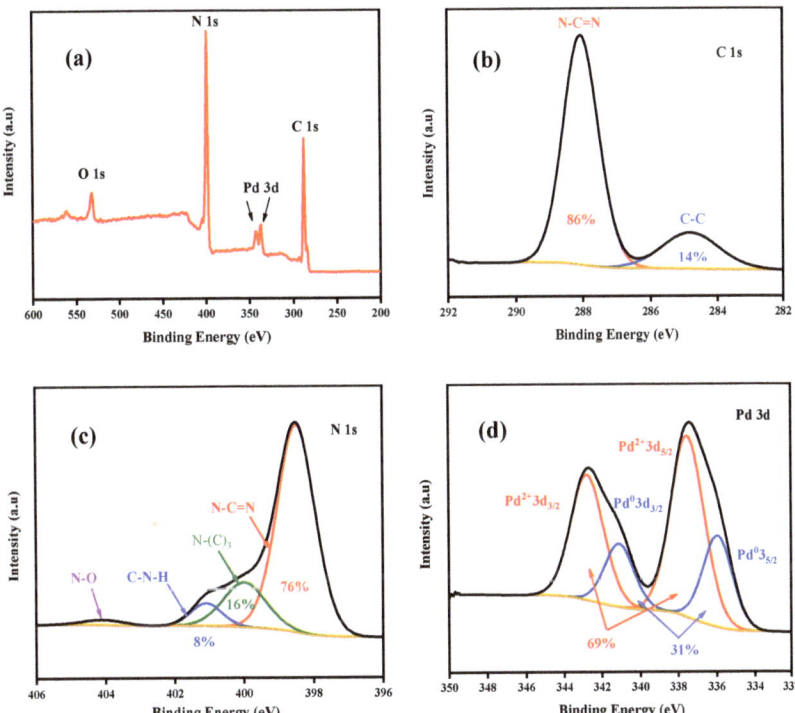

Figure 3. XPS spectra of 1.1 wt% $Pd(OH)_2/g\text{-}C_3N_4$: (**a**) the survey spectrum; (**b**) the core-level spectrum of C1s; (**c**) the core-level spectrum of N 1s; (**d**) the core-level spectrum of Pd $3d_{5/2}$ and $3d_{3/2}$ doublet region.

In order to identify the morphology of the Pd species on $g\text{-}C_3N_4$, scanning electron microscopy (SEM) and transmission electron microscopy (TEM) were performed for $Pd(OH)_2/g\text{-}C_3N_4$. The SEM image (Figure 4a) demonstrates that the $Pd(OH)_2/g\text{-}C_3N_4$ composite possesses a fluffy, wrinkled, and porous microstructure. Meanwhile, it can be seen from the TEM images (Figure 4c) that the Pd species were distributed throughout the $g\text{-}C_3N_4$ architecture uniformly without obvious aggregation, disclosing that the nanosheets are layered flakes and possess some defects and voids which provide abundant anchoring sites for Pd species. The average particle size of the Pd nanoparticles in Pd(OH)$_2$/g-C$_3$N$_4$ is 2.11 nm, as determined by statistical evaluation of 100 particles in the TEM images. The Pd content of the Pd(OH)$_2$/g-C$_3$N$_4$ and Pd/g-C$_3$N$_4$ sample determined by ICP-OES is calculated to be 1.1 wt% and 1.2 wt%.

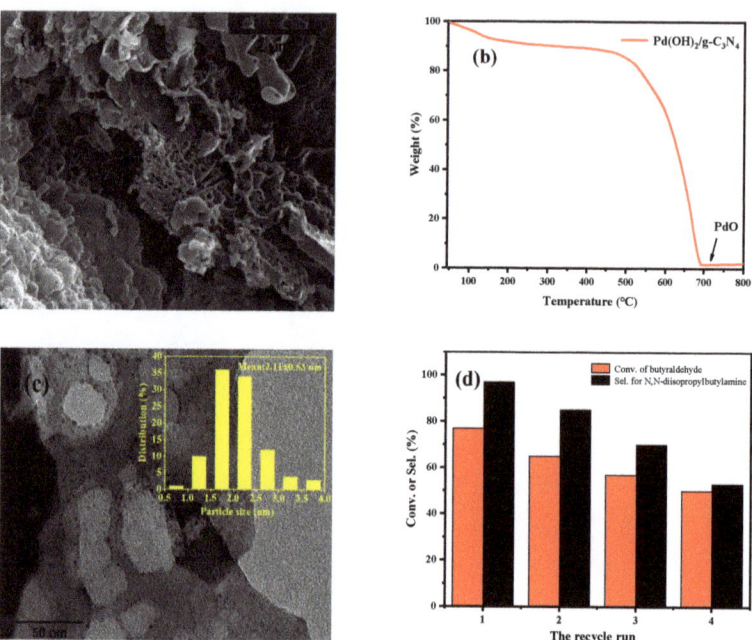

Figure 4. SEM (**a**) and TEM (**c**) images of Pd(OH)$_2$/g-C$_3$N$_4$, (**b**) Thermogravimetric plots for Pd(OH)$_2$/g-C$_3$N$_4$ and the size distribution of Pd species on Pd(OH)$_2$/g-C$_3$N$_4$. (**d**) Reuse of 1.1 wt% Pd(OH)$_2$/g-C$_3$N$_4$ catalyst at same conditions.

Subsequently, according to the thermogravimetric (TG) analysis, for Pd(OH)$_2$/g-C$_3$N$_4$, there is an initial weight loss ∼135 °C which is due to the removal of adsorbed water molecules within the system. After that, the Pd(OH)$_2$/g-C$_3$N$_4$ sample is stable up to 500 °C and starts to decompose and decomposition is complete ∼680 °C with a residual weight of 2.0% (Figure 4b); this decomposing temperature is higher compared to Pd/g-C$_3$N$_4$ [54]. Alternatively, the thermal stability of the catalyst can also be improved by introducing hydroxyl groups to the support-activated carbons with hydrothermal treatment (Figure S5). Thus, the presence of the hydroxyl groups enhances the thermal stability of the catalyst. Furthermore, for the heterogeneous catalysis reaction, it is important to test the reusability of the catalyst. Thus, the reusability of the 1.1 wt% Pd(OH)$_2$/g-C$_3$N$_4$ catalyst was investigated using the reductive amination of diisopropylamine with butyraldehyde as a model reaction. Moderate selectivity toward N,N-diisopropylbutylamine was obtained after four runs (Figure 4d), suggesting that this catalyst has receptable reusability.

2.3. Mechanism Studies

Upon addressing the structural features of different catalysts and recycling tests, the in situ ATR–FTIR spectroscopic analysis was employed to identify how the catalysts interact with the substrates during the reduction amination. The peaks of [v(C=O)=1740–1780 cm^{-1}] were firstly identified for all the spectra [55]. For commercial 1.0 wt% Pd/ACs, the imine species [v(C=N)=1625 cm^{-1}] with weak intensity was observed when acetaldehyde and ammonia were introduced to the in situ reaction cell [56]. This peak decreased but did not disappear when continuously introducing H_2 (Figure 5a) into the system. Meanwhile, the [v(C–N)] mode in the range of 1350–1280 cm^{-1} cannot be observed, probably due to its low intensity [57,58]. Note that there is no stretching vibration peak of water at 3200–3500 cm^{-1}, indicating that the dehydration process may not require a catalyst [59]. It is also confirmed by the fact that ethanimine formed without a catalyst (See Figure S6). The C=N vibration was also observed when acetone and NH_3 were introduced into the cell, and a peak around 3464 cm^{-1} was observed (Figure 5b). Considering the [v(O-H)] stretching mode adjacent to the nitrogen atom; most likely, this hydroxyl group belongs to the hemiaminal intermediate [60]. The peaks at 1200 cm^{-1} and 1130 cm^{-1} were assigned to the deformation stretching of the C–N bond (Figure 5c) [57]. The imine mechanism thus prevails as well. By contrast, for the secondary amine diethylamine, a broad peak located at 3200–3600 cm^{-1} emerged only after H_2 was introduced. This can be ascribed to the stretching vibration of adsorbed water, which was further proven by the peak in 1640 cm^{-1} (assigned to the H–O–H bending bands in product water) (Figure 5d) [59]. Meanwhile, the skeleton-stretching vibration of (-C_3)N [v((-C_3)N)=1178 cm^{-1}] is also identified [58], suggesting that the dehydration and hydrogenation processes occur simultaneously on the surface of the catalyst. As for the amination of diisopropylamine, the commercial 1.0 wt% Pd/ACs, Pd(OH)$_2$/g-C_3N_4, Pd/g-C_3N_4, and PdO/g-C_3N_4 were compared (Figure 5e–h). For all catalysts, a weak peak, as discussed above, assigned to the stretching vibration of the hydroxyl group of the hemiaminal, appeared at 3487 cm^{-1}, 3491 cm^{-1}, 3474 cm^{-1}, and 3483 cm^{-1}, respectively, after introducing diisopropylamine and acetaldehyde to the in situ reaction cell. They were found as red-shift peaks due to the difference of surface binding or adsorption strength between the different Pd catalysts since the hydroxyl group vibration of the free hemiaminal is 3628 cm^{-1}, according to DFT calculation at the B3LYP/6-31+(g) level. The peak of the water shows a completely different shape difference, which may imply a completely different catalytic performance on the home-made catalysts. Except for PdO/g-C_3N_4, all the other catalysts show the peaks of absorbed water, which is in line with the experimental results that home-made PdO/g-C_3N_4 is inactive. Moreover, no imine peak at the range 1620–1680 cm^{-1} was observed [56]. The identification of the product might be difficult because of the low intensity of the typical band at 1220–1270 cm^{-1} assigned for the skeleton-stretching vibration of (-C_3)N [58].

Based on the in situ characterization, the reaction mechanisms were investigated by theoretical calculations. Considering the different performances of Pd/g-C_3N_4 and Pd(OH)$_2$/g-C_3N_4, the valence state of Pd matters for the reductive amination of diisopropylamine. The generation of imines for different substrates was examined computationally, concerning both the hydrogen transfer and dehydration processes (see Figures S7–S10). They are all kinetically less favorable due to high reaction barriers (>40 kcal/mol). However, when a second amine is introduced into the reaction complex, the subsequent reaction barrier for both hydrogen transfer and dehydration is lowered by 2 kcal/mol and 10 kcal/mol, respectively. Further enhancement in these two steps is achieved when a third amine is introduced (see Figure S11). It seems that dehydration from aldehydes and primary amines or ammonia is a multimolecule synergistic process. This result is in line with the in situ ATR–FTIR spectroscopic findings (see Figure S6). On the contrary, for imine hydrogenation, a catalyst is required since the reaction barrier is extremely high (>60 kcal/mol; see Figure S12).

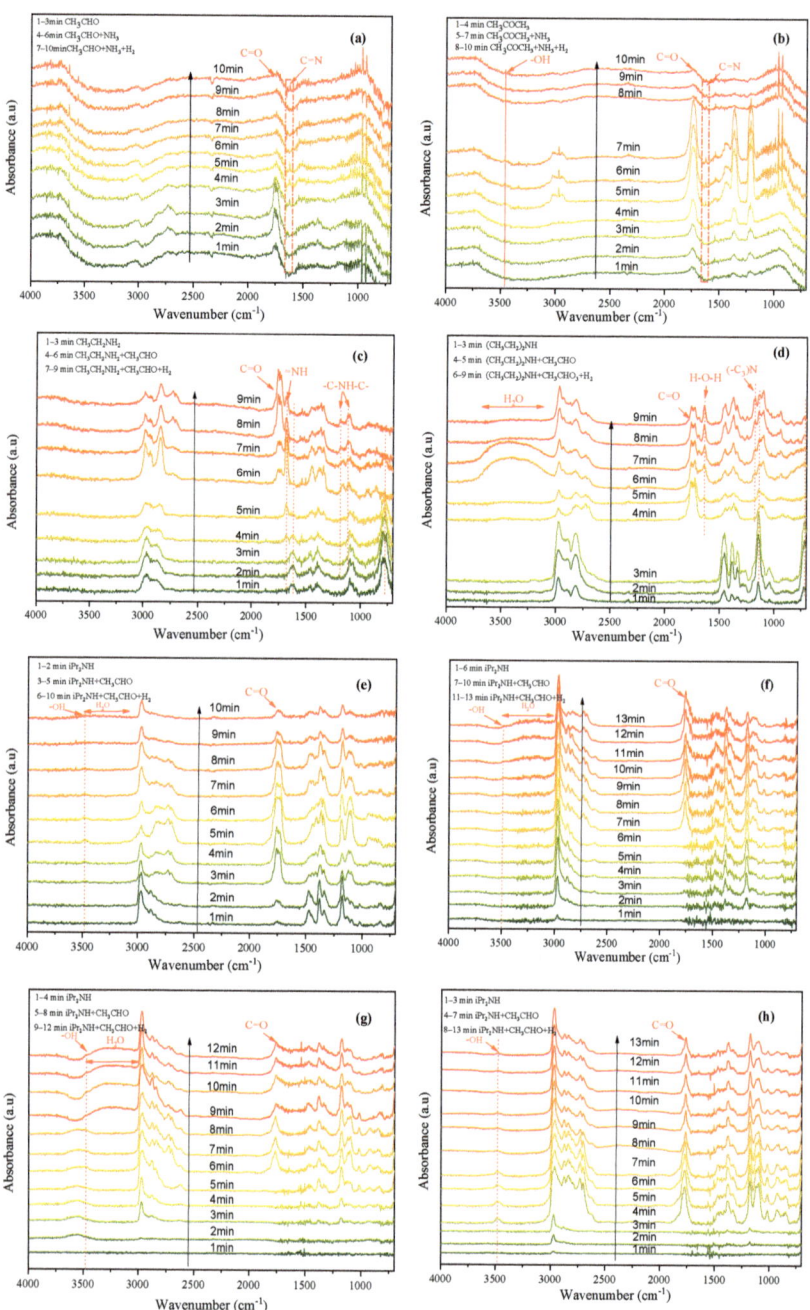

Figure 5. ATR-FTIR spectra recorded during hydrogenation of different substrates: (**a**) CH_3CHO and NH_3, (**b**) CH_3COCH_3 and NH_3, (**c**) $CH_3CH_2NH_2$ and CH_3CHO, (**d**) $(CH_3CH_2)_2NH$ and CH_3CHO, (**e**) iPr_2NH and CH_3CHO, (**f**) iPr_2NH and CH_3CHO. (**a**–**e**) commercial 1.0 wt% Pd/ACs, (**f**) 1.1 wt% $Pd(OH)_2$/g-C_3N_4, (**g**) 1.2 wt% Pd/g-C_3N_4, (**h**) 1.1 wt% PdO/g-C_3N_4. The "1.0 wt%" represents the Pd content in the catalyst, and in the other cases the Pd loading was determined by ICP-OES.

Thus, the Pd(111) plane was used for modeling the Pd/g-C$_3$N$_4$ catalyst to understand the catalytic performance on Pd^0 species. We first investigated the adsorption behavior of hydrogen on the Pd(111) surface, demonstrating that the hydrogen is preferentially absorbed on the bridge and hollow sites (see Figure S13). Furthermore, as shown in Figure 6, initially diisopropylamine and acetaldehyde are coadsorbed on the Pd(111) surface. The N–H bond activation of diisopropylamine takes place upon hydrogen transfer from the nitrogen atom to the oxygen atom via **TS1/2**, and C–N coupling takes place spontaneously (**IM1**→ → **IM2**). Species adsorption on the catalyst surface significantly lowers the reaction energy barrier, making it feasible under mild conditions (see Figure S10). Next, palladium hydride attacks the oxygen atom, thus breaking the C–O bond via **TS3/4**, which was identified as the rate-determining step (RDS). After that, **IM4** is generated, which further releases a water molecule to form **IM5**. Finally, by transferring a hydrogen atom from the Pd surface to N, an intact N,N-Diisopropylethylamine molecule is produced (**IM6**). This process is featured as the hydrogenolysis mechanism [61]. In addition to the pathways via hydrogenolysis, an alternative imine path was also investigated [62]. Specifically, after the imine intermediate is generated via **TS7/8**, hydrogen transfer from Pd to C affords the target product (**IM9** → → **IM10**). This step is followed by further hydrogenation of the hydroxyl groups, thus generating a water molecule (**IM12**). Comparing the above two pathways, the latter is energetically more favorable. Furthermore, we examined theoretically the hydrogenation of the acetaldehyde on the Pd(111) surface. As shown in Figure S14, the H atoms are successively added to O and C in a step-wise manner via the sequence **IM13** → **TS13/14** → **IM14** → **IM15** → **TS15/16** → **IM16**. Alternatively, the H atoms are successively added to C and O in a step-wise manner via the sequence **IM17** → **TS17/18** → **IM18** → **IM19** → **TS19/20** → **IM20**. Although the former is thermodynamically more favorable due to its high exothermicity (−81 kcal/mol), the latter is kinetically more preferred due to the low barrier for the hydrogenation process.

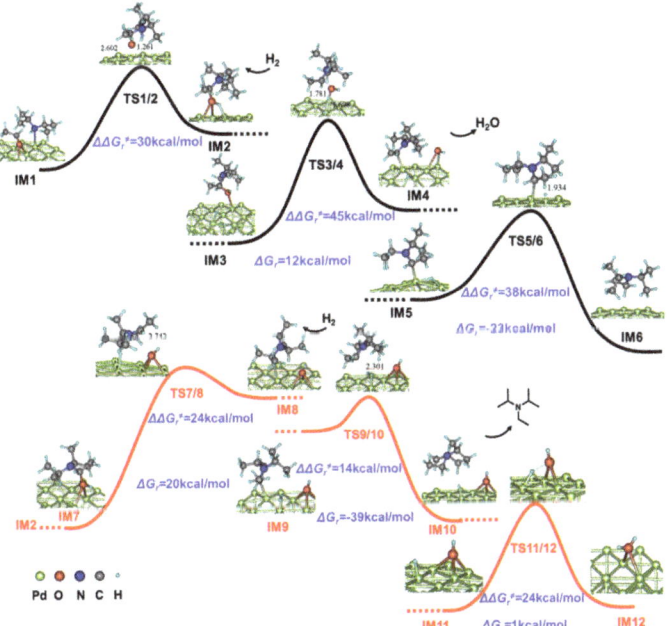

Figure 6. Simplified PES and selected structural information of alternative pathways for the generation of N,N-Diisopropylethylamine mediated by the Pd(111) plane via hydrogenolysis (black) and imine (red) mechanisms.

For Pd(OH)$_2$/g-C$_3$N$_4$, considering the structure characterized previously [51], a Pd$_8$O$_{12}$H$_8$ cluster was used to simulate the reductive amination processes. As shown in Figure 7, when diisopropylamine and acetaldehyde are coadsorbed on the clusters, the hemiaminal intermediate is formed spontaneously without a reaction barrier by simple hydrogen exchange between diisopropylamine and a hydroxyl group (**IM21** → **IM22**). On the contrary, the reaction barrier for forming a hemiaminal on the PdO cluster is extremely high (45 kcal/mol; see Figure S15), making PdO/g-C$_3$N$_4$ inactive. Here, the presence of the O–Pd–OH site is crucial. We first consider that the hydrogenolysis process occurs on the clusters which is similar to that on the Pd(111) surface. The transitional structure involving palladium hydride attacking the oxygen atom to form a water molecule was firstly located. However, the subsequent hydrogenation transition state structure could not be found (see Figure S16). After breaking the C–O bond (**IM23** → → **IM24**), the hydroxyl group on the catalyst surface serves as a "courier" to pass hydrogen. Thus, a water molecule forms first by hydrogen transfer from Pd to OH via **TS25/26**. Next, a hydrogen atom on the so-formed water molecule migrates to the imine carbon to form the product (**IM26** → → **IM27**). Here, we considered that the imine species can be polarized by a hydrogen bond formed between N and a hydrogen atom on the so-formed water, thus increasing the probability to generate a transition state for H addition. Finally, by transferring a second hydrogen atom from Pd OH, an intact water molecule forms again (**IM28**). This process is similar to phenol acting as a conduit to transfer a proton from the hydronium ion (with an accompanying charge transfer from the metal surface) to the basic carbonyl oxygen of the benzaldehyde via a PCET mechanism in the electrochemical hydrogenation (ECH) system, as reported by Udishnu Sanyal et al. [63]. Moreover, a thermodynamically more favorable pathway (−49 kcal/mol) was also found via the sequence **IM25** → **TS25/26** → **IM26** → **IM29** → **TS29/30** → **IM30**. Considering the difference in the reaction pathways presented in Pd(111) and the Pd cluster, the hydroxyl group is unique in that: (1) via barrierless (or quasi-) proton exchange, the formation of a hemiaminal is facilitated at the O–Pd–OH site; (2) the hydroxyl group serves as a hydrogen shuttle in the reduction of the imine unit concerted with the dihydrogen activation; (3) the Pd nanoclusters may be stably anchored on the basic sites of g-C$_3$N$_4$ by forming strong O–H···N or O–H···π interactions. In spite of the lower energy barrier for the hydrogenation of acetaldehyde (see Figure S17), the surface of the cluster preferentially adsorbs the hemiaminal intermediate by similarly forming O–H···N interactions which are generated spontaneously in advance. A high selectivity for reductive amination thus results, in line with the experimental findings. The probable mechanism for the generation of diisopropylethylamine is shown in Scheme 1, and the Figure S18 showed the intact outline for the generation of diisopropylethylamine catalyzed by a Pd$_8$O$_{12}$H$_8$ cluster.

To examine the versatility of the reduction amination on Pd(OH)$_2$/g-C$_3$N$_4$, various aldehydes and amines (mainly secondary ones) were used as the substrates (Table 2), especially for the ones with sterically hindered groups. For paraformaldehyde as a substrate, higher temperature (363 K) and more methanol (10 mL) were acquired to dissolve it. Other reaction conditions are similar with reported in Table 1. As for diisopropylamine and aliphatic aldehyde (**Entry 1,2**), the selectivity to the corresponding tertiary amine could reach 99 and 87%, respectively. Piperidines (**Entry 3–6**) gave a similar result to diisopropylamine (**Entry 1**). However, the selectivity of the target tertiary amine dropped as expected when acetaldehyde was employed due to its higher reactivity that can launch the reduction of the carbonyl group and aldol condensation (**Entry 7,8**) [22]. Apart from piperidines, the chain secondary amine (**Entry 9**) could also afford relatively high conversion and selectivity. Unfortunately, when aromatic amines such as diphenylamine and N-ethylaniline were employed (**Entry 10,11**), the selectivity and yield were significantly decreased due to the competing C–N activation processes [64]. On the contrary, dicyclohexylamine (**Entry 12**) could be transformed with 99% conversion and 99% selectivity. As for cyclohexylamine (**Entry 13**), the conversion and selectivity could reach 70% and 75%, respectively. The

^{1}H and ^{13}C NMR spectra of all products can be found in the Supplementary Materials (Figure S19–S46).

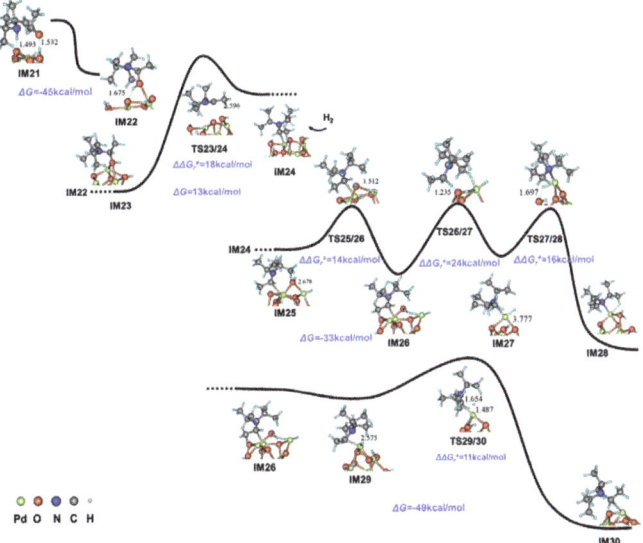

Figure 7. Simplified PES and selected structural information of alternative pathways for the generation of N,N-Diisopropylethylamine mediated by the Pd$_8$O$_{12}$H$_8$ cluster.

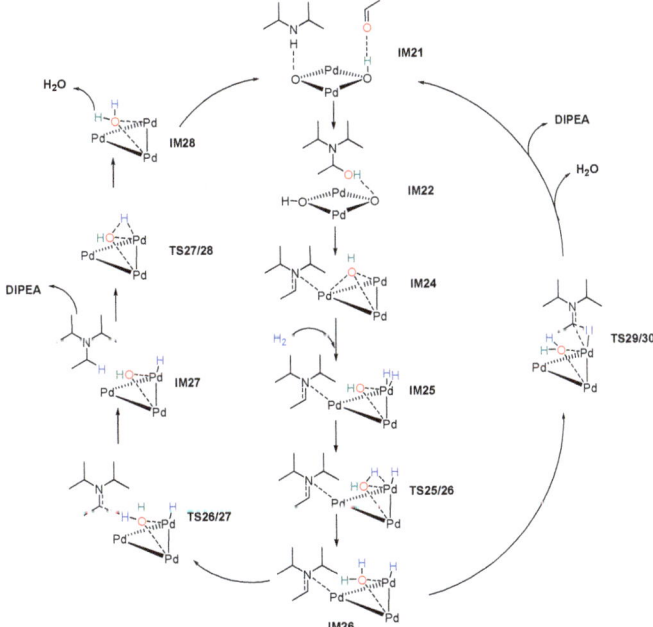

Scheme 1. Proposed mechanism for the generation of N,N-Diisopropylethylamine mediated by the Pd$_8$O$_{12}$H$_8$ cluster.

Table 2. Reductive Amination Towards Aldehydes and Amines catalyzed by Pd(OH)$_2$/g-C$_3$N$_4$.

$$R_1\text{-NH-}R_2 + R_3\text{-CHO} \xrightarrow[\text{CH}_3\text{OH}]{\text{Pd(OH)}_2/\text{g-C}_3\text{N}_4\ \text{H}_2} R_1\text{-N}(R_3)\text{-}R_2$$

Entry	Sub 1	Sub 2	Product	Temp (°C)	t (h)	Conv. (%)	Sel. (%)	Yield.[a] (%)
1	diisopropylamine	(CHO)$_n$	N-methyl diisopropylamine	90	6	90	>99	>89(83)
2	diisopropylamine	acetaldehyde	N-ethyl diisopropylamine	30	6	99	87	86(81)
3	2,2,6,6-tetramethylpiperidine	(CHO)$_n$	N-methyl-2,2,6,6-tetramethylpiperidine	90	6	>99	>99	>99(98)
4	piperidine	benzaldehyde	N-benzylpiperidine	90	6	>99	>99	>99(93)
5	morpholine	(CHO)$_n$	N-methylmorpholine	90	6	>99	>99	>99(94)
6	morpholine	benzaldehyde	N-benzylmorpholine	90	6	85	85	72(62)
7	piperidine	acetaldehyde	N-ethylpiperidine	30	6	>99	70	70(63)
8	pyrrolidine	acetaldehyde	N-ethylpyrrolidine	30	6	90	40	36(28)
9	dibutylamine	(CHO)$_n$	N-methyldibutylamine	90	6	>99	>99	>99(98)
10	diphenylamine	(CHO)$_n$	N-methyldiphenylamine	90	6	85	12	10(8)
11	N-ethylaniline	acetaldehyde	N,N-diethylaniline	30	6	40	38	15(12)
12	dicyclohexylamine	(CHO)$_n$	N-methyldicyclohexylamine	90	6	>99	>99	>99(92)
13	cyclohexylamine	(CHO)$_n$	N-methylcyclohexylamine	90	6	70	75	52(43)

Reaction conditions: catalyst loading (5.0 wt%, refer to sub 2, 1.1 wt% Pd(OH)$_2$/g-C$_3$N$_4$), sub 1 (0.1 mol), sub 2 (0.05 mol). For cases using paraformaldehyde (degree of polymerization (DP) = 3) as substrate, the usage of methanol is 10 mL, other cases are 2.5 mL. The actual molar ratio formaldehyde: amine is 1.5:1. All products were determined by GC-MS. [a] Isolated yields are reported in the brackets. All the products were separated and purified by vacuum distillation or silica gel column chromatography (see Section 3.4 for more details).

3. Materials and Methods

3.1. Chemicals and Materials

Palladium dichloride ($PdCl_2$), palladium oxide (PdO), and 10.0 wt% $Pd(OH)_2$/ACs were purchased from Alfa Aesar. Diisopropylamine, butyraldehyde, activated carbon, melamine, cyanuric acid, methanol, and molecular sieve were obtained from Sinopharm Chemical Reagent Co., Ltd. Water was purified by ion exchange and used as deionized water. Other chemicals were of analytical purity and were used as received.

3.2. Characterization Techniques

The X-ray diffraction (XRD) patterns of the samples were recorded using a Shimadzu X-ray diffractometer (MAXima XRD-7000, the Japan) with Cu Kα radiation at 40 kV and 40 mA. The X-ray photoelectron spectroscopy (XPS) measurements were performed on a Thermo Scientific K-Alpha+ system using Al Kα radiation (1486.6 eV) under a base pressure of 2×10^{-7} Torr. A FEI QUANTA FEG 650 field emission scanning electron microscope (SEM) operated at 30 kV and transmission electron microscopy (TEM) using a HT-7700 (Tokyo, Japan) at 100 kV were used for determining the morphology of the as-synthesized Pd catalysts. The Pd loading in the materials was analyzed by an Agilent 720ES type inductively coupled plasma optical emission spectroscopy (ICP-OES) instrument. The FTIR spectra were recorded using a Thermo Scientific Nicolet iS50 FTIR spectrometer, and in situ FT-IR (ATR-FT-IR) spectra were recorded using the same spectrometer with ATR appendix. The thermal gravimetric analysis (TGA) was performed by using the Pyris 1 TGA thermogravimetric analyzer with a heating rate of 5 °C/min under nitrogen atmosphere. All products were determined by 7820A-5977B GC-MS (Agilent Technologies, CA, USA) using HP-5MS column. ^{1}H NMR and ^{13}C NMR were recorded on a Bruker AVANCE III 400 MHz (101 MHz for ^{13}C), using deuterated chloroform ($CDCl_3$) and tetramethylsilane (TMS, $\delta = 0$) as internal reference. Chemical shifts are reported in parts per million (ppm) downfield and quoted to the nearest 0.01 ppm relative to the residual protons in the NMR solvent ((^{1}H NMR: δ 7.26 ppm, ^{13}C NMR: δ 77.16 ppm)), and coupling constants (J) are quoted in Hertz. GC analyses were performed on an Agilent 5973−6890 series gas chromatograph system (Agilent Technologies, CA, USA) equipped with a flame ionization detector (FID). All the separations were performed on a weakly polar capillary column SE-30.

3.3. Preparation of the Supports and Palladium Catalysts

3.3.1. Preparation of g-C_3N_4 Support

Bulk g-C_3N_4 was synthesized by a thermal treatment of a mixture of melamine and cyanuric acid with a weight ratio of 1:1 [65]. All chemicals used in the preparation were of analytical grade without further treatment. Typically, 10 g above precursor powder was placed in an alumina crucible without a cover. Then, the crucible was placed in the middle section of a tube. After vacuuming the tube, argon gas was continuously fed in with a flow rate of 50 mL min^{-1} during the thermal treatment. The sample was heated to 120 °C at a rate of 10 °C·min^{-1}, and maintained at this temperature for 20 min. The mixture was then heated to 550 °C and calcinated for 5 h. After cooling to room temperature, ~1700 mg of bulk g-C_3N_4 was obtained.

3.3.2. Pretreatment of Activated Carbon

The pretreatment of ACs has been reported elsewhere previously [38,39]. The activated carbon was pretreated with 0.4N of HNO_3 at 353 K for 4 h, was then washed with distilled water until pH = 7 after cooling to room temperature, and eventually dried in a vacuum at 323 K overnight. For hydrothermal treatment, 40 mL of deionized water was introduced into a 100 mL high-pressure autoclave; then, it was heated to 453 K and the pressure was self-generated, held at that temperature for 2 h, and subsequently cooled to room temperature. The treated ACs were filtered from the slurry and dried at 323 K overnight.

3.3.3. Preparation of the Pd/ACs Catalyst

Pd/ACs catalysts were synthesized by wetness impregnation method. First, 30 mg $PdCl_2$ was dissolved into 10 mL water with HCl mixture solution (volume/volume = 10:1). Then, the treated ACs (the ratio of ACs to water was 1 g/10 mL) were introduced into the $PdCl_2$ solution and the slurry was vigorously stirred at 353 K for 4 h; then, the slurry solution pH of 8–9 was reached by dropwise addition of KOH aqueous solution (10%). Eventually, the precipitated $Pd(OH)_2$ supported on ACs was reduced by 50% hydrazine hydrate at room temperature, filtered, and dried in vacuum at 323 K overnight.

3.3.4. Preparation of the $Pd(OH)_2$/g-C_3N_4 and $Pd(OH)_2$/ACs Catalysts

The $Pd(OH)_2/C_3N_4$ and $Pd(OH)_2$/ACs catalysts were synthesized with similar method without further reduction. First, 30 mg $PdCl_2$ was dissolved into 10 mL water with HCl mixture solution (volume/volume = 10:1). The $PdCl_2$ solution pH of 8–9 was reached by dropwise addition of KOH aqueous solution (10%). Then, the slurry was mixed with 1 g g-C_3N_4 and vigorously stirred at 323 K in a three-neck flask for 4 h, then filtered and dried in vacuum at 323 K overnight. This was the same procedure as $Pd(OH)_2$/ACs catalyst. PdO/ACs was prepared using PdO as precursor instead of the $Pd(OH)_2$ deposition.

3.4. General Procedure for the Preparation of Sterically Hindered Amine and Recycling Experiments

Diisopropylamine (0.1 mol) and butyraldehyde (0.05 mol), 5.0 wt% catalyst (0.2 g), and methanol (2.5 mL) were mixed into 100 mL high-pressure autoclave at 30 °C. Then, 1.5 MPa H_2 was fed into the reaction mixture. The mixture was stirred at 1000 rpm at 30 °C for 4 h. After the reaction, the rest of H_2 was discharged. The conversion of the butyraldehyde and the sterically hindered amine yields were determined by GC with triethylamine as the internal standard. After the reaction, the catalyst was filtered and exhaustively washed with methanol, and dried in vacuum at 323 K for 12 h. The collected catalyst was used for the next run under the same conditions. Other cycles were repeated following a similar procedure. All the products were separated and purified by vacuum distillation or silica gel column chromatography. For **Entry 12** in Table 1 and **Entry 1–3, 5, 9, 12** in Table 2, the products were separated via vacuum distillation in a rotary evaporator; for **Entry 7,8, 13** in Table 2, the products were separated via vacuum distillation in a rotary evaporator to remove the solvent and then vacuum rectification with a home-made rectification column using spring packing; for **Entry 4, 6, 10, and 11** in Table 2, the products were purified by silica gel column chromatography.

The efficiency of the catalyst was characterized by calculating the conversion (*Conv.*%) of butyraldehyde (*BA*) based on the following equation (Equation (1)):

$$Conv.\% = \frac{n_{consumed}\ BA}{n_{initial}\ BA} \cdot 100 \qquad (1)$$

Sterically hindered amine (*AN*) yield (*Y*%) was also calculated as follows (Equation (2)):

$$Yield\% = \frac{n_{formed}\ AN}{n_{theoretical}\ AN} \cdot 100 \qquad (2)$$

Furthermore, *AN* selectivity (*Sel.*%) was calculated according to the following equation (Equation (3)):

$$Sel.\% = \frac{n_{formed}\ AN}{\sum n_{products}} \cdot 100 \qquad (3)$$

3.5. General Operation Procedure for the In Situ FT-IR Spectra

In situ FT-IR spectra to investigate the adsorption of different acetaldehyde and diisopropylamine on Pd catalyst were recorded on a Nicolet iS50 spectrometer equipped with a cell fitted with BaF_2 windows and an MCT-A detector cooled with liquid nitrogen.

The spectrum was collected at a resolution of 4 cm^{-1} with an accumulation of 32 scans in the range of 4000–700 cm^{-1}. Here, the interaction of acetaldehyde and diisopropylamine was taken as an example (Figure 5f), and the catalyst sample (25 mg) was filled in priorly into a self-supported wafer. Then, the sample wafer was pretreated under a flow of N_2 for 30 min at room temperature to remove the impurities absorbed on the surface. The background was collected at room temperature under a flow of N_2. Then, liquid diisopropylamine (2 mL) filled into a glass tube was firstly vaporized by heating belt flowing by N_2 to sample wafer. Considering that pursuing a spectrum with acceptable adsorption intensity and liquefaction of substrates on the surface of the catalyst should be avoided, the temperature of the sample wafer was chosen as 323 K for all the catalysts. The spectra of adsorption of diisopropylamine were recorded per minute until stable spectra were obtained. Then, acetaldehyde (2 mL) was vaporized by heating belt to sample wafer as well. The spectra of coadsorption of diisopropylamine and acetaldehyde were recorded per minute until stable spectra were obtained. Finally, the system was purged with a flow of hydrogen instead of N_2 to obtain the spectra of aldehyde/amine/H_2 mixture. After a stable spectrum was obtained, the system was purged with a flow of N_2 again, and the desorption spectra were recorded toward the desorption time until there was no change in the band intensity. Other catalysts were then tested in a similar manner.

3.6. Computational Details

The structural optimization and the frequency analysis were performed at the GFN-xTB level of the xTB package [66,67]. Stationary points were optimized without symmetry constraint, and their nature was confirmed by vibrational frequency analysis. Unscaled vibrational frequencies were used to correct the relative energies for zero-point vibrational energy (ZPVE) contributions. Intrinsic reaction coordinate (IRC) [68–70] calculations were also performed to link transition structures with the respective intermediates; to achieve this, a Gaussian interface to the xTB code "gau_xtb" was employed [71].

4. Conclusions

In summary, a highly selective Pd catalyst designed specifically for the preparation of sterically hindered amine was reported. The selectivity of diisopropylbutylamine was up to 97%, and the yield of product amounted to 73%. The catalyst was characterized using FTIR, XRD, XPS, SEM, TEM, and TG, indicating that the active metal was well dispersed on g-C_3N_4 and stabilized by the hydroxyl group. The in situ ATR-FTIR measurement, together with the theoretical calculations, reveal that the generation of a key intermediate hemiaminal was facilitated by the proton exchange process, and the preferential adsorption of the hemiaminal enabled the selective reduction amination of diisopropylamine to sterically hindered amine. Here, the bifunctionality of the hydroxyl group mattered. An imine mechanism was thus justified for the generation of sterically hindered tertiary amines. Our work shed light on the catalytic mechanisms of reductive amination; in particular, the critical role of the hydroxyl group in the generation of sterically hindered tertiary amine was revealed. Hopefully, this may be instructive for the construction of amine moieties in specific molecules.

Supplementary Materials: The following supporting information can be downloaded at: https://www.mdpi.com/article/10.3390/ijms23147621/s1.

Author Contributions: Conceptualization and supervision, X.G. and S.Z.; methodology, software, writing—original draft preparation, Z.H. All authors have read and agreed to the published version of the manuscript.

Funding: Generous financial support by the National Natural Science Foundation of China (21878265, 22078130).

Institutional Review Board Statement: Not applicable.

Informed Consent Statement: Not applicable.

Data Availability Statement: Additional figures are available in the Supplementary Materials.

Conflicts of Interest: The authors declare no conflict of interest.

References

1. Afanasyev, O.I.; Kuchuk, E.; Usanov, D.L.; Chusov, D. Reductive Amination in the Synthesis of Pharmaceuticals. *Chem. Rev.* **2019**, *119*, 11857–11911. [CrossRef] [PubMed]
2. Smith, A.M.; Whyman, R. Review of Methods for the Catalytic Hydrogenation of Carboxamides. *Chem. Rev.* **2014**, *114*, 5477–5510. [CrossRef] [PubMed]
3. Shamim, T.; Kumar, V.; Paul, S. Silica-Functionalized CuI: An Efficient and Selective Catalyst for N-Benzylation, Allylation, and Alkylation of Primary and Secondary Amines in Water. *Synth. Commun.* **2014**, *44*, 620–632. [CrossRef]
4. Coleman, G.H. The Reaction of Alkylchloroamines with Grignard Reagents. *J. Am. Chem. Soc.* **1933**, *55*, 3001–3005. [CrossRef]
5. Ruiz-Castillo, P.; Buchwald, S.L. Applications of Palladium-Catalyzed C–N Cross-Coupling Reactions. *Chem. Rev.* **2016**, *116*, 12564–12649. [CrossRef]
6. Sambiagio, C.; Marsden, S.P.; Blacker, A.J.; McGowan, P.C. Copper catalysed Ullmann type chemistry: From mechanistic aspects to modern development. *Chem. Soc. Rev.* **2014**, *43*, 3525–3550. [CrossRef]
7. Irrgang, T.; Kempe, R. 3d-Metal Catalyzed N- and C-Alkylation Reactions via Borrowing Hydrogen or Hydrogen Autotransfer. *Chem. Rev.* **2019**, *119*, 2524–2549. [CrossRef]
8. Murugesan, K.; Senthamarai, T.; Chandrashekhar, V.G.; Natte, K.; Kamer, P.C.J.; Beller, M.; Jagadeesh, R.V. Catalytic reductive aminations using molecular hydrogen for synthesis of different kinds of amines. *Chem. Soc. Rev.* **2020**, *49*, 6273–6328. [CrossRef]
9. Irrgang, T.; Kempe, R. Transition-Metal-Catalyzed Reductive Amination Employing Hydrogen. *Chem. Rev.* **2020**, *120*, 9583–9674. [CrossRef]
10. Dong, C.; Wu, Y.; Wang, H.; Peng, J.; Li, Y.; Samart, C.; Ding, M. Facile and Efficient Synthesis of Primary Amines via Reductive Amination over a Ni/Al2O3 Catalyst. *ACS Sustain. Chem. Eng.* **2021**, *9*, 7318–7327. [CrossRef]
11. Jagadeesh, R.V.; Murugesan, K.; Alshammari, A.S.; Neumann, H.; Pohl, M.-M.; Radnik, J.; Beller, M. MOF-derived cobalt nanoparticles catalyze a general synthesis of amines. *Science* **2017**, *358*, 326–332. [CrossRef] [PubMed]
12. Lator, A.; Gaillard, Q.G.; Mérel, D.S.; Lohier, J.-F.; Gaillard, S.; Poater, A.; Renaud, J.-L. Room-Temperature Chemoselective Reductive Alkylation of Amines Catalyzed by a Well-Defined Iron(II) Complex Using Hydrogen. *J. Org. Chem.* **2019**, *84*, 6813–6829. [CrossRef] [PubMed]
13. Kapnang, H.; Charles, G.; Sondengam, B.L.; Hemo, J.H. Bis (alcoxymethyl) amines: Preparation et reduction; N-dimethylation selective d'amines primaires. *Tetrahedron Lett.* **1977**, *18*, 3469–3472. [CrossRef]
14. Tadanier, J.; Hallas, R.; Martin, J.R.; Stanaszek, R.S. Observations relevant to the mechanism of the reductive aminations of ketones with sodium cyanoborohydride and ammonium acetate. *Tetrahedron* **1981**, *37*, 1309–1316. [CrossRef]
15. Gomez, S.; Peters, J.A.; Maschmeyer, T. The Reductive Amination of Aldehydes and Ketones and the Hydrogenation of Nitriles: Mechanistic Aspects and Selectivity Control. *Adv. Synth. Catal.* **2002**, *344*, 1037–1057. [CrossRef]
16. Yagafarov, N.Z.; Kolesnikov, P.N.; Usanov, D.L.; Novikov, V.V.; Nelyubina, Y.V.; Chusov, D. The synthesis of sterically hindered amines by a direct reductive amination of ketones. *Chem. Commun.* **2016**, *52*, 1397–1400. [CrossRef]
17. Hünig, S.; Kiessel, M. Spezifische Protonenacceptoren als Hilfsbasen bei Alkylierungs- und Dehydrohalogenierungsreaktionen. *Chem. Ber.* **1958**, *91*, 380–392. [CrossRef]
18. Step, E.N.; Turro, N.J.; Gande, M.E.; Klemchuk, P.P. Mechanism of Polymer Stabilization by Hindered-Amine Light Stabilizers (HALS). Model Investigations of the Interaction of Peroxy Radicals with HALS Amines and Amino Ethers. *Macromolecules* **1994**, *27*, 2529–2539. [CrossRef]
19. Chelucci, G. Metal-complexes of optically active amino- and imino-based pyridine ligands in asymmetric catalysis. *Coord. Chem. Rev.* **2013**, *257*, 1887–1932. [CrossRef]
20. Stephan, D.W. Frustrated Lewis pairs: From concept to catalysis. *Acc. Chem. Res.* **2015**, *48*, 306–316. [CrossRef]
21. Luo, D.; He, Y.; Yu, X.; Wang, F.; Zhao, J.; Zheng, W.; Jiao, H.; Yang, Y.; Li, Y.; Wen, X. Intrinsic mechanism of active metal dependent primary amine selectivity in the reductive amination of carbonyl compounds. *J. Catal.* **2021**, *395*, 293–301. [CrossRef]
22. Podyacheva, E.; Afanasyev, O.I.; Tsygankov, A.A.; Makarova, M.; Chusov, D. Hitchhiker's Guide to Reductive Amination. *Synthesis-Stuttgart* **2019**, *51*, 2667–2677.
23. Karageorge, G.N.; Macor, J.E. Synthesis of novel serotonergics and other N-alkylamines using simple reductive amination using catalytic hydrogenation with Pd/C. *Tetrahedron Lett.* **2011**, *52*, 5117–5119. [CrossRef]
24. Guo, M.; Jayakumar, S.; Luo, M.; Kong, X.; Li, C.; Li, H.; Chen, J.; Yang, Q. The promotion effect of pi-pi interactions in Pd NPs catalysed selective hydrogenation. *Nat. Commun.* **2022**, *13*, 1770. [CrossRef]
25. Neri, G.; Musolino, M.G.; Milone, C.; Pietropaolo, D.; Galvagno, S. Particle size effect in the catalytic hydrogenation of 2,4-dinitrotoluene over Pd/C catalysts. *Appl. Catal. A* **2001**, *208*, 307–316. [CrossRef]
26. Wang, Q.; Cheng, H.; Liu, R.; Hao, J.; Yu, Y.; Zhao, F. Influence of metal particle size on the hydrogenation of maleic anhydride over Pd/C catalysts in scCO2. *Catal. Today* **2009**, *148*, 368–372. [CrossRef]
27. Suo, Y.; Hsing, I.M. Size-controlled synthesis and impedance-based mechanistic understanding of Pd/C nanoparticles for formic acid oxidation. *Electrochim. Acta* **2009**, *55*, 210–217. [CrossRef]

28. Li, J.; Chen, W.; Zhao, H.; Zheng, X.; Wu, L.; Pan, H.; Zhu, J.; Chen, Y.; Lu, J. Size-dependent catalytic activity over carbon-supported palladium nanoparticles in dehydrogenation of formic acid. *J. Catal.* **2017**, *352*, 371–381. [CrossRef]
29. Chinthaginjala, J.K.; Villa, A.; Su, D.S.; Mojet, B.L.; Lefferts, L. Nitrite reduction over Pd supported CNFs: Metal particle size effect on selectivity. *Catal. Today* **2012**, *183*, 119–123. [CrossRef]
30. Sun, W.; Wu, S.; Lu, Y.; Wang, Y.; Cao, Q.; Fang, W. Effective Control of Particle Size and Electron Density of Pd/C and Sn-Pd/C Nanocatalysts for Vanillin Production via Base-Free Oxidation. *ACS Catal.* **2020**, *10*, 7699–7709. [CrossRef]
31. Edwards, J.K.; Thomas, A.; Solsona, B.E.; Landon, P.; Carley, A.F.; Hutchings, G.J. Comparison of supports for the direct synthesis of hydrogen peroxide from H2 and O2 using Au–Pd catalysts. *Catal. Today* **2007**, *122*, 397–402. [CrossRef]
32. Xu, X.; Luo, J.; Li, L.; Zhang, D.; Wang, Y.; Li, G. Unprecedented catalytic performance in amine syntheses via Pd/g-C3N4 catalyst-assisted transfer hydrogenation. *Green Chem.* **2018**, *20*, 2038–2046. [CrossRef]
33. Wang, Z.; Wei, J.; Liu, G.; Zhou, Y.; Han, K.; Ye, H. G-C3N4-coated activated carbon-supported Pd catalysts for 4-CBA hydrogenation: Effect of nitrogen species. *Catal. Sci. Technol.* **2015**, *5*, 3926–3930. [CrossRef]
34. Wang, Y.; Yao, J.; Li, H.; Su, D.; Antonietti, M. Highly Selective Hydrogenation of Phenol and Derivatives over a Pd@Carbon Nitride Catalyst in Aqueous Media. *J. Am. Chem. Soc.* **2011**, *133*, 2362–2365. [CrossRef] [PubMed]
35. Heinen, A.W.; Peters, J.A.; Bekkum, H. The Reductive Amination of Benzaldehyde Over Pd/C Catalysts: Mechanism and Effect of Carbon Modifications on the Selectivity. *Eur. J. Org. Chem.* **2000**, *2000*, 2501–2506. [CrossRef]
36. Zhang, N.; Xie, J.; Varadan, V.K. Functionalization of carbon nanotubes by potassium permanganate assisted with phase transfer catalyst. *Smart Mater. Struct.* **2002**, *11*, 962–965. [CrossRef]
37. Carmo, M.; Linardi, M.; Rocha Poco, J.G. H_2O_2 treated carbon black as electrocatalyst support for polymer electrolyte membrane fuel cell applications. *Int. J. Hydrogen Energy* **2008**, *33*, 6289–6297. [CrossRef]
38. Li, J.; Ma, L.; Li, X.; Lu, C.; Liu, H. Effect of Nitric Acid Pretreatment on the Properties of Activated Carbon and Supported Palladium Catalysts. *Ind. Eng. Chem. Res.* **2005**, *44*, 5478–5482. [CrossRef]
39. Xu, T.-Y.; Zhang, Q.-F.; Yang, H.-F.; Li, X.-N.; Wang, J.-G. Role of Phenolic Groups in the Stabilization of Palladium Nanoparticles. *Ind. Eng. Chem. Res.* **2013**, *52*, 9783–9789. [CrossRef]
40. Song, S.; Wang, Y.Z.; Yan, N. A remarkable solvent effect on reductive amination of ketones. *Mol. Catal.* **2018**, *454*, 87–93. [CrossRef]
41. Liu, J.; Fitzgerald, A.E.; Mani, N.S. Reductive Amination by Continuous-Flow Hydrogenation: Direct and Scalable Synthesis of a Benzylpiperazine. *Synthesis-Stuttgart* **2012**, *44*, 2469–2473. [CrossRef]
42. Moro Ouma, C.N.; Modisha, P.; Bessarabov, D. Insight into the adsorption of a liquid organic hydrogen carrier, perhydro-i-dibenzyltoluene (i = m, o, p), on Pt, Pd and PtPd planar surfaces. *RSC Adv.* **2018**, *8*, 31895–31904. [CrossRef] [PubMed]
43. Santarossa, G.; Iannuzzi, M.; Vargas, A.; Baiker, A. Adsorption of Naphthalene and Quinoline on Pt, Pd and Rh: A DFT Study. *ChemPhysChem* **2008**, *9*, 401–413. [CrossRef] [PubMed]
44. Mittendorfer, F.; Thomazeau, C.; Raybaud, P.; Toulhoat, H. Adsorption of Unsaturated Hydrocarbons on Pd(111) and Pt(111): A DFT Study. *J. Phys. Chem. B* **2003**, *107*, 12287–12295. [CrossRef]
45. Han, M.; Wang, H.; Zhao, S.; Hu, L.; Huang, H.; Liu, Y. One-step synthesis of CoO/g-C3N4 composites by thermal decomposition for overall water splitting without sacrificial reagents. *Inorg. Chem. Front.* **2017**, *4*, 1691–1696. [CrossRef]
46. Qi, F.; Li, Y.; Wang, Y.; Wang, Y.; Liu, S.; Zhao, X. Ag-Doped g-C3N4 film electrode: Fabrication, characterization and photoelectro-catalysis property. *RSC Adv.* **2016**, *6*, 81378–81385. [CrossRef]
47. Wang, M.; Shen, S.; Li, L.; Tang, Z.; Yang, J. Effects of sacrificial reagents on photocatalytic hydrogen evolution over different photocatalysts. *J. Mater. Sci.* **2017**, *52*, 5155–5164. [CrossRef]
48. Baranova, E.A.; Le Page, Y.; Ilin, D.; Bock, C.; MacDougall, B.; Mercier, P.H.J. Size and composition for 1–5nm Ø PtRu alloy nano-particles from Cu Kα X-ray patterns. *J. Alloys Compd.* **2009**, *471*, 387–394. [CrossRef]
49. Bakker, J.J.W.; Neut, A G.v.d.; Kreutzer, M.T.; Moulijn, J.A.; Kapteijn, F. Catalyst performance changes induced by palladium phase transformation in the hydrogenation of benzonitrile. *J. Catal.* **2010**, *274*, 176–191. [CrossRef]
50. Amorim, C.; Keane, M.A. Palladium supported on structured and nonstructured carbon: A consideration of Pd particle size and the nature of reactive hydrogen. *J. Colloid Interface Sci.* **2008**, *322*, 196–208. [CrossRef]
51. Albers, P.W.; Möbus, K.; Wieland, S.D.; Parker, S.F. The fine structure of Pearlman's catalyst. *Phys. Chem. Chem. Phys.* **2015**, *17*, 5274–5278. [CrossRef] [PubMed]
52. Jaleel, A.; Haider, A.; Nguyen, C.V.; Lee, K.R.; Choung, S.; Han, J.W.; Baek, S.-H.; Shin, C.-H.; Jung, K.-D. Structural effect of Nitrogen/Carbon on the stability of anchored Ru catalysts for CO2 hydrogenation to formate. *Chem. Eng. J.* **2022**, *433*, 133571. [CrossRef]
53. Bulushev, D.A.; Yuranov, I.; Suvorova, E.I.; Buffat, P.A.; Kiwi-Minsker, L. Highly dispersed gold on activated carbon fibers for low-temperature CO oxidation. *J. Catal.* **2004**, *224*, 8–17. [CrossRef]
54. Das, T.K.; Banerjee, S.; Vishwanadh, B.; Joshi, R.; Sudarsan, V. On the nature of interaction between Pd nanoparticles and C3N4 support. *Solid State Sci.* **2018**, *83*, 70–75. [CrossRef]
55. García, G.; Silva-Chong, J.; Rodríguez, J.L.; Pastor, E. Spectroscopic elucidation of reaction pathways of acetaldehyde on platinum and palladium in acidic media. *J. Solid State Electrochem.* **2014**, *18*, 1205–1213. [CrossRef]
56. Knöpke, L.R.; Nemati, N.; Köckritz, A.; Brückner, A.; Bentrup, U. Reaction Monitoring of Heterogeneously Catalyzed Hydrogenation of Imines by Coupled ATR-FTIR, UV/Vis, and Raman Spectroscopy. *ChemCatChem* **2010**, *2*, 273–280. [CrossRef]

57. Tammer, M.G. Sokrates: Infrared and Raman characteristic group frequencies: Tables and charts. *Colloid Polym. Sci.* **2004**, *283*, 235. [CrossRef]
58. Aguirre, A.; Collins, S.E. Insight into the mechanism of acetonitrile hydrogenation in liquid phase on Pt/Al$_2$O$_3$ by ATR-FTIR. *Catal. Today* **2019**, *336*, 22–32. [CrossRef]
59. Litvak, I.; Anker, Y.; Cohen, H. On-line in situ determination of deuterium content in water via FTIR spectroscopy. *RSC Adv.* **2018**, *8*, 28472–28479. [CrossRef]
60. Raskó, J.; Kiss, J. Adsorption and surface reactions of acetonitrile on Al2O3-supported noble metal catalysts. *Appl. Catal. A* **2006**, *298*, 115–126. [CrossRef]
61. Wu, X.; Ge, Q.; Zhu, X. Vapor phase hydrodeoxygenation of phenolic compounds on group 10 metal-based catalysts: Reaction mechanism and product selectivity control. *Catal. Today* **2021**, *365*, 143–161. [CrossRef]
62. Ge, X.; Luo, C.; Qian, C.; Yu, Z.; Chen, X. RANEY® nickel-catalyzed reductive N-methylation of amines with paraformaldehyde: Theoretical and experimental study. *RSC Adv.* **2014**, *4*, 43195–43203. [CrossRef]
63. Sanyal, U.; Yuk, S.F.; Koh, K.; Lee, M.S.; Stoerzinger, K.; Zhang, D.; Meyer, L.C.; Lopez-Ruiz, J.A.; Karkamkar, A.; Holladay, J.D.; et al. Hydrogen Bonding Enhances the Electrochemical Hydrogenation of Benzaldehyde in the Aqueous Phase. *Angew. Chem. Int. Ed.* **2021**, *60*, 290–296. [CrossRef] [PubMed]
64. Brigas, A.F.; Clegg, W.; Dillon, C.J.; Fonseca, C.F.C.; Johnstone, R.A.W. Metal-assisted reactions. Part 29.1 Structure and hydrogenolysis of C–N bonds in derivatives of aromatic amines. Bond length and electronegativity changes from X-ray crystallographic data. *J. Chem. Soc. Perkin Trans. 2* **2001**, *8*, 1315–1324. [CrossRef]
65. Ong, W.-J.; Tan, L.-L.; Ng, Y.H.; Yong, S.-T.; Chai, S.-P. Graphitic Carbon Nitride (g-C3N4)-Based Photocatalysts for Artificial Photosynthesis and Environmental Remediation: Are We a Step Closer To Achieving Sustainability? *Chem. Rev.* **2016**, *116*, 7159–7329. [CrossRef] [PubMed]
66. Bannwarth, C.; Ehlert, S.; Grimme, S. GFN2-xTB-An Accurate and Broadly Parametrized Self-Consistent Tight-Binding Quantum Chemical Method with Multipole Electrostatics and Density-Dependent Dispersion Contributions. *J. Chem. Theory Comput.* **2019**, *15*, 1652–1671. [CrossRef]
67. Grimme, S.; Bannwarth, C.; Shushkov, P. A Robust and Accurate Tight-Binding Quantum Chemical Method for Structures, Vibrational Frequencies, and Noncovalent Interactions of Large Molecular Systems Parametrized for All spd-Block Elements (Z = 1–86). *J. Chem. Theory Comput.* **2017**, *13*, 1989–2009. [CrossRef]
68. Hratchian, H.P.; Schlegel, H.B. Using Hessian updating to increase the efficiency of a Hessian based predictor-corrector reaction path following method. *J. Chem. Theory Comput.* **2005**, *1*, 61–69. [CrossRef]
69. Truhlar, D.G.; Kilpatrick, N.J.; Garrett, B.C. Reaction-path interpolation models for variational transition-state theory. *J. Chem. Phys.* **1983**, *78*, 2438–2442. [CrossRef]
70. Fukui, K. The path of chemical-reactions—The IRC approach. *Acc. Chem. Res.* **1981**, *14*, 363–368. [CrossRef]
71. Lu, T. Gau_xtb: A Gaussian Interface for xtb Code. Available online: http://sobereva.com/soft/gau_xtb (accessed on 17 May 2021).

Article

Experimental and Theoretical Study of N_2 Adsorption on Hydrogenated $Y_2C_4H^-$ and Dehydrogenated $Y_2C_4^-$ Cluster Anions at Room Temperature

Min Gao, Yong-Qi Ding and Jia-Bi Ma *

Key Laboratory of Cluster Science of Ministry of Education, Beijing Key Laboratory of Photoelectronic/Electrophotonic Conversion Materials, School of Chemistry and Chemical Engineering, Beijing Institute of Technology, Beijing 102488, China; 3120201255@bit.edu.cn (M.G.); 3120195628@bit.edu.cn (Y.-Q.D.)
* Correspondence: majiabi@bit.edu.cn

Abstract: The adsorption of atmospheric dinitrogen (N_2) on transition metal sites is an important topic in chemistry, which is regarded as the prerequisite for the activation of robust N≡N bonds in biological and industrial fields. Metal hydride bonds play an important part in the adsorption of N_2, while the role of hydrogen has not been comprehensively studied. Herein, we report the N_2 adsorption on the well-defined $Y_2C_4H_{0,1}^-$ cluster anions under mild conditions by using mass spectrometry and density functional theory calculations. The mass spectrometry results reveal that the reactivity of N_2 adsorption on $Y_2C_4H^-$ is 50 times higher than that on $Y_2C_4^-$ clusters. Further analysis reveals the important role of the H atom: (1) the presence of the H atom modifies the charge distribution of the $Y_2C_4H^-$ anion; (2) the approach of N_2 to $Y_2C_4H^-$ is more favorable kinetically compared to that to $Y_2C_4^-$; and (3) a natural charge analysis shows that two Y atoms and one Y atom are the major electron donors in the $Y_2C_4^-$ and $Y_2C_4H^-$ anion clusters, respectively. This work provides new clues to the rational design of TM-based catalysts by efficiently doping hydrogen atoms to modulate the reactivity towards N_2.

Keywords: N_2 adsorption; mass spectrometry; density functional theory calculations

1. Introduction

More than 99% of the global nitrogen exists in the shape of gaseous dinitrogen (N_2) in the atmosphere, yet most organisms can only metabolize nitrogen-containing substances such as NH_3 rather than N_2 directly. Although N_2 is the main nitrogen source for most natural and artificial nitrogen-containing compounds, the high bond dissociation energy (9.75 eV) and the large HOMO–LUMO gap (10.8 eV) render its adsorption and activation an enormous challenge in chemistry [1–4]. Scientists regularly rely on transition metal (TM) centers to catalyze the nitrogen conversion processes [5–7]. The initial and critical step in the complicated reduction of dinitrogen is the adsorption of N_2 molecules at the TM center [8,9]. The fixation of nitrogen in industry is carried out at metal-based (Fe^- or Ru^-) catalysts under extremely high temperatures (300–500 °C) and high pressures (100–300 atm), involving the disadvantages of large energy consumption and greenhouse gas emission [10–12]. Thus, it is vital to develop mild, energy-saving, and environment–friendly catalytic systems for N_2 fixation at ambient conditions. The activation of nitrogen by transition metal compounds with the involvement of hydrogen atoms is of particular interest, while the most common feature of N_2 hydrogenative cleavage is the participation of metal hydride bonds [13–15]. A literature survey [13] shows that metal hydride bonds have several important roles: (1) as a hydrogen source; (2) as an electron source for N_2 reduction; (3) as a powerful reducing agent for the removal of activated nitrogen atoms; and so on.

As an ideal model of condensed-phase systems, gas-phase clusters can study chemical reactions and reveal related mechanisms at the strictly molecular level by simulating active sites. [16–19]. Several theoretical and experimental studies have reported the reactivity of metal species with nitrogen, however, only a few metal species such as, Sc_2 [20], Ta_2^+ [21], $V_3C_4^-$ [22], $Ta_2C_4^-$ [23], NbH_2^- [24], $Ta_3N_3H_{0,1}^-$ [25], $Sc_3NH_2^+$ [26], $FeTaC_2^-$ [27], and $AuNbBO^-$ [28] have been characterized to cleave the N≡N triple bond completely. It can be seen that for the studies on N_2 adsorption in the gas phase, there are few metal species, and they mainly focus on the early transition metals. In the previous work, we found that a suitable number of hydrogen atoms has an influence on the reactivity of transition metal-containing clusters with N_2 [24–26,29,30]. $Sc_3NH_2^+$ [26] can effectively realize the activation of N_2 by H_2, which is based on the regulation of N_2 reduction by two H atoms. $Ta_3N_3H_{0,1}^-$ is an example that highlights the importance of the assisted reactivity of a single hydrogen atom, and the reactivity of $Ta_3N_3H^-$ is higher by a factor of five compared with that of $Ta_3N_3^-$ due to the hydrogen atom changing the charge distribution and geometry [25]. How can hydrogen atoms be efficiently doped to modulate the reactivity of TM-containing systems towards N_2 at the molecular scale? Considering the previous exploration of the Sc systems and the fact that Sc and Y belong to the same group, $Y_2C_4^-$ and $Y_2C_4H^-$ cluster anions were synthesized, and the reactivity towards N_2 was investigated by mass spectrometry and DFT calculations, to answer this question. This work clearly revealed that $Y_2C_4H_{0,1}^-$ anions can adsorb N_2, and the hydrogen atom greatly enhances the reactivity of $Y_2C_4H^-$ towards N_2.

2. Results and Discussion

The time-of-flight (TOF) mass spectra of laser ablation-generated, further mass-elected $Y_2C_4^-$ and $Y_2C_4H^-$ cluster anions reacting with N_2 under thermal collision conditions in a linear ion trap (LIT) reactor are shown in Figure 1. The mass spectra for the generation of $Y_2C_4H_{0,1}^-$ clusters has been given (Supplementary Figure S1). Upon the interactions of $Y_2C_4^-$ and $Y_2C_4H^-$ with N_2, two adsorbed complexes that are assigned as $Y_2C_4N_2^-$ and $Y_2C_4HN_2^-$ are observed (Figure 1b,d), suggesting the following channels in Equations (1) and (2):

$$Y_2C_4^- + N_2 \rightarrow Y_2C_4N_2^- \quad (1)$$

$$Y_2C_4H^- + N_2 \rightarrow Y_2C_4HN_2^- \quad (2)$$

Compared with $Y_2C_4^-$, $Y_2C_4H^-$ shows a higher reactivity towards N_2 under the same reaction conditions in Figure 1f. Besides the major products, two weak peaks in Figure 1 are assigned to $Y_2C_4OH^-$ and $Y_2C_4O_2H^-$, generated from the reaction of $Y_2C_4H_{0,1}^-$ anions with water impurities in the LIT. The pseudo-first-order rate constants (k_1) for the reactions one and two are estimated to be $(3.7 \pm 0.8) \times 10^{-12}$ cm^3 $molecule^{-1}$ s^{-1} and $(6.2 \pm 1.3) \times 10^{-14}$ cm^3 $molecule^{-1}$ s^{-1}, which are based on a least-square fitting procedure, corresponding to reaction efficiencies (Φ) [31,32] of 0.6% and 0.01%, respectively. Additionally, the signal dependence of product $Y_2C_4H_{0,1}N_2^-$ ions on N_2 pressures was obtained, which are derived and fitted with the mass spectrometry experimental data (Supplementary Figure S2).

BPW91 calculations are performed to investigate the structures of reactant $Y_2C_4H_{0,1}^-$ anion clusters (Supplementary Figure S3), as well as the reaction mechanisms between $Y_2C_4H_{0,1}^-$ and N_2. The lowest-energy isomer of $Y_2C_4^-$ (doublet, ^{2}**IA1**, Supplementary Figure S3), which is 0.08 eV lower than its quartet isomer, is a C_s-symmetric six-membered ring, with the Y-Y bond as the symmetry axis and two C_2 ligands bonded to the two Y atoms. Moreover, the most stable isomer of $Y_2C_4H^-$ (^{1}**IA2**) has a hydrogen atom binding to the Y1 atom in the six-membered ring, similar to the $Y_2C_4^-$ (^{2}**IA1**), and it is 0.07 eV lower than the triplet state in energy (Supplementary Figure S3). Since the energies of the isomers are very close, their reaction paths are calculated. The results show that, in the reaction coordinates, the energies of the doublet and singlet stationary points and the products in the $Y_2C_4^-/N_2$ and $Y_2C_4H^-/N_2$ systems are lower than those of the corresponding quartet and triplet analogues, respectively (Supplementary Figure S4). Enthalpy and Gibbs free energies along with electronic and zero-point correction energies are added (Supplementary Table S1). The concentration of dinitrogen adducts in the gas phase is relatively low, so it is difficult to collect and continue to measure Raman spectra. Currently, it is difficult to characterize structures due to technical and instrumental limitations. Infrared multiple photon dissociation may be applied to reveal such types of anions. We have added the calculated infrared spectra (Supplementary Figure S5), and the vibrational frequencies may be used for future experimental identification of these clusters.

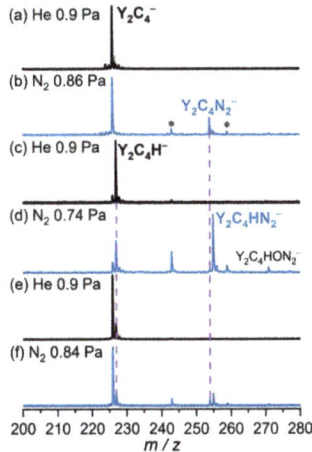

Figure 1. TOF mass spectra for the reactions of (**a**) mass-selected $Y_2C_4^-$ with He and (**b**) N_2 for 6 ms, (**c**) mass-selected $Y_2C_4H^-$ with He and (**d**) N_2 for 14 ms, and (**e**) the coexisting $Y_2C_4^-$ and $Y_2C_4H^-$ clusters with (**f**) N_2 for 10 ms, respectively. The effective reactant gas pressures are shown. The asterisked peaks (*) are $Y_2C_4OH^-$ and $Y_2C_4O_2H^-$, due to the reactions with residual water in the LIT. Black bold, blue bold and black font represent reactants, products and impurities, respectively.

The potential energy surfaces (PESs) of the most favorable reaction pathways are given in Figure 2. The N_2 molecule is initially captured by the Y1 atom in both $Y_2C_4^-$ and $Y_2C_4H^-$ to form the end-on-coordinated complexes ^{2}**I1** and ^{1}**I4**. Notably, ^{2}**I1** (−0.71 eV) in Figure 2a is as stable as ^{1}**I4** (−0.70 eV) in Figure 2b, suggesting that the N_2-adsorbed intermediates ^{2}**I1** and ^{1}**I4** are not the final products in the $Y_2C_4^-/N_2$ and $Y_2C_4H^-/N_2$ systems. As for the $Y_2C_4^-/N_2$ system, the coordination mode of N_2 is further changed from η^1 in ^{2}**I1** to η^2 in ^{2}**I2** via ^{2}**TS1**. During this process, the N-N bond length is elongated from 110 pm in free N_2 to 119 pm in ^{2}**I2**. Subsequently, the adsorbed N_2 unit is anchored by two Y atoms via ^{2}**TS2**, forming a Y-N-N-Y bridge; at the same time, a longer N-N bond of 123 pm is generated in ^{2}**P1**. Note that the rupture of the N-N bonds encounters a high energy barrier (^{2}**TS3**, +2.46 eV with respect to the separated reactants), so that further activation of N_2 is hampered in this system.

The reaction of $Y_2C_4H^-/N_2$ (Figure 2b) follows the similar mechanism. The complex is coordinated laterally to form a Y-N-N-Y bridge like ^{2}**P1** by overcoming a negligible barrier ^{1}**TS4**, and the activation energy (ΔE_a, i.e., the energy difference between the encounter complex and the transition state) is lower than that of ^{2}**I2** → ^{2}**TS2** (ΔE_a = 0.23 eV) in $Y_2C_4^-$. In the step of ^{1}**I4** → ^{1}**P2**, an elongation of the N–N bond from 115 to 121 pm occurs. Further cleavage of N–N is also hindered due to the positive energy barrier of 4.89 eV (^{1}**TS5**). In addition, another adsorption of N_2 on the Y2 atom (Supplementary Figure S6) that is not bonded with the hydrogen atom can be eventually trapped in ^{1}**P2** by generating the η^2-mode intermediate ^{1}**I7**. In conclusion, the reactions of $Y_2C_4H^-$ and $Y_2C_4^-$ with N_2 result in the formation of bridging adsorption products ^{2}**P1** and ^{1}**P2**, and the adsorbed N_2

molecules are in the η^1:η^2 mode. As shown in Figure 3, the potential energy curves reveal that the adsorption process of $Y_2C_4H^-/N_2$ is more favorable kinetically compared to that of $Y_2C_4^-/N_2$, since it is barrier−free for $Y_2C_4H^-/N_2$. A small barrier exists in the shallow entrance channels when N_2 approaches $Y_2C_4^-$, which further explains the experimental observed low reaction rate constant for the dehydrogenated $Y_2C_4^-/N_2$.

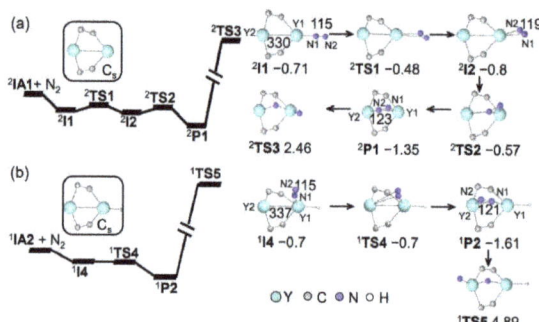

Figure 2. BPW91-D3-calculated potential energy surfaces for the reactions of $Y_2C_4^-$ (**a**) and $Y_2C_4H^-$ (**b**) with N_2. The zero-point vibration-corrected energies (ΔH_{0K} in eV) of the reaction intermediates (**I1–I4**), transition states (**TS1–TS4**), and products (**P1, P2**), with respect to the separated reactants, are given. The bond lengths are given in pm. The green, blue, grey and white atoms represent Y, N, C and H atoms, respectively. Spin multiplicity is located in superscript.

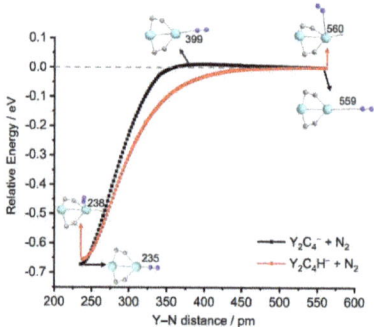

Figure 3. The BPW91-calculated relaxed potential energy curves of N_2 approaching $Y_2C_4^-$ and $Y_2C_4H^-$ anions.

Frontier orbital analysis shows that the immobilization of the N_2 ligand, as well as the formation of ^{2}**P1** and ^{1}**P2**, involve d-electrons transfer from the single-occupied molecular orbital-1 (SOMO-1) of $Y_2C_4^-$ and the HOMO orbital of $Y_2C_4H^-$ to the antibonding π^*-orbitals of N_2 (Supplementary Figure S7). The presence of hydrogen atoms enhances the reactivity of the cluster cations toward N_2 since it changes the charge distribution. As shown in Figure 4a, the Y1 linked to the hydrogen atom on the $Y_2C_4H^-$ cluster has more negative charges compared to $Y_2C_4^-$, and it promotes π-back-donation. Note that the energy differences between the transition states and the separated reactants, which is the apparent barrier ($\Delta E^\ddagger$), matters in gas−phase studies. The apparent barrier for $Y_2C_4H^-/N_2$ ($\Delta E^\ddagger = -0.70$ eV) is lower than that of $Y_2C_4^-/N_2$ ($\Delta E^\ddagger = -0.48$ eV), and the energy of ^{1}**P2** is lower than that of ^{2}**P1** (-1.61 eV vs. -1.35 eV). According to the Rice-Ramsperger-Kassel-Marcus (RRKM) theory [33], the internal conversion rate of **I4** → **TS4** (8.49×10^{11} s^{-1}) is 32 times larger than that of **I2** → **TS2** (2.65×10^{10} s^{-1}). These theoretical results are consistent with the experiments.

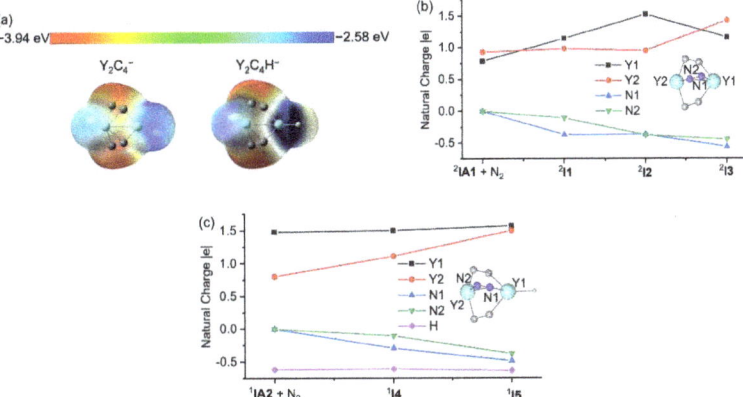

Figure 4. (a) Electrostatic potentials of the $Y_2C_4H_{0,1}^-$. Charges on atoms of stationary points along reaction coordinates of N_2 absorption on (b) $Y_2C_4^-$ and (c) $Y_2C_4H^-$ clusters.

To further improve the understanding of $Y_2C_4H_{0,1}^-/N_2$ systems, NBO analysis along reaction coordinates was performed (Figure 4b,c). The charge details were added (Supplementary Table S2). In the adsorption processes **IA1** → **I1** and **IA2** → **I4** of $Y_2C_4H_{0,1}^-/N_2$, the yttrium atoms transfer 0.37 e and 0.29 e to the N1 atom, respectively, leading to the formation of the Y-N1 bonds, while two N2 atoms in $Y_2C_4^-$ and $Y_2C_4H^-$ only increase by 0.11 e. In the subsequent steps **I2** → **P1** and **I4** → **P2** for the formation of the N2-Y2 bonds, more electrons are stored in the two nitrogen atoms, resulting in the gradual elongation of the N-N bonds. Overall, the electrons required for the N_2 adsorption by the $Y_2C_4^-$ and $Y_2C_4H^-$ clusters are mainly provided by Y atoms with total transferred amounts of 0.88 e and 0.78 e, respectively. Differently, two and one Y atoms are the electron donors in $Y_2C_4^-$ and $Y_2C_4H^-$, respectively. The active Y1 atom in $Y_2C_4^-$ (**IA1**) has more 5s electron occupancies (5s$^{1.10}$ 4d$^{1.03}$), which causes an unfavorable approach and a high σ-repulsion on the N_2 molecule. When one hydrogen atom on the $Y_2C_4^-$ (^{2}**IA1**) cluster bonds to form $Y_2C_4H^-$ (^{1}**IA2**), the natural charge on the Y1 increases from 0.79 e to 1.48 e; at the same time, more 4d and less 5s electron occupancies are located (5s$^{0.38}$ 4d$^{1.12}$), which can make N_2 more accessible to the $Y_2C_4H^-$ cluster anions. The values of bond orders of Y-Y bond in $Y_2C_4H_{0,1}^-$ anions are an important indicator for the ability of storing electrons, which increases from 0.55 in $Y_2C_4^-$ (^{2}**IA1**) to 0.66 in $Y_2C_4H^-$ (^{1}**IA2**). Therefore, although hydrogen appears to be a bystander in N_2 adsorption, its presence indeed stores more electrons in the Y-Y bond and facilitates N_2 adsorption. It can be concluded that the hydrogen atom in the $Y_2C_4H^-$ cluster significantly affects the charge distribution and electronic structure, and a suitable number of hydrogen atoms can enhance the reactivity towards N_2.

3. Methods

3.1. Experimental Methods

The metal carbide clusters were generated by laser ablation metal target (made of pure yttrium powder) (Jiangxi Ketai New Materials Co. Ltd, Jiangxi, China) seeded at 2‰ CH$_4$ (Beijing Huatong Jingke Gas Chemical Co. Ltd, Beijing, China) in a helium carrier gas (backing pressure 4 atm). The pulsed laser is a 532 nm laser with 5–8 mJ/energy pulses and 10 Hz repetition rate (140 Baytech Drive, San Jose, CA, USA). $Y_2C_4^-$ and $Y_2C_4H^-$ anion clusters were mass-selected by a quadrupole mass filter (QMF) (China Academy of Engineering Physics, Mianyang, Sichuan, China) [34] and subsequently entered into a linear ion trap (LIT) reactor (homemade) [35]. After being confined and thermalized by the pulsed gas He for about 2 ms, they interacted with N_2 for about 6 ms and 14 ms, at room temperature, respectively. The anion clusters were ejected from the LIT and then detected by a reflection time-of-flight mass spectrometer (TOF-MS) [36]. The rate constants of the reactions between $Y_2C_4H_{0,1}^-$ cluster anions and N_2 were described [37]. A schematic diagram of the experimental apparatus is shown in ref [34].

3.2. Computational Methods

All DFT [38] calculations were formed using the Gaussian 09 [39] program package to explore the structures of reactant clusters $Y_2C_4H_{0,1}^-$ and the mechanistic details of $Y_2C_4H_{0,1}^-$ with N_2. To give the best interpretation of the experimental data, we calculated the dissociation energies of the Y-Y Y-C, N-N and C-C (Supplementary Table S3) bonds using 20 methods. The results show that BPW91 functional [40–42] performs very well. For application of basis sets in reaction systems, the def2-TZVP [43] basis set was used for the Y atom, and the 6 − 311 + G * basis sets [44,45] were selected for the C, H, and N atoms, which were applied in other systems containing these elements [24,27,46]. The zero-point vibration corrected energies (ΔH_{0K} in eV) in unit of eV are reported. Vibrational frequency calculations must be performed for the geometric optimization of the reaction intermediates (IMs) and transition states (**TS**s) [47]. Intrinsic reaction coordinate [48] calculations were employed to ensure whether each TS was connected to two appropriate local minima. DFT-D3 correction for the complexes were contained in the system. Natural population analysis was performed using NBO 6.0 [49], and the orbital composition was analyzed by the method of natural atomic orbitals employing the Multiwfn program [50].

4. Conclusions

In summary, the reactions of $Y_2C_4H^-$ and dehydrogenated $Y_2C_4^-$ cluster anions with N_2 have been investigated experimentally and theoretically. The experimental results indicate that the reaction rate constant of $Y_2C_4H^-/N_2$ is higher by a factor of 50 compared with that of $Y_2C_4^-/N_2$. DFT calculations indicate that the differences are caused by the different charge distributions and the bonding of the additional hydrogen atom to the yttrium atom in the $Y_2C_4H^-$ cluster, resulting in more 4d electron occupancies and thus more efficient π-back-donation bonding with N_2 molecules. The electron donor atoms of $Y_2C_4^-$ and $Y_2C_4H^-$ anion clusters are different, for $Y_2C_4^-$, two Y atoms donate electrons, while only one Y atom donates electrons in $Y_2C_4H^-$. Storing more electrons in the Y-Y bond is also an important influence of the hydrogen atom on the reactivity of $Y_2C_4H^-$ to N_2. This study clearly reveals the significance of hydrogen-assisted reactions in N_2 adsorption processes. Attaching an appropriate number of hydrogen atoms on active sites can enhance the N_2 adsorption rates, providing a new strategic direction for the rational design of TM-based energy-efficient nitrogen fixation catalysts.

Supplementary Materials: The following supporting information can be downloaded at: https://www.mdpi.com/article/10.3390/ijms23136976/s1. References [51–55] are cited in the supplementary materials.

Author Contributions: Data curation, M.G. and Y.-Q.D.; writing—original draft preparation, M.G.; writing—review and editing, M.G. and J.-B.M.; supervision, J.-B.M.; project administration, J.-B.M.; funding acquisition, J.-B.M. All authors have read and agreed to the published version of the manuscript.

Funding: This research was funded by National Natural Science Foundation of China (No. 91961122) and the Beijing Natural Science Foundation (No. 2222023).

Institutional Review Board Statement: Not applicable.

Informed Consent Statement: Not applicable.

Data Availability Statement: The data presented in this study are available on request from the corresponding author.

Conflicts of Interest: The authors declare no conflict of interest.

References

1. Burford, R.J.; Fryzuk, M.D. Examining the relationship between coordination mode and reactivity of dinitrogen. *Nat. Rev. Chem.* **2017**, *1*, 0026. [CrossRef]
2. Chen, J.G.; Crooks, R.M.; Seefeldt, L.C.; Bren, K.L.; Bullock, R.M.; Darensbourg, M.Y.; Holland, P.L.; Hoffman, B.; Janik, M.J.; Jones, A.K.; et al. Beyond fossil fuel–driven nitrogen transformations. *Science* **2018**, *360*, eaar6611. [CrossRef] [PubMed]
3. Légaré, A.; Rang, M.; Bélanger-Chabot, G.; Schweizer, J.I.; Krummenacher, I.; Bertermann, R.; Arrowsmith, M.; Holthausen, M.C.; Braunschweig, H. The reductive coupling of dinitrogen. *Science* **2019**, *363*, 1329–1332. [CrossRef] [PubMed]
4. Tomaszewski, R. Citations to chemical resources in scholarly articles: CRC handbook of chemistry and physics and the merck index. *Scientometrics* **2017**, *112*, 1865–1879. [CrossRef]
5. Avenier, P.; Taoufik, M.; Lesage, A.; Solans-Monfort, X.; Baudouin, A.; de Mallmann, A.; Veyre, L.; Basset, J.M.; Eisenstein, O.; Emsley, L.; et al. Dinitrogen dissociation on an isolated surface tantalum atom. *Science* **2007**, *317*, 1056–1060. [CrossRef]
6. Shima, T.; Hu, S.; Luo, G.; Kang, X.; Luo, Y.; Hou, Z. Dinitrogen cleavage and hydrogenation by a trinuclear titanium polyhydride complex. *Science* **2013**, *340*, 1549–1552. [CrossRef]
7. Qiu, P.Y.; Wang, J.W.; Liang, Z.Q.; Xue, Y.J.; Zhou, Y.L.; Zhang, X.L.; Cui, H.Z.; Cheng, G.Q.; Tian, J. The metallic 1T-WS2 as cocatalysts for promoting photocatalytic N2 fixation performance of Bi5O7Br nanosheets. *Chin. Chem. Lett.* **2021**, *32*, 3501–3504. [CrossRef]
8. Deng, G.; Pan, S.; Wang, G.; Zhao, L.; Zhou, M.; Frenking, G. Beryllium atom mediated dinitrogen activation via coupling with carbon monoxide. *Angew. Chem. Int. Ed.* **2020**, *59*, 18201–18207. [CrossRef]
9. Wang, Y.Y.; Ding, X.L.; Israel Gurti, J.; Chen, Y.; Li, W.; Wang, X.; Wang, W.J.; Deng, J.J. Non-dissociative activation of chemisorbed dinitrogen on one or two vanadium atoms supported by a Mo_6S_8 cluster. *Chem. Phys. Chem.* **2021**, *22*, 1645–1654. [CrossRef]
10. Qing, G.; Ghazfar, R.; Jackowski, S.T.; Habibzadeh, F.; Ashtiani, M.M.; Chen, C.P.; Smith, M.R., III; Hamann, T.W. Recent advances and challenges of electrocatalytic N_2 reduction to ammonia. *Chem. Rev.* **2020**, *120*, 5437–5516. [CrossRef]
11. Cherkasov, N.; Ibhadon, A.O.; Fitzpatrick, P. A review of the existing and alternative methods for greener nitrogen fixation. *Chem. Eng. Process.* **2015**, *90*, 24–33. [CrossRef]
12. van der Ham, C.J.M.; Koper, M.T.M.; Hetterscheid, D.G.H. Challenges in reduction of dinitrogen by proton and electron transfer. *Chem. Soc. Rev.* **2014**, *43*, 5183–5191. [CrossRef] [PubMed]
13. Jia, H.P.; Quadrelli, E.A. Mechanistic aspects of dinitrogen cleavage and hydrogenation to produce ammonia in catalysis and organometallic chemistry: Relevance of metal hydride bonds and dihydrogen. *Chem. Soc. Rev.* **2014**, *43*, 547–564. [CrossRef] [PubMed]
14. Li, J.; Li, S. Energetics and mechanism of dinitrogen cleavage at a mononuclear surface tantalum center: A new way of dinitrogen reduction. *Angew. Chem. Int. Ed.* **2008**, *47*, 8040–8043. [CrossRef]
15. Chow, C.; Taoufik, M.; Quadrelli, E.A. Cheminform abstract: Ammonia and dinitrogen activation by surface organometallic chemistry on silica-grafted tantalum hydrides. *Eur. J. Inorg. Chem.* **2011**, *2011*, 1349–1359. [CrossRef]
16. Lang, S.M.; Bernhardt, T.M. Gas phase metal cluster model systems for heterogeneous catalysis. *Phys. Chem. Chem. Phys.* **2012**, *14*, 9255–9269. [CrossRef]
17. O'Hair, R.A.J.; Khairallah, G.N. Gas phase ion chemistry of transition metal clusters: Production, reactivity, and catalysis. *J. Clust. Sci.* **2004**, *15*, 331–363. [CrossRef]
18. Schwarz, H. Menage-a-Trois: Single-atom catalysis, mass spectrometry, and computational chemistry. *Catal. Sci. Technol.* **2017**, *7*, 4302–4314. [CrossRef]
19. Schwarz, H. How and why do cluster size, charge state, and ligands affect the course of metal-mediated gas-phase activation of methane? *Isr. J. Chem.* **2014**, *54*, 1413–1431. [CrossRef]
20. Gong, Y.; Zhao, Y.Y.; Zhou, M.F. Formation and characterization of the tetranuclear scandium nitride: Sc_4N_4. *J. Phys. Chem. A* **2007**, *111*, 6204–6207. [CrossRef]
21. Geng, C.; Li, J.L.; Weiske, T.; Schwarz, H. Ta^{2+}-Mediated ammonia synthesis from N_2 and H_2 at ambient temperature. *Proc. Natl. Acad. Sci. USA* **2018**, *115*, 11680–11687. [CrossRef] [PubMed]
22. Li, Z.Y.; Li, Y.; Mou, L.H.; Chen, J.J.; Liu, Q.Y.; He, S.G.; Chen, H. A facile N≡N bond cleavage by the trinuclear metal center in vanadium carbide cluster anions $V_3C_4^-$. *J. Am. Chem. Soc.* **2020**, *142*, 10747–10754. [CrossRef] [PubMed]
23. Li, Z.Y.; Mou, L.H.; Wei, G.P.; Ren, Y.; Zhang, M.Q.; Liu, Q.Y.; He, S.G. C–N coupling in N_2 fixation by the ditantalum carbide cluster anions $Ta_2C_4^-$. *Inorg. Chem.* **2019**, *58*, 4701–4705. [CrossRef] [PubMed]
24. Wang, M.; Chu, L.Y.; Li, Z.Y.; Messinis, A.M.; Ding, Y.Q.; Hu, L.R.; Ma, J.B. Dinitrogen and carbon dioxide activation to form C–N bonds at room temperature: A new mechanism revealed by experimental and theoretical studies. *J. Phys. Chem. Lett.* **2021**, *12*, 3490–3496. [CrossRef]
25. Zhao, Y.; Cui, J.T.; Wang, M.; Valdivielso, D.Y.; Fielicke, A.; Hu, L.R.; Ma, J.B. Dinitrogen fixation and reduction by $Ta_3N_3H_{0,1}^-$ cluster anions at room temperature: Hydrogen-assisted enhancement of reactivity. *J. Am. Chem. Soc.* **2019**, *141*, 12592–12600. [CrossRef]
26. Wang, M.; Zhao, C.Y.; Zhou, H.Y.; Zhao, Y.; Li, Y.K.; Ma, J.B. The sequential activation of H_2 and N_2 mediated by the gas-phase Sc_3N^+ clusters: Formation of amido unit. *J. Chem. Phys.* **2021**, *154*, 054307. [CrossRef]
27. Mou, L.H.; Li, Y.; Li, Z.Y.; Liu, Q.Y.; Chen, H.; He, S.G. Dinitrogen activation by heteronuclear metal carbide cluster anions $FeTaC_2^-$: A 5d early and 3d late transition metal strategy. *J. Am. Chem. Soc.* **2021**, *143*, 19224–19234. [CrossRef]

28. Li, Y.; Ding, Y.Q.; Zhou, S.D.; Ma, J.B. Dinitrogen activation by dihydrogen and quaternary cluster anions AuNbBO⁻: Nb− and B−Mediated N_2 activation and Au-assisted nitrogen transfer. *J. Phys. Chem. Lett.* **2022**, *13*, 4058–4063. [CrossRef]
29. Mou, L.H.; Li, Z.Y.; Liu, Q.Y.; He, S.G. Size-dependent association of cobalt deuteride cluster anions $Co_3D_n^-$ (n = 0–4) with dinitrogen. *J. Am. Soc. Mass Spectrom.* **2019**, *30*, 1956. [CrossRef]
30. Cheng, X.; Li, Z.Y.; Mou, L.H. Size-dependent reactivity of rhodium deuteride cluster anions $Rh_3D_n^-$ (n = 0–3) toward dinitrogen: The prominent role of σ donation. *J. Chem. Phys.* **2022**, *156*, 064303. [CrossRef]
31. Gioumousis, G.; Stevenson, D.P. Reactions of gaseous molecule ions with gaseous molecules. *J. Chem. Phys.* **1958**, *29*, 294–299. [CrossRef]
32. Kummerlöwe, G.; Beyer, M.K. Rate estimates for collisions of ionic clusters with neutral reactant molecules. *Int. J. Mass Spectrom.* **2005**, *244*, 84–90. [CrossRef]
33. Steinfeld, J.I.; Francisco, J.S.; Hase, W.L. *Chemical Kinetics and Dynamics*; Prentice-Hall: Hobboken, NJ, USA, 1999; p. 231.
34. Yuan, Z.; Zhao, Y.X.; Li, X.N.; He, S.G. Reactions of $V_4O_{10}^+$ cluster ions with simple inorganic and organic molecules. *Int. J. Mass Spectrom.* **2013**, *354–355*, 105–112. [CrossRef]
35. Jiang, L.X.; Liu, Q.Y.; Li, X.N.; He, S.G. Design and application of a high-temperature linear ion trap reactor. *J. Am. Soc. Mass Spectrom.* **2018**, *29*, 78–84. [CrossRef]
36. Wu, X.N.; Xu, B.; Meng, J.H.; He, S.G. C−H bond activation by nanosized scandium oxide clusters in gas-phase. *Int. J. Mass Spectrom.* **2012**, *310*, 57–64. [CrossRef]
37. Li, Z.Y.; Yuan, Z.; Li, X.N.; Zhao, Y.X.; He, S.G. CO oxidation catalyzed by single gold atoms supported on aluminum oxide clusters. *J. Am. Chem. Soc.* **2014**, *136*, 14307–14313. [CrossRef]
38. Chan, B.; Gill, P.M.W.; Kimura, M. Assessment of DFT methods for transition metals with the TMC151 compilation of data sets and comparison with accuracies for main group chemistry. *J. Chem. Theory. Comput.* **2019**, *15*, 3610–3622. [CrossRef]
39. Frisch, M.J.; Trucks, G.W.; Schlegel, H.B.; Scuseria, G.E.; Robb, M.A.; Cheeseman, J.R.; Scalmani, G.; Barone, V.; Mennucci, B.; Petersson, G.A.; et al. *Gaussian 09, revision A.1*; Gaussian, Inc.: Wallingford, CT, USA, 2009.
40. Lee, C.T.; Yang, W.T.; Parr, R.G. Development of the collesalvetti correlation-energy formula into a functional of the electrondensity. *Phys. Rev. B* **1988**, *37*, 785–789. [CrossRef]
41. Becke, A.D. Density-functional exchange-energy approximation with correct asymptotic-behavior. *Phys. Rev. A* **1988**, *38*, 3098–3100. [CrossRef]
42. Becke, A.D. Density-functional Thermochemistry III. The role of exact exchange. *J. Chem. Phys.* **1993**, *98*, 5648–5652. [CrossRef]
43. Gonzalez, C.; Schlegel, H.B. Reaction path following in mass-weighted internal coordinates. *J. Chem. Phys.* **1990**, *94*, 5523–5527. [CrossRef]
44. Krishnan, R.; Binkley, J.S.; Seeger, R.; Pople, J.A. Self-consistent molecular-orbital methods 0.20. basis set for correlated wave-functions. *J. Chem. Phys.* **1980**, *72*, 650–654. [CrossRef]
45. Clark, T.; Chandrasekhar, J.; Spitznagel, G.W.; Schleyer, P.V.R. Efficient diffuse function-augmented basis sets for anion calculations. III.* The 3-21+G Basis set for first-row elements, Li-F. *J. Comput. Chem.* **1983**, *4*, 294–301. [CrossRef]
46. Ma, J.B.; Wang, Z.C.; Schlangen, M.; He, S.G.; Schwarz, H. On the Origin of the Surprisingly Sluggish Redox Reaction of the N_2O/CO Couple Mediated by $[Y_2O_2]^{+\bullet}$ and $[YAlO_2]^{+\bullet}$ Cluster Ions in the Gas Phase. *Angew. Chem. Int. Ed.* **2013**, *52*, 1226–1230. [CrossRef] [PubMed]
47. Berente, I.; Náray-Szabó, G. Multicoordinate driven method for approximating enzymatic reaction paths: Automatic definition of the reaction coordinate using a subset of chemical coordinates. *J. Phys. Chem. A.* **2006**, *110*, 772–778. [CrossRef] [PubMed]
48. Gonzalez, C.; Schlegel, H.B. An improved algorithm for reaction path following. *J. Chem. Phys.* **1989**, *90*, 2154–2161. [CrossRef]
49. Glendening, E.D.; Badenhoop, J.K.; Reed, A.E.; Carpenter, J.E.; Bohmann, J.A.; Morales, C.M.; Landis, C.R.; Weinhold, F. *NBO 6.0*; Theoretical Chemistry Institute, University of Wisconsin: Madison, WI, USA, 2013. Available online: http://nbo6.chem.wisc.edu/ (accessed on 20 January 2022).
50. Lu, T.; Chen, F.W. Multiwfn: A multifunctional wavefunction analyzer. *J. Comput. Chem.* **2012**, *33*, 580–592. [CrossRef]
51. Beyer, T.; Swinehart, D.F. Algorithm 448: Number of Multiply-Restricted Partitions. *Commun. ACM* **1973**, *16*, 379. [CrossRef]
52. Simoes, J.A.M.; Beauchamp, J.L. Transition metal-hydrogen and metal-carbon bond strengths: The keys to catalysis. *Chem. Rev.* **1990**, *90*, 629–688. [CrossRef]
53. Mallard, W.G. (Ed.) *NIST Chemistry Webbook*; August 1990. Available online: http://webbook.nist.gov (accessed on 20 January 2022).
54. Gurvich, L.V.; Karachevtsev, G.V. *Bond Energies of Chemical Bonds, Ionization Potentials and Electron Affinities*; Nauka: Moscow, Russia, 1974.
55. Tang, X.N.; Hou, Y.; Ng, C.Y.; Ruscic, B. Pulsed field-ionization photoelectronphotoion coincidence study of the process $N_2 + h\nu \rightarrow N^+ + N + e^-$: Bond dissociation energies of N_2 and N_2^+. *J. Chem. Phys.* **2011**, *123*, 074330. [CrossRef]

Article

Solid-State Construction of $CuO_x/Cu_{1.5}Mn_{1.5}O_4$ Nanocomposite with Abundant Surface CuO_x Species and Oxygen Vacancies to Promote CO Oxidation Activity

Baolin Liu [1,2,†], Hao Wu [3,†], Shihao Li [3], Mengjiao Xu [2], Yali Cao [2,*] and Yizhao Li [1,3,*]

1. Yangtze Delta Region Institute (Huzhou), University of Electronic Science and Technology of China, Huzhou 313001, China; liubaolin6977@163.com
2. State Key Laboratory of Chemistry and Utilization of Carbon Based Energy Resources, College of Chemistry, Xinjiang University, Urumqi 830017, China; xmj_1117@163.com
3. College of Chemical Engineering, Xinjiang University, Urumqi 830046, China; wuhaode123@eyou.com (H.W.); lsh1531@126.com (S.L.)
* Correspondence: caoyali@xju.edu.cn (Y.C.); yizhao@csj.uestc.edu.cn (Y.L.); Tel./Fax: +86-572-2370780 (Y.L.)
† These authors contributed equally to this work.

Abstract: Carbon monoxide (CO) oxidation performance heavily depends on the surface-active species and the oxygen vacancies of nanocomposites. Herein, the $CuO_x/Cu_{1.5}Mn_{1.5}O_4$ were fabricated via solid-state strategy. It is manifested that the construction of $CuO_x/Cu_{1.5}Mn_{1.5}O_4$ nanocomposite can produce abundant surface CuO_x species and a number of oxygen vacancies, resulting in substantially enhanced CO oxidation activity. The CO is completely converted to carbon dioxide (CO_2) at 75 °C when $CuO_x/Cu_{1.5}Mn_{1.5}O_4$ nanocomposites were involved, which is higher than individual CuO_x, MnO_x, and $Cu_{1.5}Mn_{1.5}O_4$. Density function theory (DFT) calculations suggest that CO and O_2 are adsorbed on $CuO_x/Cu_{1.5}Mn_{1.5}O_4$ surface with relatively optimal adsorption energy, which is more beneficial for CO oxidation activity. This work presents an effective way to prepare heterogeneous metal oxides with promising application in catalysis.

Keywords: solid-state synthesis; $CuO_x/Cu_{1.5}Mn_{1.5}O_4$ nanocomposites; surface CuO_x species; oxygen vacancies; CO oxidation

1. Introduction

Transition metal oxide catalysts for eliminating carbon monoxide (CO) at lower temperatures have attracted enormous attention in the past decades for their inexpensive cost and wide applications in catalytic applications and environmental protection [1–3]. Many techniques, such as morphology control [4–7], engineering defects [8–10], and construction of composite oxides [11–13] have been developed to improve the CO oxidation performance of transition metal oxide catalysts. Particularly, heterogeneous metal oxides that have exhibited excellent performances in CO oxidation fields [14–16] are the most widely studied because of their interactions of components [17]. Previous works have gradually demonstrated that heterogeneous metal oxides with synergistic interactions between two components are crucial for promoting catalytic performances [12,18,19]. The catalytic activity of nanocomposites depends significantly on surface active species and oxygen vacancies [20]. Therefore, the manipulation of the surface-active species and the oxygen vacancies of nanocomposites by simple strategy to optimize their catalytic performance is of great importance to meet the application in practice.

In the past few decades, many transition metal oxide catalysts have been developed, mainly including CeO_2 [13,21,22], MnO_2 [23–25], Co_3O_4 [26–28], CuO [29–31], Fe_2O_3 [32–34] et al., Co_3O_4-CeO_{2-x} [16,35], Cu-Mn [36], Ce-Cu [37,38], and Ce-Mn [39,40] composite oxides. Different active metals and carriers will result in different interactions between metals and carriers and different exposed active sites, thus making them have different reactivity for CO catalytic oxidation. Yu [16] synthesized the catalyst of Co_3O_4-CeO_2 nanocomposite, which showed good catalytic activity due to its special hollow multishell structure and the interaction between the two components. Chen [41] constructed CuO_x-CeO_2 nanorods and explained the relationship between reduction treatment and catalytic activity, indicating that reduction treatment accelerates the generation of active sites. In our previous works, we fabricated a CuO_x-CeO_2 catalyst via the solid-state method [22], and investigated the influence of heating rate on catalytic performance. It was demonstrated that the heating rate can regulate the surface dispersion of CuO_x on CeO_2 surface, resulting in enhanced catalytic performance. Copper-manganese mixed oxide catalyst is a typical transition metal-based catalyst in the CO oxidation reaction, which is known for its high activity at high temperature and low cost [36,42]. However, current commercial copper–manganese catalysts exhibit relatively low catalytic activity at low temperatures for CO oxidation. Furthermore, specific deactivation frequently occurs during the catalytic process [43,44]. The catalytic performance of heterogeneous catalysts is closely associated with their synergistic interactions and oxygen vacancies. In addition, many controllable synthetic strategies, such as direct calcination [30] and hydrothermal/solvothermal synthesis [45,46], have been developed for the fabrication of heterogeneous catalysts with the active two-phase interface, controllable size, shape, and composition in view of the purpose of improving catalytic performance. These synthetic routes are usually low-producing, time-consuming, and high-energy-consuming [47,48]. Solid-state synthesis integrated the advantages of low cost, eco-friendly and large-scale and have aroused wide concern in recent years [49–53].

Herein, a solid-state synthesis was developed to fabricate the $CuO_x/Cu_{1.5}Mn_{1.5}O_4$, which was implemented by the straightforward grinding of copper salt, manganese salt, and potassium hydroxide at ambient conditions. The metal oxide catalysts fabricated by solid-state synthesis are considered a simple and economical approach because they are without complicated procedures and organic solvents. The as-prepared $CuO_x/Cu_{1.5}Mn_{1.5}O_4$ exhibit significant advantages compared to other methods. The catalytic performance was obviously promoted, which can be attributed to the surface CuO_x species and the number of oxygen vacancies. More importantly, this work presents us with an effective way to prepare heterogeneous metal oxides with outstanding catalytic performance.

2. Results and Discussion

The preparation process of $CuO_x/Cu_{1.5}Mn_{1.5}O_4$ is schematically illustrated in Scheme 1. The $CuO_x/Cu_{1.5}Mn_{1.5}O_4$ can be efficiently synthesized by the solvent-free strategy. The corresponding X-ray powder diffraction (XRD) patterns of $CuO_x/Cu_{1.5}Mn_{1.5}O_4$ are exhibited in Figure 1. The peaks detected at 2θ = 18.55, 30.51, 35.94, 37.60, 43.69, 54.23, 57.81, and 63.51 degrees were indexed to $Cu_{1.5}Mn_{1.5}O_4$ spinel solid solution (JCPDS No:70-0262). The diffraction peak at 38.92 degrees attributes to the isolated CuO phases. These results demonstrate the successful synthesis of $CuO_x/Cu_{1.5}Mn_{1.5}O_4$ nanocomposites by solid-state strategy. As shown in Figure 1b, compared with individual CuO species, the major diffraction peaks of the products with different Cu/Mn mole ratios were slightly shifted to a higher degree, which can be ascribed to the change of the lattice parameters. In addition, XRD patterns show the characteristic peaks of CuO and MnO_x, as shown in Figure S1. The ratios of Cu^{2+}/Cu^+ in Cu_2O/CuO can be controlled by adjusting the types of copper salts. The energy-dispersive X-ray spectrum (EDS) mapping analyses was implemented to identify the elementary composition of the $CuO_x/Cu_{1.5}Mn_{1.5}O_4$. The Cu/Mn molar ratio in the as-prepared samples is close to 1.0 (Figure S2), which has no significant difference compared to the theoretical value during synthesis.

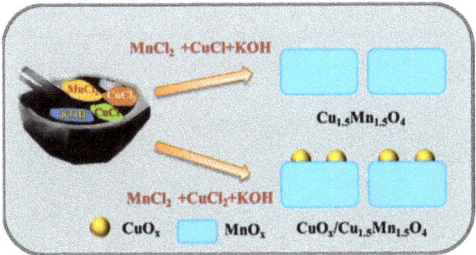

Scheme 1. Schematic illustration of the construction of $CuO_x/Cu_{1.5}Mn_{1.5}O_4$ nanocomposite with different surface CuO_x species.

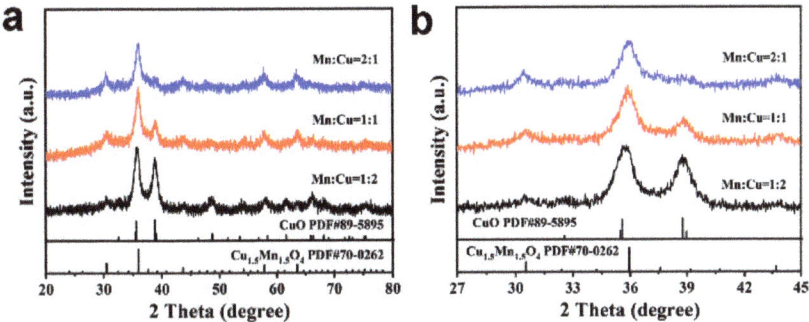

Figure 1. (a) The powder XRD patterns of $CuO_x/Cu_{1.5}Mn_{1.5}O_4$; (b) partially enlarged profiles.

The scanning electron microscope (SEM) and transmission electron microscopes (TEM) images as shown in Figure 2 further prove that the nanoparticles with 11–16 nm (Figure 2e) existed in $CuO_x/Cu_{1.5}Mn_{1.5}O_4$. Additionally, the interplanar spacing of 0.250 nm and 0.232 nm, revealed by the high-resolution transmission electron microscopes (HRTEM) image (Figure 2i), correspond to (311) and (111) planes of $Cu_{1.5}Mn_{1.5}O_4$ and CuO, respectively. The element mapping results are exhibited in Figure 2, confirming that the Cu and Mn elements are uniformly dispersed on the surface of $CuO_x/Cu_{1.5}Mn_{1.5}O_4$. These results further confirm the successful synthesis of $CuO_x/Cu_{1.5}Mn_{1.5}O_4$ by solid-state strategy. Furthermore, the morphologies of $CuO_x/Cu_{1.5}Mn_{1.5}O_4$ prepared by adjusting the molar ratios of Cu/Mn were also acquired (Figure S3), displaying agglomerated nanoparticles. The morphologies of CuO, Cu_2O-CuO, and MnO_x are shown in Figures S4–S6, which also exhibit ir-regular nanoparticles.

The CO catalytic performance of the as-obtained $CuO_x/Cu_{1.5}Mn_{1.5}O_4$ were firstly evaluated. As shown in Figure S7a,b, the best CO catalytic activity is $CuO_x/Cu_{1.5}Mn_{1.5}O_4$ with Cu/Mn molar ratio of 1:1 calcined at 400 °C. The $CuO_x/Cu_{1.5}Mn_{1.5}O_4$ can completely convert CO to CO_2 at 75 °C, especially at low temperatures. T_{50} (the temperature of 50% of CO conversion) is only 41 °C. The CO_2 yield in the CO oxidation has been shown in Figure S8a, which presents nearly 100% yield. Moreover, they have been compared with previous works (Table S1), which also presents a better catalytic property. Other samples exhibit a relatively lower catalytic activity performance with higher T_{100} (the temperature of 100% of CO conversion) and T_{50} (the temperature of 50% of CO conversion) in Figure 3. The 100% CO conversion was accomplished for individual CuO, Cu_2O-CuO, and MnO_x samples at 140 °C, 130 °C, and 200 °C, respectively. The individual CuO_x and MnO_x particles show poorer performance than $CuO_x/Cu_{1.5}Mn_{1.5}O_4$, implying that the synergistic effect between CuO_x and MnO_x may promote its catalytic activity that is not presented in the individual components. As shown in Figure S8b, the sample of physical mixing of $CuO_x + MnO_x$ was also prepared, which exhibits the poor catalytic performance for CO oxidation. The stability of the $CuO_x/Cu_{1.5}Mn_{1.5}O_4$ was also tested at 60 °C. The negligible decline of activity can be observed during 30 h testing from Figure S9, which implies the excellent stability of $CuO_x/Cu_{1.5}Mn_{1.5}O_4$ for CO oxidation reaction.

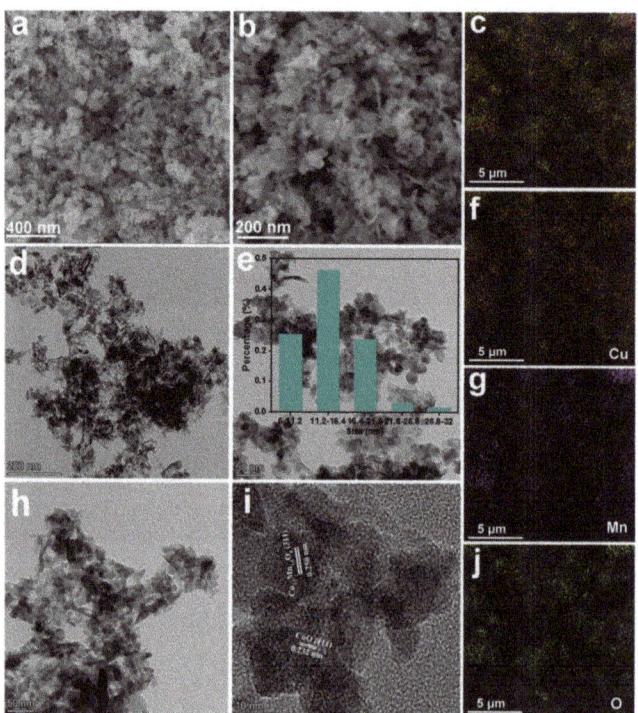

Figure 2. (**a,b**) the SEM images; (**d,e,h**) TEM images; (**i**) HRTEM image; and (**c,f,g,j**) the corresponding element mapping patterns of $CuO_x/Cu_{1.5}Mn_{1.5}O_4$.

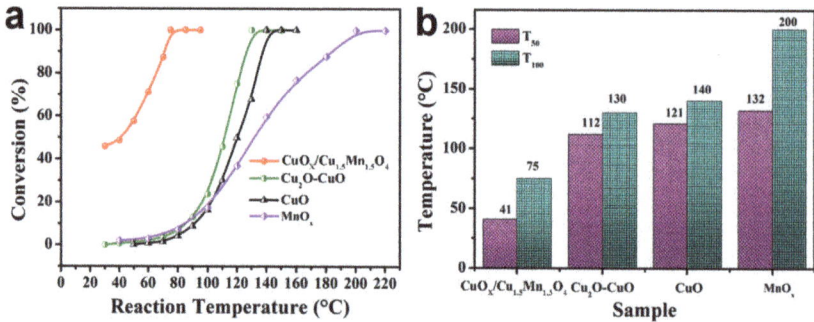

Figure 3. (**a,b**) The CO conversion performances on various catalysts.

The X-ray photoelectron spectra (XPS) were used to investigate the chemical states of samples. The XPS spectrum in Figure 4 and Figure S10 indicate the coexistence of the Cu, Mn, and O elements. Two peaks at about 931.1 and 950.9 eV, respectively, shown in Figure 4a refer to the Cu^+ or Cu^0 due to the fact that their binding energies are basically the same [31,54]. Cu^0 is unstable at room temperature and easily oxidized to copper oxide. The $CuO_x/Cu_{1.5}Mn_{1.5}O_4$ nanocomposites were acquired after being calcined at high temperature in air. Therefore, the peak is assigned to Cu^+ because of the successful synthesis of $CuO_x/Cu_{1.5}Mn_{1.5}O_4$ nanocomposites. The two main peaks have small shoulder peaks that appeared at 933.3 and 953.4 eV, relating to the Cu^{2+} [55–58]. The XPS analysis results imply that Cu^+ and Cu^{2+} coexist on the surfaces of the $CuO_x/Cu_{1.5}Mn_{1.5}O_4$. As shown in Figure 4b, the asymmetrical Mn 2p spectra of individual MnO_x catalysts could be fitted into four components based on their binding energies. The binding energies of 640.2 eV, 641.2 eV, 642.5 eV, and 646.0 eV correspond to Mn^{2+}, Mn^{3+}, Mn^{4+} species, and the satellite peak, respectively [59]. The O 1s XPS spectrum of samples can be divided into three single peaks (Figure 4c), corresponding to surface lattice oxygen (O_α), surface adsorbed oxygen (O_β), and adsorbed molecular water species (O_γ), respectively [14,60–62]. The $CuO_x/Cu_{1.5}Mn_{1.5}O_4$ demonstrates the highest surface adsorbed oxygen, which is beneficial to the adsorption of O_2 molecules, and thus help to improve catalytic performance.

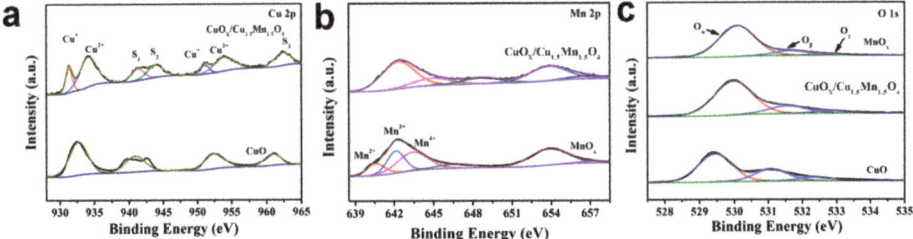

Figure 4. (**a**) Cu 2p; (**b**) Mn 2p; (**c**) O 1s spectra of MnO_x, $CuO_x/Cu_{1.5}Mn_{1.5}O_4$, and CuO catalysts, respectively.

As shown in Figure 5a, the reduction property of prepared samples was investigated by hydrogen temperature-programmed reduction (H_2-TPR). A peak at 200 °C–400 °C was presented for an individual CuO sample, attributing to the gradual reduction of copper oxide [30]. In addition, two H_2 reduction peaks at 245 °C and 364 °C occurred for an MnO_x sample, which correspond to the gradual reduction of $MnO_2 \rightarrow Mn_3O_4 \rightarrow MnO$ [23,39,59]. For $CuO_x/Cu_{1.5}Mn_{1.5}O_4$ nanocomposite, the first peak at below 158 °C refers to the reduction of fine CuO to Cu or MnO_2 to Mn_3O_4. Other peaks at 174 °C and 201 °C correspond to the gradual reduction of $Cu_{1.5}Mn_{1.5}O_4$ oxides [63]. The lower reduction temperature for $CuO_x/Cu_{1.5}Mn_{1.5}O_4$ compared with individual CuO_x and MnO_x indicates the strong synergistic effect between CuO_x and MnO_x. The strong interactions in $CuO_x/Cu_{1.5}Mn_{1.5}O_4$ are often related to the abundant oxygen vacancies, which can promote catalytic performance. The oxygen storage capacity (OSC) of catalysts was assessed by oxygen temperature-programmed desorption (O_2-TPD). In Figure 5b, the temperature below 200 °C is due to the desorption of surface oxygen species (O_β) [64]. The second peak, appearing at 250–550 °C, corresponds to the overflow of surface lattice oxygen (O_α). The high-temperature zone at above 550 °C is related to the bulk lattice oxygen species [63]. The $CuO_x/Cu_{1.5}Mn_{1.5}O_4$ shows the highest amount of adsorbed oxygen species compared with other samples, confirming the higher oxygen capacity that is conducive to the promotion of catalytic performance.

To elucidate the effects of surface CuO_x species and the oxygen vacancies in $CuO_x/Cu_{1.5}Mn_{1.5}O_4$ in detail, the $Cu_{1.5}Mn_{1.5}O_4$ was fabricated by changing the types of copper salt in the synthesis process to investigate the structure-activity relationships. The Cu^{2+} and Mn^{4+} proportion was evaluated by XPS (Figure 6d). It indicates that the ratio of Cu^{2+} and Mn^{4+} in $CuO_x/Cu_{1.5}Mn_{1.5}O_4$ is higher than $Cu_{1.5}Mn_{1.5}O_4$ sample. The higher contents of Cu^{2+} and Mn^{4+} are beneficial to the formation of Cu^{2+}-O^{2-}-Mn^{4+} entities at the two-phase interface [36]. As reported in the studies [63,65], the presence of abundant Mn^{4+} proportion can create many adsorbed oxygen species [63]. In addition, the ratios of Cu^+/Cu^{2+} in different samples show a change by altering the types of copper salts in the synthetic process (Table S2). The $CuO_x/Cu_{1.5}Mn_{1.5}O_4$ exhibits a higher Cu^{2+} ratio than $Cu_{1.5}Mn_{1.5}O_4$, which corresponds to XRD results that the $CuO_x/Cu_{1.5}Mn_{1.5}O_4$ shows the higher intensity of the CuO diffraction peaks. The XPS results indicate that the Cu^{2+} and Mn^{4+} proportion can be engineered by changing the component of Cu-based oxides in the synthetic process.

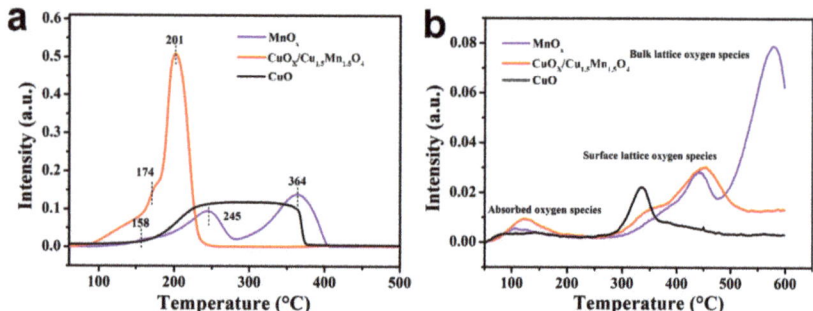

Figure 5. (a) H_2-TPR; (b) O_2-TPD profiles of MnO_x, $CuO_x/Cu_{1.5}Mn_{1.5}O_4$, and CuO catalysts, respectively.

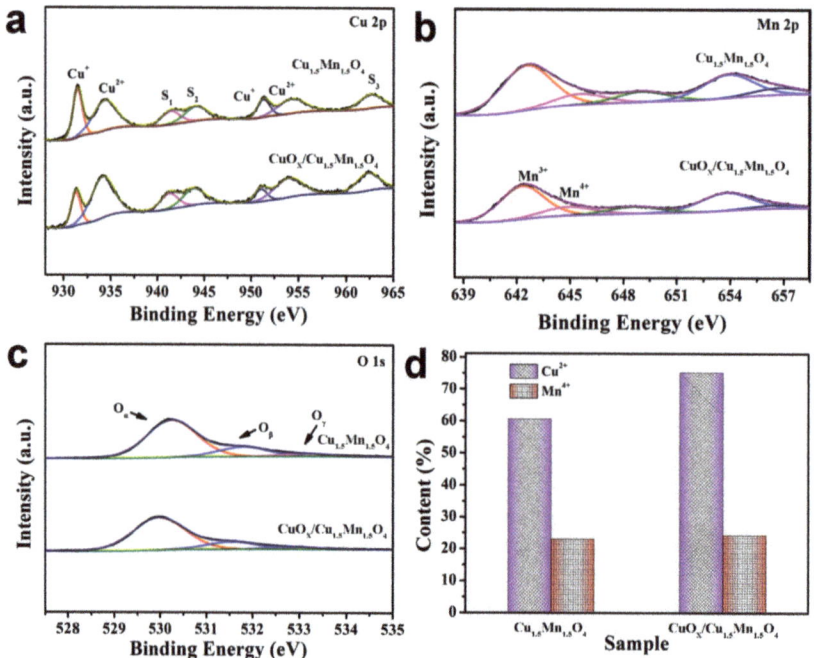

Figure 6. (a) Cu 2p; (b) Mn 2p; (c) O 1s spectra; (d) the contents of Cu^{2+} and Mn^{4+} in $Cu_{1.5}Mn_{1.5}O_4$, and $CuO_x/Cu_{1.5}Mn_{1.5}O_4$, respectively.

As shown in Figure 7a, there are obvious differences in catalytic performance after changing the copper salts. The $Cu_{1.5}Mn_{1.5}O_4$ exhibits poor catalytic activity compared with $CuO_x/Cu_{1.5}Mn_{1.5}O_4$ at the same condition. Herein, the performance of catalysts can be meaningfully boosted by altering the surface CuO_x species and the oxygen vacancies in $CuO_x/Cu_{1.5}Mn_{1.5}O_4$. As shown in Figure S11a,b, the $Cu_{1.5}Mn_{1.5}O_4$ exhibited ir-regular nanoparticles morphology. The Cu and Mn elements are homogeneously dispersed on the surface of $Cu_{1.5}Mn_{1.5}O_4$, and their molar ratio is close to 1.0 (Figure S2), which has no significant difference with $CuO_x/Cu_{1.5}Mn_{1.5}O_4$. The XRD diffraction patterns indicate the formation of $Cu_{1.5}Mn_{1.5}O_4$ containing some CuO_x (Figure 7b), while the catalytic performance showed obvious differences. The $CuO_x/Cu_{1.5}Mn_{1.5}O_4$ exhibits the higher intensity of the CuO diffraction peaks than $Cu_{1.5}Mn_{1.5}O_4$. In our previous work [31], the individual Cu_2O/CuO nanocomposites and CuO can be fabricated by tuning the types of copper salt in the synthesis. The $Cu_{1.5}Mn_{1.5}O_4$ with different surface CuO_x types were fabricated by altering the types of copper salts (+1, +2 valence state) in the synthetic process. Therefore, the different surface types in $Cu_{1.5}Mn_{1.5}O_4$ depend on the types of copper salts (+1, +2 valence state) in the synthetic process used. The higher content of CuO in $CuO_x/Cu_{1.5}Mn_{1.5}O_4$ can significantly enhance redox reaction between Cu and Mn species, promoting charge transfer in nanocomposites, and thus achieving a stronger interaction. From Figure 7c, the $CuO_x/Cu_{1.5}Mn_{1.5}O_4$ shows a lower reduction temperature than $Cu_{1.5}Mn_{1.5}O_4$, implying the better reducibility. The peak areas for different samples were estimated from the H_2-TPR results. In Table S3, the higher peak areas of first peak α is presented for $CuO_x/Cu_{1.5}Mn_{1.5}O_4$, indicating the ratio of Cu^{2+} and Mn^{4+} in $CuO_x/Cu_{1.5}Mn_{1.5}O_4$, which is consistent with XPS results. Therefore, we can confirm that CO oxidation activity is heavily dependent on the surface CuO_x species in $CuO_x/Cu_{1.5}Mn_{1.5}O_4$. In addition, as shown in Figure 7d, the amounts of oxygen desorption ($O_β$) over the obtained $CuO_x/Cu_{1.5}Mn_{1.5}O_4$ are greatly changed by adjusting the surface CuO_x species. The $CuO_x/Cu_{1.5}Mn_{1.5}O_4$ shows the highest amount of adsorbed oxygen species compared with $Cu_{1.5}Mn_{1.5}O_4$, confirming the higher oxygen capacity that is conducive to the promotion of catalytic performance. The H_2-TPR analysis results indicate that the interaction of nanocomposites could be manipulated by changing the surface compositions of $CuO_x/Cu_{1.5}Mn_{1.5}O_4$. The O_2-TPD results further identify the presence of abundant surface-adsorbed oxygen on the surface of $CuO_x/Cu_{1.5}Mn_{1.5}O_4$. Therefore, the construction of $CuO_x/Cu_{1.5}Mn_{1.5}O_4$ with abundant surface CuO_x species not only strengthens the interactions in $CuO_x/Cu_{1.5}Mn_{1.5}O_4$, but also facilitates the absorption and activation of surface oxygen species.

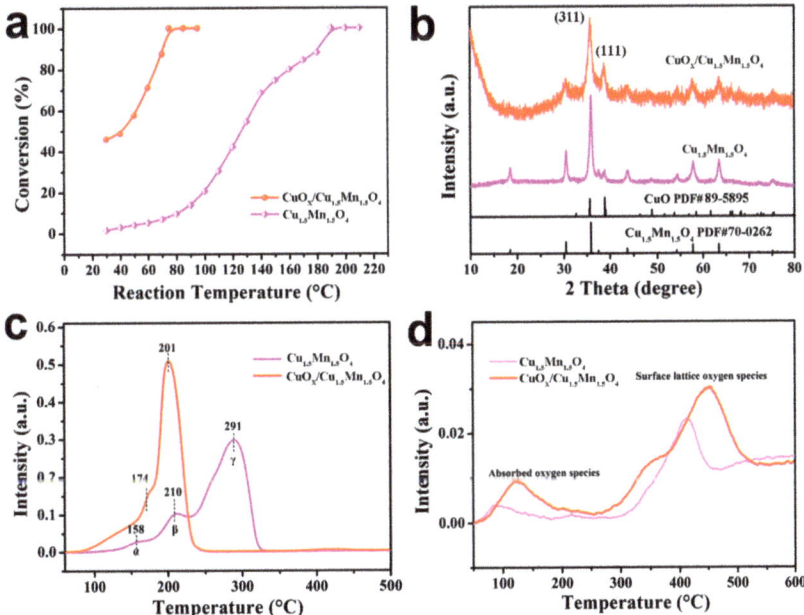

Figure 7. (a) The CO conversion performances; (b) the powder XRD patterns; (c) H_2-TPR; (d) O_2-TPD profiles of $Cu_{1.5}Mn_{1.5}O_4$ and $CuO_x/Cu_{1.5}Mn_{1.5}O_4$, respectively.

Density functional theory (DFT) calculations were implemented to understand the intrinsic reason about the mechanism of CO and O_2 adsorption and the subsequent oxidation process for $CuO_x/Cu_{1.5}Mn_{1.5}O_4$. The adsorption configurations of CO and O_2 molecules on the CuO, $Cu_{1.5}Mn_{1.5}O_4$, and $CuO_x/Cu_{1.5}Mn_{1.5}O_4$ are shown in Figure 8. It is found that CO is adsorbed on the surface of CuO and $Cu_{1.5}Mn_{1.5}O_4$ with the adsorption energy of −0.316 eV and −0.164 eV, respectively. The adsorption energy of CO molecules adsorbed at the $CuO_x/Cu_{1.5}Mn_{1.5}O_4$ surface is −1.346 eV, which are lower than that on CuO and $Cu_{1.5}Mn_{1.5}O_4$. In addition, the adsorption energy of O_2 molecules on $CuO_x/Cu_{1.5}Mn_{1.5}O_4$ surface is −1.018 eV, which is much lower than the 0.026 eV and −0.962 eV for CuO and $Cu_{1.5}Mn_{1.5}O_4$. The lower adsorption energy indicates that gas molecules are easier to adsorb on the surface of $CuO_x/Cu_{1.5}Mn_{1.5}O_4$. In summary, DFT calculation showed that construction of $CuO_x/Cu_{1.5}Mn_{1.5}O_4$ nanocomposite with abundant surface CuO_x species and oxygen vacancies significantly improved the adsorption capacity of CO and O_2 molecules, and thus is more beneficial for CO oxidation activity.

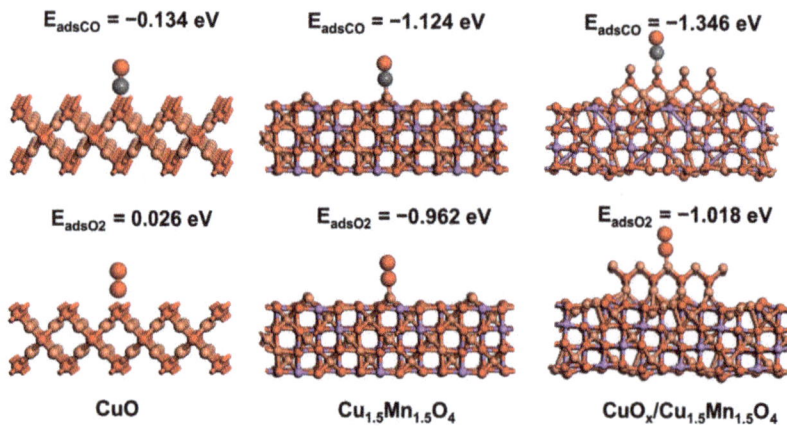

Figure 8. The side views of CO and O_2 adsorption on the surfaces of CuO, $Cu_{1.5}Mn_{1.5}O_4$, and $CuO_x/Cu_{1.5}Mn_{1.5}O_4$, respectively.

The effects of surface CuO_x species and the oxygen vacancies in composite oxide can be clarified based on the above results. In Figure 9, the catalytic property of the $CuO_x/Cu_{1.5}Mn_{1.5}O_4$ is improved compared to $Cu_{1.5}Mn_{1.5}O_4$, which confirms the important role of surface CuO_x species and oxygen vacancies. After construction of $CuO_x/Cu_{1.5}Mn_{1.5}O_4$ nanocomposite with abundant surface CuO_x species and oxygen vacancies, the abundant Cu^{2+} and Mn^{4+} proportions in $CuO_x/Cu_{1.5}Mn_{1.5}O_4$ are higher than in $Cu_{1.5}Mn_{1.5}O_4$, which facilitated the formation of more (Cu^{2+}-O^{2-}-Mn^{4+}) entities at the two interfaces. In addition, the construction of $CuO_x/Cu_{1.5}Mn_{1.5}O_4$ nanocomposites is beneficial for enhancing the synergetic interaction between MnO_x species and CuO_x species, which promotes the massive production of surface adsorbed oxygen species [36]. In CO oxidation, surface CuO_x species and oxygen vacancies play significant roles in catalytic activity. The abundant surface CuO_x species and oxygen vacancies could preferentially adsorb CO and O_2 molecules [22], and the adsorbed O_2 reacts with CO to form CO_2, which ultimately enhances catalytic activity.

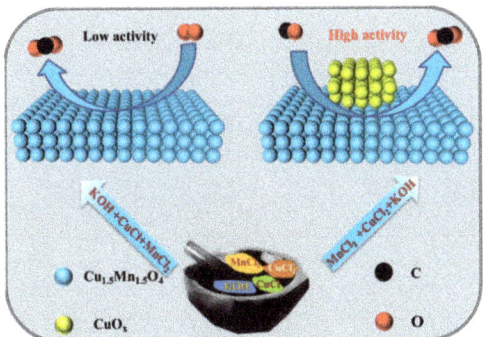

Figure 9. Reaction mechanisms of $CuO_x/Cu_{1.5}Mn_{1.5}O_4$ toward CO oxidation.

3. Materials and Methods

3.1. Materials

Copper (II) chloride ($CuCl_2$, AR), cuprous (I) chloride (CuCl, AR), manganese (II) chloride ($MnCl_2$, AR), and potassium hydroxide (KOH, AR) were purchased from Tianjin Zhiyuan Chemical Reagents Co., Ltd. (Tianjin, China), which were used without further refinement.

3.2. The Preparation of $CuO_x/Cu_{1.5}Mn_{1.5}O_4$ Nanocomposite with Various Surface CuO_x Species and the Oxygen Vacancies

As shown in Scheme 1, in a typical procedure, 1.70 g of $CuCl_2$ (10 mmol) and 1.25 g of $MnCl_2$ (10 mmol) were mixed well in an agate mortar by grinding. Then 4.49 g of KOH (60 mmol) was added into the agate mortar. After continuous grinding for about 1 h, the resulting solid products were sufficiently washed with deionized water and anhydrous ethanol to clear the residual Cl or K species, and then dried at ambient temperature overnight. The final $CuO_x/Cu_{1.5}Mn_{1.5}O_4$ nanocomposites were acquired after calcining the mixtures in the air at 400 °C for 2 h (5 °C/min). In addition, the $CuO_x/Cu_{1.5}Mn_{1.5}O_4$ with different Cu/Mn mole ratios (Cu/Mn = 1:2 and 2:1) were calcined at 300 °C or 500 °C.

The $Cu_{1.5}Mn_{1.5}O_4$ nanocomposite (containing some CuO) was also obtained, and only the $CuCl_2$ was replaced by CuCl during the solvent-free synthesis route.

3.3. The Preparation of MnO_x

As a comparison, the individual MnO_x particles were also fabricated by straightforward grinding $MnCl_2$ with KOH under a similar process.

3.4. The Preparation of Cu_2O/CuO and CuO

The Cu_2O/CuO nanocomposite was fabricated according to our previous work [31]. The 0.99 g of CuCl (10 mmol) and 1.68 g of KOH (30 mmol) were ground in the agate mortar for 1 h. The other parameters are consistent with the $CuO_x/Cu_{1.5}Mn_{1.5}O_4$ nanocomposite above.

In addition, the CuO was also prepared by mixing $CuCl_2$ and KOH in the agate mortar. The sample of physical mixing of $CuO_x + MnO_x$ was also prepared by straightforward grinding CuO and MnO_x, and then calcining the mixtures in the air at 400 °C for 2 h (5 °C/min).

3.5. The Characterization and Testing Processes of Catalyst

XRD, SEM, HRTEM, EDS, XPS, H_2-TPR, and O_2-TPD were implemented to investigate the morphology and structure of $CuO_x/Cu_{1.5}Mn_{1.5}O_4$ nanocomposites. Detailed characterization and testing processes are presented in the Supplementary Materials.

4. Conclusions

In summary, the $Cu_{1.5}Mn_{1.5}O_4$ with different surface CuO_x types were fabricated by altering the types of copper salts (+1, +2 valence state) in the synthetic process. The higher content of CuO in $CuO_x/Cu_{1.5}Mn_{1.5}O_4$ can significantly enhance redox reaction between Cu and Mn species, promoting charge transfer in nanocomposites, thus achieving a stronger interaction. In addition, the higher ratio of Cu^{2+} and Mn^{4+} is beneficial to the formation of Cu^{2+}-O^{2-}-Mn^{4+} entities at the two-phase interface, which produced abundant surface CuO_x species and oxygen vacancies. DFT calculations suggest that CO and O_2 molecules are adsorbed on the $CuO_x/Cu_{1.5}Mn_{1.5}O_4$ surface with relatively optimal adsorption energy, resulting in the highest CO oxidation activity. The as-synthesized $CuO_x/Cu_{1.5}Mn_{1.5}O_4$ delivers excellent CO catalytic performance compared with individual CuO_x and MnO_x particles. The CO is completely converted to CO_2 at 75 °C when $CuO_x/Cu_{1.5}Mn_{1.5}O_4$ is involved. This work opens new avenues for the efficient and sustainable production of heterogeneous metal oxides with an outstanding catalytic performance.

Supplementary Materials: The following supporting information can be downloaded at: https://www.mdpi.com/article/10.3390/ijms23126856/s1. References [66–68] are cited in Supplementary Materials.

Author Contributions: Conceptualization, B.L. and Y.L.; methodology, B.L.; software, H.W.; validation, B.L., Y.L. and S.L.; formal analysis, B.L.; investigation, B.L.; resources, Y.L. and Y.C.; data curation, M.X.; writing—original draft preparation, B.L.; writing—review and editing, B.L. and Y.L.; visualization, B.L.; supervision, Y.L. and Y.C.; project administration, Y.L. and Y.C.; funding acquisition, Y.L. and Y.C. All authors have read and agreed to the published version of the manuscript.

Funding: This work was financially supported by the National Natural Science Foundation of China (Nos. 21766036 and 52100166), the Natural Science Foundation of Xinjiang Province (No. 2019D04005), the Graduate Innovation Project of Xinjiang Province (No. XJ2021G037), and the Key Research and Development Project of Xinjiang Province (No. 2020B02008).

Institutional Review Board Statement: This study did not require ethical approval.

Informed Consent Statement: This study did not involve humans.

Data Availability Statement: The study did not report any data, we choose to exclude this statement.

Conflicts of Interest: The authors declare no conflict of interest.

References

1. Hu, Z.; Liu, X.; Meng, D.; Guo, Y.; Guo, Y.; Lu, G. Effect of Ceria ceria crystal plane on the physicochemical and catalytic properties of Pd/Ceria for CO and propane oxidation. *ACS Catal.* **2016**, *6*, 2265–2279. [CrossRef]
2. Du, P.-P.; Hu, X.-C.; Wang, X.; Ma, C.; Du, M.; Zeng, J.; Jia, C.-J.; Huang, Y.-Y.; Si, R. Synthesis and metal–support interaction of subnanometer copper–palladium bimetallic oxide clusters for catalytic oxidation of carbon monoxide. *Inorg. Chem. Front.* **2017**, *4*, 668–674. [CrossRef]
3. Guo, X.; Li, J.; Zhou, R. Catalytic performance of manganese doped CuO–CeO$_2$ catalysts for selective oxidation of CO in hydrogen-rich gas. *Fuel* **2016**, *163*, 56–64. [CrossRef]
4. Zou, W.X.; Liu, L.C.; Zhang, L.; Li, L.; Cao, Y.; Wang, X.B.; Tang, C.J.; Gao, F.; Dong, L. Crystal-plane effects on surface and catalytic properties of Cu$_2$O nanocrystals for NO reduction by CO. *Appl. Catal. A Gen.* **2015**, *505*, 334–343. [CrossRef]
5. Tao, L.; Shi, Y.; Huang, Y.-C.; Chen, R.; Zhang, Y.; Huo, J.; Zou, Y.; Yu, G.; Luo, J.; Dong, C.-L.; et al. Interface engineering of Pt and CeO$_2$ nanorods with unique interaction for methanol oxi-dation. *Nano Energy* **2018**, *53*, 604–612. [CrossRef]
6. Sun, L.; Zhan, W.W.; Li, Y.A.; Wang, F.; Zhang, X.L.; Han, X.G. Understanding the facet-dependent catalytic performance of hematite microcrystals in a CO oxidation reaction. *Inorg. Chem. Front.* **2018**, *5*, 2332–2339. [CrossRef]
7. Singhania, N.; Anumol, E.A.; Ravishankar, N.; Madras, G. Influence of CeO$_2$ morphology on the catalytic activity of CeO$_2$–Pt hybrids for CO oxidation. *Dalton Trans.* **2013**, *42*, 15343–15354. [CrossRef]
8. Yang, J.; Hu, S.; Fang, Y.; Hoang, S.; Li, L.; Yang, W.; Liang, Z.; Wu, J.; Hu, J.; Xiao, W.; et al. Oxygen Vacancy promoted O$_2$ activation over perovskite oxide for low-temperature CO oxidation. *ACS Catal.* **2019**, *9*, 9751–9763. [CrossRef]
9. Lykaki, M.; Pachatouridou, E.; Carabineiro, S.A.; Iliopoulou, E.; Andriopoulou, C.; Kallithrakas-Kontos, N.; Boghosian, S.; Konsolakis, M. Ceria nanoparticles shape effects on the structural defects and surface chemistry: Implications in CO oxidation by Cu/CeO$_2$ catalysts. *Appl. Catal. B Environ.* **2018**, *230*, 18–28. [CrossRef]
10. Lawrence, N.J.; Brewer, J.R.; Wang, L.; Wu, T.-S.; Wells-Kingsbury, J.; Ihrig, M.M.; Wang, G.; Soo, Y.-L.; Mei, W.-N.; Cheung, C.L. Defect engineering in cubic cerium oxide nanostructures for catalytic oxidation. *Nano Lett.* **2011**, *11*, 2666–2671. [CrossRef]

11. Li, W.; Feng, X.L.; Zhang, Z.; Jin, X.; Liu, D.P.; Zhang, Y. A controllable surface etching strategy for well-defined spiny yolk@shell CuO@CeO$_2$ cubes and their catalytic performance boost. *Adv. Funct. Mater.* **2018**, *28*, 1802559. [CrossRef]
12. Ma, B.; Kong, C.; Lv, J.; Zhang, X.; Yang, S.; Yang, T.; Yang, Z. Cu–Cu$_2$O heterogeneous architecture for the enhanced CO catalytic oxidation. *Adv. Mater. Interfaces* **2020**, *7*, 1901643. [CrossRef]
13. Liu, B.L.; Li, Y.Z.; Qing, S.J.; Wang, K.; Xie, J.; Cao, Y.L. Engineering CuO$_x$–ZrO$_2$–CeO$_2$ nanocat-alysts with abundant surface Cu species and oxygen vacancies toward high catalytic performance in CO oxidation and 4-nitrophenol reduction. *CrystEngComm* **2020**, *22*, 4005–4013. [CrossRef]
14. Fang, Y.; Chi, X.; Li, L.; Yang, J.; Liu, S.; Lu, X.; Xiao, W.; Wang, L.; Luo, Z.; Yang, W.; et al. Elucidating the nature of the Cu(I) active site in CuO/TiO$_2$ for excellent low-temperature CO oxidation. *ACS Appl. Mater. Interfaces* **2020**, *12*, 7091–7101. [CrossRef]
15. Jin, X.; Duan, Y.; Liu, D.P.; Feng, X.L.; Li, W.; Zhang, Z.; Zhang, Y. CO oxidation catalyzed by two-dimensional Co$_3$O$_4$/CeO$_2$ nanosheets. *ACS Appl. Nano Mater.* **2019**, *2*, 5769–5778. [CrossRef]
16. Wang, H.; Mao, D.; Qi, J.; Zhang, Q.H.; Ma, X.H.; Song, S.Y.; Gu, L.; Yu, R.B.; Wang, D. Hollow multishelled structure of heterogeneous Co$_3$O$_4$-CeO$_2$−x nanocomposite for CO catalytic oxidation. *Adv. Funct. Mater.* **2019**, *29*, 1806588. [CrossRef]
17. Yuan, C.; Wang, H.-G.; Liu, J.; Wu, Q.; Duan, Q.; Li, Y. Facile synthesis of Co$_3$O$_4$-CeO$_2$ composite oxide nanotubes and their multifunctional applications for lithium ion batteries and CO oxidation. *J. Colloid Interface Sci.* **2017**, *494*, 274–281. [CrossRef]
18. Chen, K.; Ling, J.L.; Wu, C.D. In situ generation and stabilization of accessible Cu/Cu$_2$O hetero-junctions inside organic frameworks for highly efficient catalysis. *Angew. Chem. Int. Ed. Engl.* **2020**, *59*, 1925–1931. [CrossRef]
19. Liu, A.; Liu, L.; Cao, Y.; Wang, J.; Si, R.; Gao, F.; Dong, L. Controlling dynamic structural transformation of atomically dispersed CuO$_x$ species and influence on their catalytic performances. *ACS Catal.* **2019**, *9*, 9840–9851. [CrossRef]
20. Jampaiah, D.; Velisoju, V.K.; Devaiah, D.; Singh, M.; Mayes, E.L.H.; Coyle, V.E.; Reddy, B.M.; Bansal, V.; Bhargava, S.K. Flower-like Mn$_3$O$_4$/CeO$_2$ microspheres as an efficient catalyst for diesel soot and CO oxidation: Synergistic effects for enhanced catalytic performance. *Appl. Surf. Sci.* **2019**, *473*, 209–221. [CrossRef]
21. Peng, H.; Rao, C.; Zhang, N.; Wang, X.; Liu, W.; Mao, W.; Han, L.; Zhang, P.; Dai, S. Confined ultrathin Pd-Ce nanowires with outstanding moisture and SO$_2$ tolerance in methane combustion. *Angew. Chem. Int. Ed.* **2018**, *57*, 8953–8957. [CrossRef]
22. Liu, B.L.; Li, Y.Z.; Cao, Y.L.; Wang, L.; Qing, S.J.; Wang, K.; Jia, D.Z. Optimum balance of Cu$^+$ and oxygen vacancies of CuO$_x$-CeO$_2$ composites for CO oxidation based on thermal treatment. *Eur. J. Inorg. Chem.* **2019**, *2019*, 1714–1723. [CrossRef]
23. Bai, B.Y.; Li, J.H.; Hao, J.M. 1D-MnO$_2$, 2D-MnO$_2$ and 3D-MnO$_2$ for low-temperature oxidation of ethanol. *Appl. Catal. B Environ.* **2015**, *164*, 241–250. [CrossRef]
24. Wang, M.; Liu, H.; Huang, Z.-H.; Kang, F. Activated carbon fibers loaded with MnO$_2$ for removing NO at room temperature. *Chem. Eng. J.* **2014**, *256*, 101–106. [CrossRef]
25. Chen, H.; Wang, Y.; Lv, Y.-K. Catalytic oxidation of NO over MnO$_2$ with different crystal struc-tures. *RSC Adv.* **2016**, *6*, 54032–54040. [CrossRef]
26. Wang, K.; Cao, Y.L.; Hu, J.D.; Li, Y.Z.; Xie, J.; Jia, D.Z. Solvent-free chemical approach to synthesize various morphological Co$_3$O$_4$ for CO oxidation. *ACS Appl. Mater. Interfaces* **2017**, *9*, 16128–16137. [CrossRef] [PubMed]
27. Wang, K.; Liu, B.L.; Cao, Y.L.; Li, Y.Z.; Jia, D.Z. V-modified Co$_3$O$_4$ nanorods with superior cata-lytic activity and thermostability for CO oxidation. *CrystEngComm* **2018**, *20*, 5191–5199. [CrossRef]
28. Zhang, C.; Zhang, L.; Xu, G.C.; Ma, X.; Xu, J.L.; Zhang, L.; Qi, C.L.; Xie, Y.Y.; Sun, Z.P.; Jia, D.Z. Hollow and core–shell nanostructure Co$_3$O$_4$ derived from a metal formate framework toward high catalytic activity of CO oxidation. *ACS Appl. Nano Mater.* **2018**, *1*, 800–806. [CrossRef]
29. Zhang, R.R.; Hu, L.; Bao, S.X.; Li, R.; Gao, L.; Li, R.; Chen, Q.W. Surface polarization enhancement: High catalytic performance of Cu/CuO$_x$/C nanocomposites derived from Cu-BTC for CO ox-idation. *J. Mater. Chem. A* **2016**, *4*, 8412–8420. [CrossRef]
30. Wei, B.; Yang, N.T.; Pang, F.; Ge, J.P. Cu$_2$O–CuO hollow nanospheres as a heterogeneous catalyst for synergetic oxidation of CO. *J. Phys. Chem. C* **2018**, *122*, 19524–19531. [CrossRef]
31. Liu, B.L.; Li, Y.Z.; Wang, K.; Cao, Y.L. The solid-state in situ construction of Cu$_2$O/CuO hetero-structures with adjustable phase compositions to promote CO oxidation activity. *CrystEngComm* **2020**, *22*, 7808–7815. [CrossRef]
32. Li, Y.; Shen, W. Morphology-dependent nanocatalysts: Rod-shaped oxides. *Chem. Soc. Rev.* **2014**, *43*, 1543–1574. [CrossRef]
33. Zhang, J.; Cao, Y.; Wang, C.-A.; Ran, R. Design and preparation of MnO$_2$/CeO$_2$–MnO$_2$ double-shelled binary oxide hollow spheres and their application in CO oxidation. *ACS Appl. Mater. Interfaces* **2016**, *8*, 8670–8677. [CrossRef]
34. Elias, J.S.; Artrith, N.; Bugnet, M.; Giordano, L.; Botton, G.A.; Kolpak, A.M.; Shao-Horn, Y. Elucidating the nature of the active phase in copper/ceria catalysts for CO oxidation. *ACS Catal.* **2016**, *6*, 1675–1679. [CrossRef]
35. He, Y.; Chen, D.; Li, N.; Xu, Q.; Li, H.; He, J.; Lu, J. Hollow mesoporous Co$_3$O$_4$-CeO$_2$ composite nanotubes with open ends for efficient catalytic CO oxidation. *ChemSusChem* **2019**, *12*, 1084–1090. [CrossRef]
36. Wang, Y.; Yang, D.; Li, S.; Zhang, L.; Zheng, G.; Guo, L. Layered copper manganese oxide for the efficient catalytic CO and VOCs oxidation. *Chem. Eng. J.* **2018**, *357*, 258–268. [CrossRef]
37. Zeng, S.H.; Wang, Y.; Ding, S.P.; Sattler, J.J.H.B.; Borodina, E.; Zhang, L.; Weckhuysen, B.M.; Su, H.Q. Active sites over CuO/CeO$_2$ and inverse CeO$_2$/CuO catalysts for preferential CO oxidation. *J. Power Sources* **2014**, *256*, 301–311. [CrossRef]
38. Chen, A.; Yu, X.; Zhou, Y.; Miao, S.; Li, Y.; Kuld, S.; Sehested, J.; Liu, J.; Aoki, T.; Hong, S.; et al. Structure of the catalytically active copper–ceria interfacial perimeter. *Nat. Catal.* **2019**, *2*, 334–341. [CrossRef]

39. Chen, G.; Song, G.; Zhao, W.; Gao, D.; Wei, Y.; Li, C. Carbon sphere-assisted solution combustion synthesis of porous/hollow structured CeO$_2$-MnO$_x$ catalysts. *Chem. Eng. J.* **2018**, *352*, 64–70. [CrossRef]
40. Pahalagedara, L.; Kriz, D.A.; Wasalathanthri, N.; Weerakkody, C.; Meng, Y.; Dissanayake, S.; Pahalagedara, M.; Luo, Z.; Suib, S.L.; Nandi, P.; et al. Benchmarking of manganese oxide materials with CO oxidation as catalysts for low temperature selective oxidation. *Appl. Catal. B Environ.* **2017**, *204*, 411–420. [CrossRef]
41. Chen, G.Z.; Xu, Q.H.; Yang, Y.; Li, C.C.; Huang, T.Z.; Sun, G.X.; Zhang, S.X.; Ma, D.L.; Li, X. Facile and mild strategy to construct mesoporous CeO$_2$-CuO nanorods with enhanced catalytic activity toward CO oxidation. *ACS Appl. Mater. Interfaces* **2015**, *7*, 23538–23544. [CrossRef]
42. May, Y.A.; Wei, S.; Yu, W.-Z.; Wang, W.-W.; Jia, C.-J. Highly efficient CuO/α-MnO$_2$ catalyst for low-temperature CO oxidation. *Langmuir* **2020**, *36*, 11196–11206. [CrossRef]
43. Hasegawa, Y.-I.; Maki, R.-U.; Sano, M.; Miyake, T. Preferential oxidation of CO on cop-per-containing manganese oxides. *Appl. Catal. A Gen.* **2009**, *371*, 67–72. [CrossRef]
44. Elmhamdi, A.; Pascual, L.; Nahdi, K.; Martínez-Arias, A. Structure/redox/activity relationships in CeO$_2$/CuMn$_2$O$_4$ CO-PROX catalysts. *Appl. Catal. B Environ.* **2017**, *217*, 1–11. [CrossRef]
45. Wang, W.-W.; Yu, W.-Z.; Du, P.-P.; Xu, H.; Jin, Z.; Si, R.; Ma, C.; Shi, S.; Jia, C.-J.; Yan, C.-H. Crystal plane effect of ceria on supported copper oxide cluster catalyst for CO oxidation: Importance of metal–support interaction. *ACS Catal.* **2017**, *7*, 1313–1329. [CrossRef]
46. Liu, Z.Q.; Li, J.H.; Wang, R.G. CeO$_2$ nanorods supported M–Co bimetallic oxides (M = Fe, Ni, Cu) for catalytic CO and C$_3$H$_8$ oxidation. *J. Colloid Interface Sci.* **2020**, *560*, 91–102. [CrossRef] [PubMed]
47. Diao, J.X.; Qiu, Y.; Liu, S.Q.; Wang, W.; Chen, K.; Li, H.L.; Yuan, W.; Qu, Y.T.; Guo, X.H. Inter-facial engineering of W$_2$N/WC heterostructures derived from solid-state synthesis: A highly effi-cient trifunctional electrocatalyst for ORR, OER, and HER. *Adv. Mater.* **2019**, *32*, 1905679. [CrossRef] [PubMed]
48. Konopatsky, A.S.; Firestein, K.L.; Leybo, D.V.; Sukhanova, E.V.; Popov, Z.I.; Fang, X.; Manakhov, A.M.; Kovalskii, A.M.; Matveev, A.T.; Shtansky, D.V.; et al. Structural evolution of Ag/BN hybrids via a polyol-assisted fabrication process and their catalytic activity in CO oxidation. *Catal. Sci. Technol.* **2019**, *9*, 6460–6470. [CrossRef]
49. Xie, Y.H.; Liu, B.L.; Li, Y.Z.; Chen, Z.X.; Cao, Y.L.; Jia, D.Z. Cu/Cu$_2$O/rGO nanocomposites: Solid-state self-reduction synthesis and catalytic activity for p-nitrophenol reduction. *New J. Chem.* **2019**, *43*, 12118–12125. [CrossRef]
50. Rong, Y.; Cao, Y.; Guo, N.; Li, Y.; Jia, W.; Jia, D. A simple method to synthesize V$_2$O$_5$ nanostructures with controllable morphology for high performance Li-ion batteries. *Electrochim. Acta* **2016**, *222*, 1691–1699. [CrossRef]
51. Xie, J.; Cao, Y.; Jia, D.; Li, Y.; Wang, K.; Xu, H. In situ solid-state fabrication of hybrid AgCl/AgI/AgIO$_3$ with improved UV-to-visible photocatalytic performance. *Sci. Rep.* **2017**, *7*, 12365. [CrossRef]
52. Wang, Q.; Ming, M.; Niu, S.; Zhang, Y.; Fan, G.Y.; Hu, J.S. Scalable solid-state synthesis of highly dispersed uncapped metal (Rh, Ru, Ir) nanoparticles for efficient hydrogen evolution. *Adv. Energy Mater.* **2018**, *8*, 1801698. [CrossRef]
53. Xiong, H.; Zhou, H.; Sun, G.; Liu, Z.; Zhang, L.; Zhang, L.; Du, F.; Qiao, Z.; Dai, S. Solvent-free self-assembly for scalable preparation of highly crystalline mesoporous metal oxides. *Angew. Chem. Int. Ed.* **2020**, *59*, 11053–11060. [CrossRef]
54. Chawla, S.K.; Sankarraman, N.; Payer, J.H. Diagnostic spectra for XPS analysis of Cu-O-S-H com-pounds. *J. Electron. Spectrosc.* **1992**, *61*, 1–18. [CrossRef]
55. Rastegarpanah, A.; Rezaei, M.; Meshkani, F.; Dai, H. 3D ordered honeycomb-shaped CuO·Mn$_2$O$_3$: Highly active catalysts for CO oxidation. *Mol. Catal.* **2020**, *485*, 110820. [CrossRef]
56. Chang, H.-W.; Chen, S.-C.; Chen, P.-W.; Liu, F.-J.; Tsai, Y.-C. Constructing morphologically tunable copper oxide-based nanomateri-als on cu wire with/without the deposition of manganese oxide as bifunctional materials for glucose sensing and supercapacitors. *Int. J. Mol. Sci.* **2022**, *23*, 3299. [CrossRef]
57. Liu, B.; Cao, Y.; Zhang, H.; Wang, S.; Geng, Q.; Li, Y.; Dong, F. Constructing ultrafine Cu nanoparticles encapsulated by N-doped carbon nanosheets with fast kinetics for high-performance lithium/sodium storage. *Chem. Eng. J.* **2022**, *446*, 136918. [CrossRef]
58. Skinner, W.M.; Prestidge, C.A.; Smart, R.S.C. Irradiation effects during XPS studies of Cu(I1) ac-tivation of zinc sulphide. *Surf. Interface Anal.* **1996**, *24*, 620–626. [CrossRef]
59. Qian, K.; Qian, Z.; Hua, Q.; Jiang, Z.; Huang, W. Structure–activity relationship of CuO/MnO$_2$ catalysts in CO oxidation. *Appl. Surf. Sci.* **2013**, *273*, 357–363. [CrossRef]
60. Xu, X.L.; Li, L.; Huang, J.; Jin, H.; Fang, X.Z.; Liu, W.M.; Zhang, N.; Wang, H.M.; Wang, X. Engineering Ni^{3+} cations in NiO lattice at the atomic level by Li+ doping: The roles of Ni^{3+} and oxygen species for CO oxidation. *ACS Catal.* **2018**, *8*, 8033–8045. [CrossRef]
61. Wang, X.Y.; Li, X.Y.; Mu, J.C.; Fan, S.Y.; Chen, X.; Wang, L.; Yin, Z.F.; Tade, M.; Liu, S.M. Ox-ygen vacancy-rich porous Co$_3$O$_4$ nanosheets toward boosted NO reduction by CO and CO oxidation: Insights into the structure-activity relationship and performance enhancement mechanism. *ACS Appl. Mater. Interfaces* **2019**, *11*, 41988–41999. [CrossRef] [PubMed]
62. Liu, B.; Li, Y.; Wu, H.; Ma, F.; Cao, Y. Room-temperature solid-state preparation of CoFe$_2$O$_4$@Coal composites and their catalytic performance in direct coal liquefaction. *Catalysts* **2020**, *10*, 503. [CrossRef]
63. Mo, S.P.; Zhang, Q.; Li, J.; Sun, Y.H.; Ren, Q.M.; Zou, S.B.; Zhang, Q.; Lu, J.H.; Fu, M.L. Highly efficient mesoporous MnO$_2$ catalysts for the total toluene oxidation: Oxygen-vacancy defect engineering and involved intermediates using in situ DRIFTS. *Appl. Catal. B Environ.* **2020**, *264*, 1184641–11846416. [CrossRef]

64. Zou, Q.; Zhao, Y.; Jin, X.; Fang, J.; Li, D.; Li, K.; Lu, J.; Luo, Y. Ceria-nano supported copper ox-ide catalysts for CO preferential oxidation: Importance of oxygen species and metal-support inter-action. *Appl. Surf. Sci.* **2019**, *494*, 1166–1176. [CrossRef]
65. Wu, P.; Dai, S.Q.; Chen, G.X.; Zhao, S.Q.; Xu, Z.; Fu, M.L.; Chen, P.R.; Chen, Q.; Jin, X.J.; Qiu, Y.C.; et al. Interfacial effects in hierarchically porous α-MnO_2/Mn_3O_4 heterostruc-tures promote photocatalytic oxidation activity. *Appl. Catal. B Environ.* **2019**, *268*, 118418. [CrossRef]
66. Lin, J.; Guo, Y.; Chen, X.; Li, C.; Lu, S.; Liew, K.M. CO Oxidation over Nanostructured Ceria Supported Bimetallic Cu-Mn Oxides Catalysts: Effect of Cu/Mn Ratio and Calcination Temperature. *Catal. Lett.* **2018**, *148*, 181–193. [CrossRef]
67. Zhou, Y.; Liu, X.; Wang, K.; Li, J.; Zhang, X.; Jin, X.; Tang, X.; Zhu, X.; Zhang, R.; Jiang, X.; et al. Porous Cu-Mn-O catalysts fabricated by spray pyrolysis method for efficient CO oxidation. *Results Phys.* **2019**, *12*, 1893–1900. [CrossRef]
68. Choi, K.-H.; Lee, D.-H.; Kim, H.-S.; Yoon, Y.-C.; Park, C.-S.; Kim, Y.H. Reaction Characteristics of Precious-Metal-Free Ternary Mn–Cu–M (M = Ce, Co, Cr, and Fe) Oxide Catalysts for Low-Temperature CO Oxidation. *Ind. Eng. Chem. Res.* **2016**, *55*, 4443–4450. [CrossRef]

Article

Development of High-Efficiency, Magnetically Separable Palladium-Decorated Manganese-Ferrite Catalyst for Nitrobenzene Hydrogenation

Viktória Hajdu [1], Gábor Muránszky [1], Miklós Nagy [1,*], Erika Kopcsik [1], Ferenc Kristály [2], Béla Fiser [1], Béla Viskolcz [1] and László Vanyorek [1,*]

[1] Institute of Chemistry, University of Miskolc, Miskolc-Egyetemváros, 3515 Miskolc, Hungary; kemviki@uni-miskolc.hu (V.H.); kemmug@uni-miskolc.hu (G.M.); ria.toth1@gmail.com (E.K.); kemfiser@uni-miskolc.hu (B.F.); bela.viskolcz@uni-miskolc.hu (B.V.)

[2] Institute of Mineralogy and Geology, University of Miskolc, Miskolc-Egyetemváros, 3515 Miskolc, Hungary; askkf@uni-miskolc.hu

* Correspondence: nagy.miklos@uni-miskolc.hu (M.N.); kemvanyi@uni-miskolc.hu (L.V.)

Citation: Hajdu, V.; Muránszky, G.; Nagy, M.; Kopcsik, E.; Kristály, F.; Fiser, B.; Viskolcz, B.; Vanyorek, L. Development of High-Efficiency, Magnetically Separable Palladium-Decorated Manganese-Ferrite Catalyst for Nitrobenzene Hydrogenation. *Int. J. Mol. Sci.* 2022, 23, 6535. https://doi.org/10.3390/ijms23126535

Academic Editor: Shaodong Zhou

Received: 16 May 2022
Accepted: 10 June 2022
Published: 10 June 2022

Publisher's Note: MDPI stays neutral with regard to jurisdictional claims in published maps and institutional affiliations.

Copyright: © 2022 by the authors. Licensee MDPI, Basel, Switzerland. This article is an open access article distributed under the terms and conditions of the Creative Commons Attribution (CC BY) license (https://creativecommons.org/licenses/by/4.0/).

Abstract: Aniline (AN) is one of the most important compounds in the chemical industry and is prepared by the catalytic hydrogenation of nitrobenzene (NB). The development of novel, multifunctional catalysts which are easily recoverable from the reaction mixture is, therefore, of paramount importance. Compared to conventional filtration, magnetic separation is favored because it is cheaper and more facile. For satisfying these requirements, we developed manganese ferrite ($MnFe_2O_4$)–supported, magnetically separable palladium catalysts with high catalytic activity in the hydrogenation of nitrobenzene to aniline. In addition to high NB conversion and AN yield, remarkable aniline selectivity (above 96 n/n%) was achieved. Surprisingly, the magnetic support alone also shows moderate catalytic activity even without noble metals, and thus, up to 94 n/n% nitrobenzene conversion, along with 47 n/n% aniline yield, are attainable. After adding palladium nanoparticles to the support, the combined catalytic activity of the two nanomaterials yielded a fast, efficient, and highly selective catalyst. During the test of the $Pd/MnFe_2O_4$ catalyst in NB hydrogenation, no by-products were detected, and consequently, above 96 n/n% aniline yield and 96 n/n% selectivity were achieved. The activity of the $Pd/MnFe_2O_4$ catalyst was not particularly sensitive to the hydrogenation temperature, and reuse tests indicate its applicability in at least four cycles without regeneration. The remarkable catalytic activity and other favorable properties can make our catalyst potentially applicable to both NB hydrogenation and other similar or slightly different reactions.

Keywords: manganese ferrite; spinel; aniline; nanoparticles; $Pd/MnFe_2O_4$; nitrobenzene

1. Introduction

The global aniline market reached a volume of 8.93 million tons in 2021 and is expected to reach 12.47 million tons by 2027. This huge production volume is a clear indicator that aniline is one of the most important bulk chemicals, with wide applications in the manufacture of herbicides, dyes, pigments, pharmaceuticals, and polymers (e.g., polyurethanes, ~75% of world production) [1–5]. Aniline is produced industrially by the catalytic hydrogenation of nitrobenzene carried out either in the gas or liquid phase. In most cases, the liquid phase process is performed in an organic solvent, such as methanol [6], ethanol [7], or isopropanol [8]. Heterogeneous catalysts, such as Raney nickel [9], copper [10], gold [11], platinum [12], and palladium [13], are commonly used in the hydrogenation of nitrobenzene. Recently, a number of palladium-based catalysts have been developed and used in industrial processes due to their high activity and selectivity. Recovery of homogeneous Pd catalysts from the reaction medium is cumbersome, time-consuming, and limits recycling; therefore, the development of heterogeneous catalysts has received much attention [14].

Among the various substrates, clays with high thermal and mechanical stability were one of the most facile [15]. Large specific surface area ($a = A/V$, m^2/m^3) is a key factor in heterogeneous catalysis since the reaction happens on the catalyst surface. Larger surface yields higher rate of reaction, hence the application of nanoparticles is favored. Moreover, nanoparticles with different geometries can also be combined. For example placing palladium on poly-dopamine-decorated halloysite nanotubes hybridized with N-doped porous carbon monolayer (Pd@Hal-pDA-NPC) yielded a hybrid catalyst which proved to be very effective in C–C coupling (Sonogashira, Heck and Suzuki reactions) and the hydrogenation of nitro compounds with high recyclability [16]. The adhesion between the catalyst and the clay support can be further enhanced by decoration and support with various ligands, such as dendrimer (PAMAM) [17], different diamines [18], or multi-dentate ligands [19]. Instead of clay, natural polymers, such as ligand-decorated chitosan, can also be used as support [20]. Using these supports has the advantages of easy catalyst recovery and reusability without losing catalytic activity and with little Pd dissolution; however, they are difficult to make, and therefore cheaper alternatives, such as activated carbon, are still in use.

Palladium catalysts on activated carbon are the main types, although their separation from the reaction medium is difficult and time-consuming due to their very small (several 10–100 nm) particle size. The use of magnetic materials as catalyst supports is a novel approach in the hydrogenation of NB [21–24]. Magnetic materials allow easy and efficient separation of the catalyst using an external magnet or magnetic field, avoiding complicated and time-consuming additional separation operations such as filtration and centrifugation [25]. Spinel ferrites (MFe_2O_4, M = Mn, Co, Cu, Ni, etc.) appear as new and promising heterogeneous catalysts due to their ferromagnetic properties and stable mineral structures. They are widely used in information storage, ferrofluid technology, electronic devices, biomedical applications, and catalysis [26–29]. Manganese(II) ferrite ($MnFe_2O_4$) has an inverse spinel structure, high adsorption capacity, and exceptional chemical stability. $MnFe_2O_4$ nanoparticles can be prepared by a variety of methods, such as hydrothermal [30,31], solvothermal [32,33], coprecipitation [34,35], sol-gel [36], microemulsion [37], combustion [38], and sonochemical synthesis [39].

Sonochemical synthesis has a few advantages compared to conventional methods. Catalysts are generally prepared in several steps. During the activation step, metal ions (e.g., palladium ions), oxides, and complex ions are reduced with hydrogen gas to a catalytically active form, namely the metal phase; however, this reduction step is generally time- and energy-consuming. In contrast, using ultrasonic cavitation, the catalyst preparation process is simplified, and no post-activation in hydrogen is required as the precious metals are present in the elemental state on the support.

In our work, we used a simple process to prepare novel, magnetic $MnFe_2O_4$-supported Pd catalysts, the performance of which was tested in nitrobenzene hydrogenation at four different temperatures. Important catalytic properties, such as conversion, yield, selectivity, ease of separation (magnetically), and reusability were also investigated in detail.

2. Results and Discussion

2.1. Synthesis and Characterization of the Developed Magnetic Catalysts

For the economical production of aniline, the price of the catalyst is a key factor. One of the most facile and cheapest methods for the preparation of the magnetic support nanoparticles is sonochemical activation combined with combustion. During the first step, upon the action of ultrasound, highly dispersed metal hydroxide nanoparticles formed from the metal nitrate precursors in polyethylene-glycol-based dispersion. PEG400 was used in this study since it is a viscous liquid at room temperature, and it yielded the desired small (metal oxide/hydroxide) particle size. The chemical effects of ultrasound irradiation are, primarily, attributed to its acoustic cavitation: the formation, growth, and implosive collapse of bubbles in the irradiated liquid. During the collapse, there is a high energy density from the conversion of the kinetic energy of the liquid's motion into the heating of the contents of the bubble, which can cover the energy requirements of the chemical reactions, such as the formation of metal hydroxides from their nitrate precursors in the polyol phase. Iron(III) was added to Mn(II) in a 2:1 molar ratio (16 mmol vs. 8 mmol) to obtain the favored spinel structure with the following formula: MFe_2O_4 (where M stands for Mn, Cd, Co, Ni, Cu, Zn, etc.). Deviation from the stoichiometric form of the spinel structure may result in the formation of undesirable, non-magnetic metal oxides.

In the second step, during the combustion method, the PEG-based colloid system of the iron and manganese hydroxides was heated in a furnace in the presence of an air atmosphere at 573 K, 623 K, and 673 K. The duration of the heat treatment was 3 h in each case. After burning of the PEG and the dehydration of the metal hydroxide nanoparticles, the expected spinel structures with magnetic properties formed. These $MnFe_2O_4$ samples were used as magnetic catalyst support for the preparation of palladium-decorated spinel catalysts. Palladium was also deposited on the surface of the support using a sonochemical step. During the process, elemental palladium nanoparticles formed from the Pd(II) ions and the alcohol (patosolv), which acted as the reducing agent. The synthesis, properties, and further application of the catalyst nanoparticles are summarized in Figure 1 and Table 1.

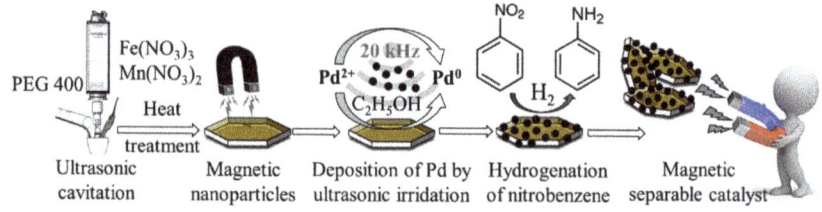

Figure 1. Scheme of the preparation and application of the magnetic $Pd/MnFe_2O_4$ catalyst.

Table 1. Summary table for the physical and chemical properties of the magnetic nanoparticles prepared: mean particle sizes and the phase composition of the manganese ferrite and magnetite based on XRD results, the palladium contents based on ICP-OES analysis, A_{DA} specific surface area determined by the Dubinin–Astakhov (DA) method. X_{max}, Y_{max} and S_{max} are the maximum conversion, yield, and selectivity values determined at the time and temperature indicated in parentheses.

	$MnFe_2O_4$		Fe_3O_4		Pd		A_{DA} m^2/g	X_{max}, (t) (%)	Y_{max}, (t) (%)	S_{max}, (t) (%)
	d (nm)	wt%	d (nm)	wt%	d (nm)	wt%				
$MnFe_2O_4$ (573 K)	11 ± 3	100	⊘	⊘	⊘	⊘	⊘	34.3 [a]	4.90 [a]	n/a
$MnFe_2O_4$ (623 K)	12 ± 2	61.1	14 ± 3	38.9	⊘	⊘	⊘	93.4 [a]	47.0 [a]	n/a
$MnFe_2O_4$ (673 K)	13 ± 2	41.8	12 ± 3	48.2	⊘	⊘	⊘	94.8 [a]	42.5 [a]	n/a
$Pd/MnFe_2O_4$ (573 K)	10 ± 3	89.1	35 ± 5	3.3	4 ± 1	4.20	74	99.9 (180 min) (303 K)	94.8 (60 min) (323 K)	94.6 (120 min) (323 K)
$Pd/MnFe_2O_4$ (623 K)	8 ± 2	66.0	15 ± 3	25.7	4 ± 1	4.64	69	99.9 (40 min) (323 K)	96.7 (240 min) (283 K)	96.8 (240 min) (283 K)
$Pd/MnFe_2O_4$ (673 K)	10 ± 3	64.3	13 ± 2	20.3	6 ± 1	4.61	78	99.9 (120 min) (323 K)	95.7 (180 min) (293 K)	95.8 (180 min) (293K)

[a] These parameters were measured at 240 min and 323 K.

Surface polarity and the presence of functional groups on the surface are important factors which determine the dispersibility of the nanoparticles in the liquid phase. Furthermore, the functional groups could serve as potential anchor points for the catalytically active metals on the supports. On the other hand, surface polarity, and the potential binding centra (functional groups, crystal defect sites etc.), play a great role in the adsorption and desorption processes of the reactant and product molecules during catalytic hydrogenation reactions. Henceforth, it is important to determine the presence or absence of certain functional groups and the corresponding Zeta potential of the system which contribute to the dispersibility and colloidal stability in water or aqueous solutions. Thus, FTIR and Zeta potential measurements were carried out in the case of the three magnetic catalyst supports which were prepared at different temperatures (Figure 2A,B).

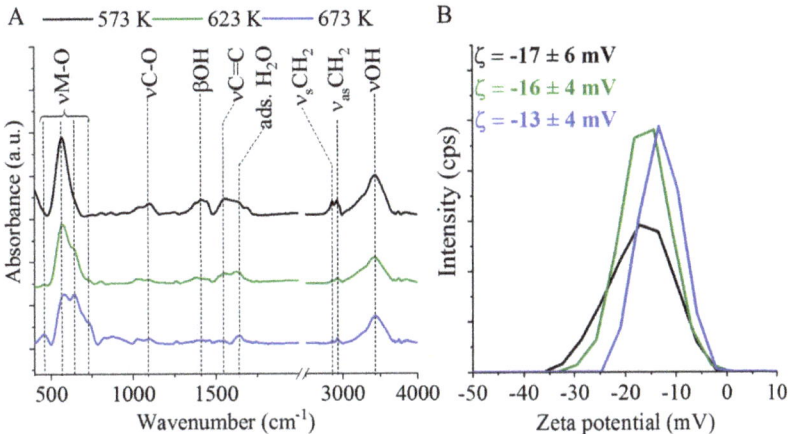

Figure 2. FTIR spectra (**A**) and Zeta potential distributions (**B**) of the magnetic catalyst supports.

In the case of the manganese ferrite support synthesized at 573 K, two absorption bands at 448 cm^{-1} and 568 cm^{-1} wavelengths were identified in the spectrum, possibly arising from the intrinsic stretching vibration modes of the metal–oxygen bond at the octahedral and tetrahedral sites of the MnFe$_2$O$_4$ [40]. Furthermore, based on the band at 1099 cm^{-1}, C–O bonds are also located on the surface, indicating the presence of alcoholic, carbonyl, and carboxyl functional groups. The band at 1557 cm^{-1} can be associated with the νC=C vibration mode of the deposited carbon from the thermal decomposition of the polyol (PEG). Additional carbon vibration bands are also visible at 2864 cm^{-1} and 2927 cm^{-1}, which correspond to the symmetric and asymmetric stretching vibration of the aliphatic and aromatic C–H bonds, respectively. The spectra of the samples synthesized at higher temperatures (623 K and 673 K) clearly differ as the intensity of νC–O, νC=C and νC–H bands decreased because of the partial oxidation of carbon as the heating occurred in the air atmosphere. The band at 1410 cm^{-1} corresponds to the βOH vibration mode. The presence of the –OH stretching vibration is further confirmed by a band at 3429 cm^{-1}. Furthermore, adsorbed water molecules are also detected as an OH vibration band at 1641 cm^{-1}. At higher temperatures, distinct shoulders can be found on the bands of the metal–oxygen vibrations at 642 cm^{-1} and 726 cm^{-1} due to the formation of a new phase, the magnetite beside MnFe$_2$O$_4$.

For the identification and quantification of the different crystalline phases in the magnetic catalyst supports and the palladium-decorated catalysts, XRD analysis was performed (Figure 3). On the diffractograms of the magnetic ferrite samples (Figure 3A), reflexions at 18.1°, 29.9°, 35.3°, 36.8°, 42.5°, 52.7°, 56.3°, and 61.7° two Theta degrees were identified, which correspond to (111), (220), (311), (222), (400), (422), (511), and (440) Miller-indexed crystal lattices of the MnFe$_2$O$_4$ phase (PDF 74–2403), respectively. In the case of the ferrite sample, which was prepared at 573 K, only one phase, manganese ferrite, was found (Figure 3A). On the diffractograms of the magnetic nanopowders, which were made at 623 K and 673 K, magnetite (Fe$_3$O$_4$) was also identified, in addition to the Mn spinel (Figure 3B,C). The reflexions, which are characteristic for the Fe$_3$O$_4$ are identified at 18.2°

(111), 30.3° (220), 35.7° (311), 43.2° (400), 53.5° (422), 57.1° (511), and 62.3° (440) two Theta degrees (PDF 19-629).

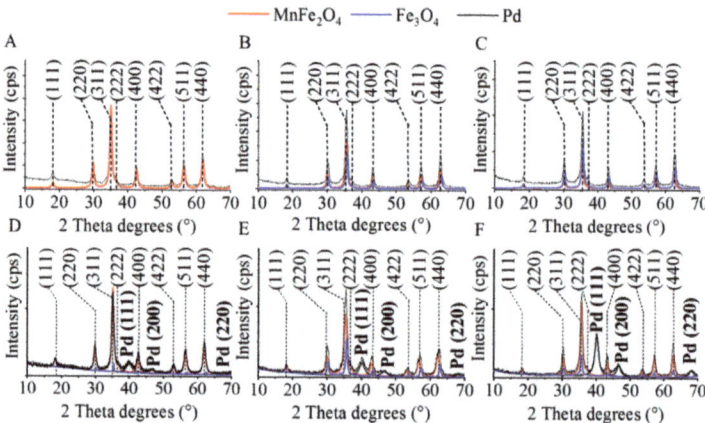

Figure 3. XRD diffractograms of the magnetic catalyst supports prepared at 573 K (**A**), 623 K (**B**), and 673 K (**C**) and the corresponding palladium-decorated catalysts (**D**) 573 K, (**E**) 623 K, and (**F**) 673 K.

Surprisingly, magnetite was also identified in the pure manganese ferrite catalyst support in the case of the palladium-decorated magnetic catalysts. (Figure 3D). This phenomenon can be explained by the decomposition of the spinel upon sonication and the subsequent phase-separation of magnetite. This phase transition could be promoted by the high energy released during the acoustic cavitation, which was applied for the palladium decomposition on the surface of the $MnFe_2O_4$ particles. Magnetite was also found in the other two catalysts (Figure 3E,F). The reflexions of the elemental palladium are located at 40.2°, 46.3°, and 68.0° two Theta degrees which belong to the Pd (111), Pd (200), and Pd (220) phases (PDF 046–1043), respectively.

Based on the Rietveld analysis, only $MnFe_2O_4$ (maximum spinel phase) was identified in the sample prepared at the lowest temperature (573 K, Table 1). The average particle size was 11 ± 3 nm. By increasing the temperature to 623 K and 673 K during the heating step, the magnetite phase also formed in increasing amounts (38.9 wt% and 48.2 wt%, respectively) in addition to the manganese spinel. The particle sizes were similar in the case of the ferrite and magnetite crystallites (between 11 nm and 14 nm). The deposition of the palladium particles onto surface of the magnetic crystallites was carried out by high-energy ultrasonic treatment. Due to the high energy, the manganese ferrite (jacobsite) partially decomposed, and magnetite formed. Nevertheless, this does not affect separability since the nanopowders contain only magnetic phases in addition to metallic palladium. The average crystallite size of the palladium nanoparticles is between 4 nm and 6 nm. The real palladium contents were determined by ICP-OES measurements, and the results were similar (between 4.20 and 4.65 wt%) for the three catalysts (Table 1).

Sorptometric investigations revealed no significant differences (approx. 10%) between the specific surface areas (A_{DA}, Table 1) of the palladium-containing magnetic catalysts prepared at different temperatures. It was found that, by increasing the synthesis temperature from 573 to 623 K, the surface is slightly decreased (74 m^2 g^{-1} for Pd/MnFe$_2$O$_4$-573 K and 69 m^2 g^{-1} for Pd/MnFe$_2$O$_4$-623 K, respectively). However, if the temperature is further increased up to 673 K, the surface area also increases and it becomes even larger than at 573 K (78 m^2 g^{-1}, Pd/MnFe$_2$O$_4$-673 K).

2.2. Catalytic Tests of the Prepared Magnetic Catalysts in Nitrobenzene Hydrogenation

To check the catalytic activity of the supports, the palladium-free manganese ferrite samples were tested in nitrobenzene hydrogenation at 323 K temperature and 20 bar hydrogen pressure. All the samples turned out to be catalytically active. The highest nitrobenzene conversion (94.8 n/n%) was reached by using the MnFe$_2$O$_4$ sample prepared at 673 K. However, the aniline yield was only 42.5 n/n% (Figure 4, Table 1). Similar results were achieved by applying the other support prepared at 623 K, but in this case, the nitrobenzene conversion was slightly lower (93.4 n/n%), while the aniline yield was a bit higher, at 47.0 n/n%. In the case of the MnFe$_2$O$_4$ prepared at 573 K, the nitrobenzene conversion and aniline yield were only 34.3 n/n% and 4.9 n/n%, respectively (Table 1).

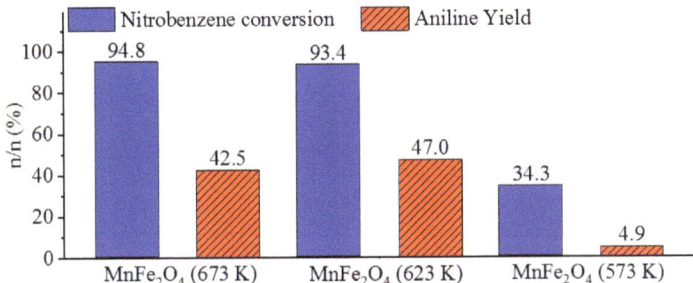

Figure 4. Nitrobenzene conversions and aniline yields achieved by using the palladium-free MnFe$_2$O$_4$ supports prepared at 673 K, 623 K, and 573 K. (T = 323 K, p = 20 bar H$_2$).

Despite the high (X > 90%) nitrobenzene conversions in two cases, the corresponding aniline yields (Y < 50%) were unsatisfactory, and thus, the application of palladium is necessary.

Next, the novel palladium-decorated ferrite nanoparticle catalysts were tested in nitrobenzene hydrogenation at four different reaction temperatures (283 K, 293 K, 303 K, and 323 K) and at constant hydrogen pressure (20 bar). Almost complete nitrobenzene conversion (99.9 n/n%) was reached for all three Pd-decorated catalysts; however, the reaction time varied significantly. For Pd/MnFe$_2$O$_4$ (623 K) it took only 40 min to reach complete conversion, while in the case of Pd/MnFe$_2$O$_4$ (573 K) and Pd/MnFe$_2$O$_4$ (673 K) 180 min and 120 min, respectively, were necessary (Figure 5A, and Table 1).

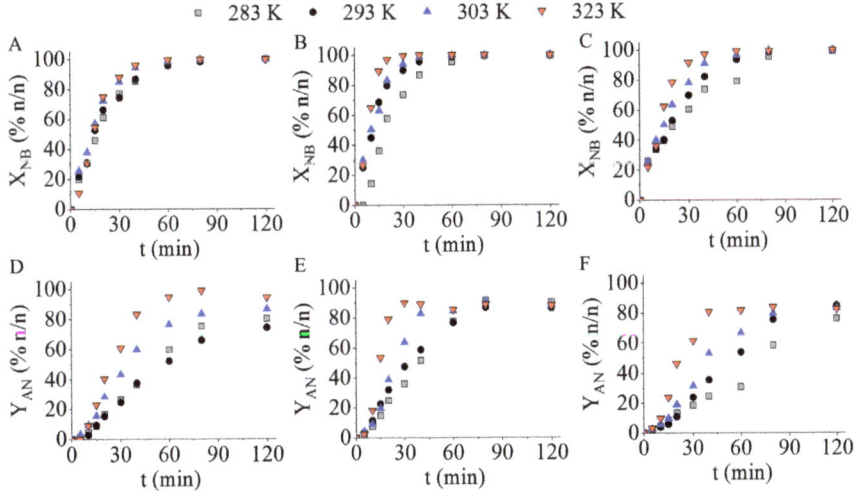

Figure 5. Nitrobenzene conversion and aniline yield as a function of hydrogenation time by using the magnetic Pd catalysts (Pd/MnFe$_2$O$_4$) prepared at 573 K (**A,D**), 623 K (**B,E**), and 673 K (**C,F**). The different marks represent different reaction temperatures.

The change of reaction temperature does not have a significant effect on the nitrobenzene conversions (X_{NB}), but it does affect the aniline yield (Y_{AN}). The highest yield (96.8 n/n%) was achieved after 240 min hydrogenation at 283 K in the case of the Pd/MnFe$_2$O$_4$ (623 K) catalyst. The formation of aniline from its intermediates (azobenzene and azoxybenzene) showed a slowing trend at lower temperatures. In contrast to the catalyst prepared at 573 K, in the case of the two other catalysts, Pd/MnFe$_2$O$_4$ (623 K) and Pd/MnFe$_2$O$_4$ (673 K), reaction temperature had a more pronounced effect on the nitrobenzene conversion. This may be explained by the significantly different magnetite loading of the three catalysts, namely the Pd/MnFe$_2$O$_4$ (573 K) contained only 3.3 wt% Fe$_3$O$_4$ phase, while the two other samples contained 25.7 wt% and 20.3 wt% (Table 1). The corresponding palladium-free manganese ferrites (38.9 wt%, 623 K and 48.2 wt%, 673 K magnetite content) showed higher catalytic activity than the pure MnFe$_2$O$_4$-523 K (Figure 4). In summary, the different magnetite content may be able to influence the catalytic activity and temperature sensitivity of the tested palladium-decorated ferrite catalysts.

The maximum aniline selectivity (S_{AN}) versus reaction temperature was also determined in the case of the three Pd/MnFe$_2$O$_4$ catalysts (Figure 6). High selectivity values were obtained for each catalyst/reaction temperature pairs. At 283 K, the Pd/MnFe$_2$O$_4$ (623 K) catalyst showed the highest selectivity. Common by-products (e.g., N-methylaniline, or o-toluidine) were not detected. During the reaction, the presence of intermediates, azobenzene and azoxybenzene, were confirmed, but these were completely converted to aniline until the end of the reaction. At lower temperatures (283 K and 293 K), the aniline selectivity was S_{AN} > 96 n/n%. There were no significant amounts of by-products detected during the catalytic test; therefore, the selectivity lower than 100% may be explained by the formation of polyaniline, which could not be detected by GC–MS measurements. In all, the prepared manganese ferrite–supported palladium catalysts are highly selective and may be applicable in aniline synthesis.

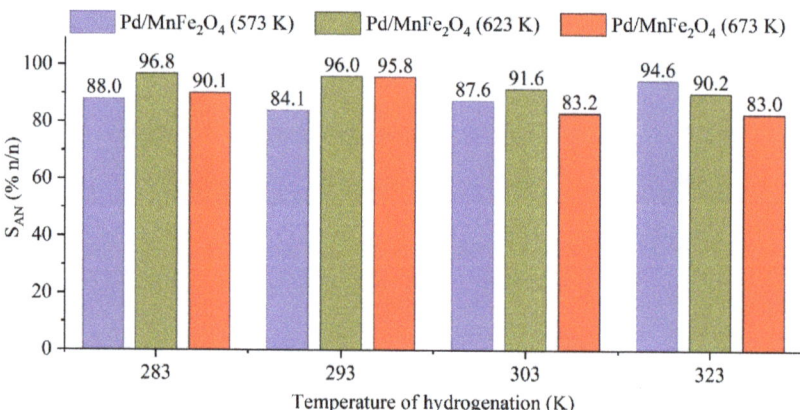

Figure 6. Aniline selectivity vs. hydrogenation temperature by using the prepared Pd/MnFe$_2$O$_4$ (573 K), Pd/MnFe$_2$O$_4$ (623 K), and Pd/MnFe$_2$O$_4$ (673 K) catalysts.

From the three developed magnetic catalysts, the one which was prepared at 573 K was selected to carry out reuse tests because its activity was less sensitive to the hydrogenation temperature, which may be related to the very low magnetite content. The reuse tests were carried at four cycles at 323 K. The nitrobenzene conversion and aniline yield did not change significantly during the time (180 min) of hydrogenation (Figure 7A,B).

During the third and fourth reuse tests, nitrobenzene conversion turned out to be lower during the initial phase of hydrogenation. However, after 180 min reaction time, the conversion and yield values were the same as those of the first two cycles. That is, the catalysts were able to convert the total amount of nitrobenzene with maximum aniline yield in each cycle. It should be noted, however, that the catalyst was not regenerated between the cycles, it was only washed with methanol. Thus, the developed Pd/MnFe$_2$O$_4$ catalyst is applicable in at least in four cycles of hydrogenation without regeneration. In addition, the magnetic feature of the catalyst further increases its applicability in the above commercially important hydrogenation system.

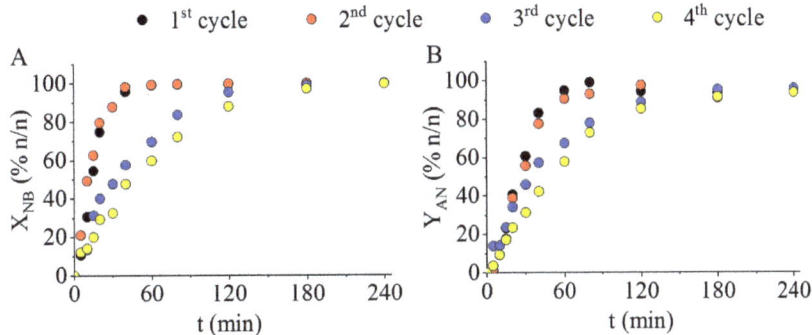

Figure 7. Reuse tests of the Pd/MnFe$_2$O$_4$ (573 K) catalyst. Nitrobenzene conversion (**A**) and aniline yield (**B**) vs. time of hydrogenation during the catalytic tests.

3. Materials and Methods

The spinel nanopowders were made from manganese(II) nitrate tetrahydrate (Mn(NO$_3$)$_2$·4H$_2$O, MW: 251.01 g/mol, Carl Roth GmbH, D-76185 Karlsruhe, Germany), iron(III) nitrate nonahydrate (Fe(NO$_3$)$_3$·9H$_2$O, MW: 404.0 g/mol, VWR Int. Ltd., B-3001 Leuven, Belgium). As reducing agent and dispersion medium, polyethylene glycol (PEG 400, Mw: ~400 Da, which corresponds to a chain length of approximately 8–9 links) from VWR Int. Ltd. (F-94126 Fontenay-sous-Bois, France), was applied. Palladium(II) nitrate dihydrate (Pd(NO$_3$)$_2$*2H$_2$O, MW: 266.46 g/mol, Thermo Fisher Ltd., D-76870 Kandel, Germany) as Pd precursor and Patosolv®, a mixture of 90 vol% ethanol and 10 vol% isopropanol (Molar Chem. Ltd., H-2314 Halásztelek, Hungary) were used during the preparation of the magnetic catalyst. Catalytic tests were carried out on nitrobenzene (NB, Merck KGaA, D-64293 Darmstadt, Germany).

3.1. Preparation of the Magnetic Spinel Nanoparticles and the Final Pd Catalyst

Manganese ferrite magnetic catalyst supports were synthesized in a two-step process, involving a sonochemical and a heat treatment step in air (Figure 1). In the sonochemical step, 6.44 g (15.9 mmol) iron(III) nitrate nonahydrate and 2.00 g (7.97 mmol) manganese(II) nitrate tetrahydrate were dissolved in 20 g polyethylene glycol, and the solution was sonicated by using a Hielscher UIP1000 Hdt homogenizer for 3 min (120 W, 17 kHz). The energy released during ultrasonic cavitation was enough to support the formation of metal oxyhydroxide nanoparticles from the metal precursors. The reddish-brown viscous dispersions were heated for 3 h at 573 K, 623 K, and 673 K, respectively. After elimination or burning of the polyethylene glycol, dehydration of the oxide-hydroxides occurred, resulting in the formation of a spinel phase. For the deposition of the palladium nanoparticles onto the surface of the ferrite crystals, a Hielscher UIP100 Hdt homogenizer was used. First, palladium(II) nitrate (0.25 g, 0.94 mmol) was dissolved in patosolv (50 mL) within which ferrite sample (2.00 g) was dispersed and then was treated by ultrasonic cavitation for two minutes.

3.2. Catalytic Tests—Nitrobenzene Hydrogenation

0.10 g magnetic spinel supported Pd catalysts were used for nitrobenzene (NB, c = 0.25 mol/dm^3 in methanolic solution) hydrogenation in an agitated (1000 rpm), Büchi Uster Picoclave reactor of 200 mL volume. The pressure of the hydrogenation was kept at 20 bar in all experiments and the reaction temperature was set to 283 K, 293 K, 303 K, and 323 K. Samples for analysis were taken after 0, 5, 10, 15, 20, 30, 40, 60, 80, 120, 180, and 240 min of hydrogenation.

3.3. Characterization Techniques

High-resolution transmission electron microscopy (HRTEM, FEI Technai G2 electron microscope, 200 kV) was used for size and morphological characterization of the nanoparticles. During sample preparation an aqueous suspension of the nanoparticles was dropped on 300 mesh copper grids (Ted Pella Inc., Redding, CA, USA). The qualitative and the quantitative analysis of the different oxide forms was carried out with X-ray diffraction (XRD) measurements by applying the Rietveld method. Bruker D8 diffractometer (Cu-Kα source) in parallel beam geometry (Göbel mirror) with Vantec detector was used. The carbon content, which remained from the polyol, was determined by Vario Macro CHNS element analyzer for all ferrite samples. Phenanthrene was used as standard (C: 93.538%, H: 5.629%, N:0.179%, S: 0.453%) from Carlo Erba Inc. Helium (99.9990%) was the carrier gas, and oxygen (99.995%) was used as oxidative atmosphere. Electrokinetic (Zeta) potential measurements were performed on the ferrite samples by using their aqueous colloids with Malvern Zetasizer Nano ZS equipment. The Zeta potentials were calculated based on electrophoretic mobility measurements by applying laser Doppler electrophoresis. The functional groups located on the surface of the prepared nanopowders were identified with Fourier transform infrared spectroscopy (FTIR) by using a Bruker Vertex 70 spectroscope. The measurements were carried out in transmission mode, and in each case, a 10 mg sample was pelletized with 250 mg potassium bromide. The palladium contents of the catalyst samples were determined by using a Varian 720 ES inductively coupled optical emission spectrometer (ICP-OES). For the ICP-OES measurements the samples were dissolved in aqua regia. The specific surface area of the catalysts was measured by CO_2 adsorption–desorption experiments at 273 K by using a Micromeritics ASAP 2020 sorptometer based on the Dubinin–Astakhov (DA) method. After the hydrogenation tests, the aniline-containing samples were quantitatively analyzed by using an Agilent 7890A gas chromatograph coupled with Agilent 5975C Mass Selective detector. An RTX-624 column (60 m × 0.25 mm × 1.4 µm) was used and the injected sample volume was 1 µL at 200:1 split ratio, while the inlet temperature was set to 473 K. Helium was used as the carrier gas (2.28 mL/min), and the oven temperature was set to 323 K for 3 min and it was heated up to 523 K with a heating rate of 10 K/min and kept there for another 3 min. The analytical standards of aniline, potential by-products, and the intermediates originated from Sigma Aldrich (Burlington, MA, USA) and Dr. Ehrenstorfer Ltd. (Wesel, Nordrhein, Germany).

The catalytic activity of the magnetic Pd catalysts was compared based on Equation (1):

$$Y = X \times S = \frac{(n_{NB,t} - n_{NB,0})}{n_{NB,0}} \times \frac{n_{AN,t}}{(n_{NB,t} - n_{NB,0})} \qquad (1)$$

where Y is the yield of aniline, X the conversion of nitrobenzene, S aniline selectivity, $n_{NB,0}$ the initial amount of nitrobenzene at $t = 0$, $(n_{NB,t} - n_{NB,0})$ the amount of nitrobenzene consumed after t reaction time and $n_{AN,t}$ the amount of aniline formed. In the main text the % values are presented, which are obtained by multiplying Y, X, and S by 100.

4. Conclusions

Manganese ferrite–supported magnetic separable palladium catalysts (Pd/MnFe$_2$O$_4$) were prepared using a simple sonochemical method starting from the corresponding metal nitrates and consecutive heat treatment at three different temperatures of the formed metal-(oxi)hydroxides. Each of these catalysts shows high catalytic activity in nitrobenzene hydrogenation even without the noble metal (Pd). Catalytic activity investigations of the Pd-free magnetic supports revealed that the noble metal–free system can reach up to 94% nitrobenzene conversion. Due to the local hotspots during sonication, magnetite also formed outside of the ferrite phase, which proved to be crucial for increased catalytic activity, since without magnetite only low (<34 n/n%) conversions were determined. However, the aniline yield was unsatisfactory in each case for the pure supports, and thus, the inclusion of palladium was inevitable. The catalytic activity of the two nanomaterials ($MnFe_2O_4$ and Pd) combined led to a fast, efficient, and selective catalyst. In addition to the nitrobenzene conversion and aniline yield, the aniline selectivity of the catalysts is also remarkable (above 96 n/n%). The increasing amount of magnetite in addition to manganese ferrite greatly affects the temperature sensitivity of the nitrobenzene conversion. The $Pd/MnFe_2O_4$ (573 K) catalyst's activity had the lowest sensitivity for the hydrogenation temperature, which may be related to its very low magnetite content. Reuse tests showed that the catalyst prepared at 573 K can give the same performance in at least four cycles without regeneration. The ease of preparation, combined with high conversion, yield and selectivity, as well as the reusability without regeneration, make the developed $Pd/MnFe_2O_4$ magnetic catalyst system successfully applicable in the industrially important hydrogenation of nitro compounds.

Author Contributions: Conceptualization, L.V. and B.V.; methodology, V.H., G.M. and F.K.; validation, B.F. and L.V.; formal analysis, V.H., G.M., F.K. and E.K.; investigation, B.V.; resources, V.H.; data curation, B.F.; writing—original draft preparation, L.V., V.H., B.F. and M.N.; writing—review and editing, V.H. and M.N.; visualization, L.V.; supervision, L.V.; project administration, B.V.; funding acquisition, B.V. All authors have read and agreed to the published version of the manuscript.

Funding: This research was supported by the European Union and the Hungarian State, co-financed by the European Regional Development Fund in the framework of the GINOP-2.3.4-15-2016-00004 project, aimed to promote the cooperation between the higher education and the industry. Prepared with the professional support of the Doctoral Student Scholarship Program of the Co-operative Doctoral Program of the Ministry of Innovation and Technology financed from the National Research, Development and Innovation Fund.

Institutional Review Board Statement: Not applicable.

Informed Consent Statement: Not applicable.

Data Availability Statement: Data is available upon request from the corresponding authors.

Conflicts of Interest: The authors declare no conflict of interest.

References

1. Downing, R.S.; Kunkeler, P.J.; Van Bekkum, H. Catalytic syntheses of aromatic amines. *Catal. Today* **1997**, *37*, 121–136. [CrossRef]
2. Grirrane, A.; Corma, A.; Garcia, H. Preparation of symmetric and asymmetric aromatic azo compounds from aromatic amines or nitro compounds using supported gold catalysts. *Nat. Protoc.* **2010**, *5*, 429–438. [CrossRef] [PubMed]
3. Fu, L.; Cai, W.; Wang, A.; Zheng, Y. Photocatalytic hydrogenation of nitrobenzene to aniline over tungsten oxide-silver nanowires. *Mater. Lett.* **2015**, *142*, 201–203. [CrossRef]
4. Xiao, Q.; Sarina, S.; Waclawik, E.R.; Jia, J.; Chang, J.; Riches, J.D.; Wu, H.; Zheng, Z.; Zhu, H. Alloying Gold with Copper Makes for a Highly Selective Visible-Light Photocatalyst for the Reduction of Nitroaromatics to Anilines. *ACS Catal.* **2016**, *6*, 1744–1753. [CrossRef]
5. Sun, X.; Olivos-Suarez, A.I.; Osadchii, D.; Romero, M.J.V.; Kapteijn, F.; Gascon, J. Single cobalt sites in mesoporous N-doped carbon matrix for selective catalytic hydrogenation of nitroarenes. *J. Catal.* **2018**, *357*, 20–28. [CrossRef]
6. Prekob, A.; Hajdu, V.; Muránszky, G.; Fiser, B.; Sycheva, A.; Ferenczi, T.; Viskolcz, B.; Vanyorek, L. Application of carbonized cellulose-based catalyst in nitrobenzene hydrogenation. *Mater. Today Chem.* **2020**, *17*, 100337. [CrossRef]
7. Turek, F.; Geike, R.; Lange, R. Liquid-phase hydrogenation of nitrobenzene in a slurry reactor. *Chem. Eng. Process. Process. Intensif.* **1986**, *20*, 213–219. [CrossRef]
8. Gelder, E.A.; Jackson, S.D.; Lok, C.M. A Study of Nitrobenzene Hydrogenation Over Palladium/Carbon Catalysts. *Catal. Lett.* **2002**, *84*, 205–208. [CrossRef]
9. Mahata, N.; Cunha, A.F.; Órfão, J.J.M.; Figueiredo, J.L. Hydrogenation of nitrobenzene over nickel nanoparticles stabilized by filamentous carbon. *Appl. Catal. A Gen.* **2008**, *351*, 204–209. [CrossRef]
10. Petrov, L.; Kumbilieva, K.; Kirkov, N. Kinetic model of nitrobenzene hydrogenation to aniline over industrial copper catalyst considering the effects of mass transfer and deactivation. *Appl. Catal.* **1990**, *59*, 31–43. [CrossRef]
11. Cano, M.; Villuendas, P.; Benito, A.M.; Urriolabeitia, E.P.; Maser, W.K. Carbon nanotube-supported gold nanoparticles as efficient catalyst for the selective hydrogenation of nitroaromatic derivatives to anilines. *Mater. Today Commun.* **2015**, *3*, 104–113. [CrossRef]

12. Li, C.H.; Yu, Z.X.; Yao, K.F.; Ji, S.F.; Liang, J. Nitrobenzene hydrogenation with carbon nanotube-supported platinum catalyst under mild conditions. *J. Mol. Catal. A Chem.* **2005**, *226*, 101–105. [CrossRef]
13. Couto, C.S.; Madeira, L.M.; Nunes, C.P.; Araújo, P. Hydrogenation of Nitrobenzene over a Pd/Al$_2$O$_3$ Catalyst–Mechanism and Effect of the Main Operating Conditions. *Chem. Eng. Technol.* **2015**, *38*, 1625–1636. [CrossRef]
14. Yu, X.; Wang, M.; Li, H. Study on the nitrobenzene hydrogenation over a Pd-B/SiO$_2$ amorphous catalyst. *Appl. Catal. A Gen.* **2000**, *202*, 17–22. [CrossRef]
15. Sadjadi, S.; Heravi, M.M.; Malmir, M. Pd@HNTs-CDNS-g-C$_3$N$_4$: A novel heterogeneous catalyst for promoting ligand and copper-free Sonogashira and Heck coupling reactions, benefits from halloysite and cyclodextrin chemistry and g-C$_3$N$_4$ contribution to suppress Pd leaching. *Carbohydr. Polym.* **2018**, *186*, 25–34. [CrossRef]
16. Sadjadi, S.; Lazzara, G.; Malmir, M.; Heravi, M.M. Pd nanoparticles immobilized on the poly-dopamine decorated halloysite nanotubes hybridized with N-doped porous carbon monolayer: A versatile catalyst for promoting Pd catalyzed reactions. *J. Catal.* **2018**, *366*, 245–257. [CrossRef]
17. Bahri-Laleh, N.; Sadjadi, S.; Poater, A. Pd immobilized on dendrimer decorated halloysite clay: Computational and experimental study on the effect of dendrimer generation, Pd valance and incorporation of terminal functionality on the catalytic activity. *J. Colloid Interface Sci.* **2018**, *531*, 421–432. [CrossRef]
18. Tabrizi, M.; Sadjadi, S.; Pareras, G.; Nekoomanesh-Haghighi, M.; Bahri-Laleh, N.; Poater, A. Efficient hydro-finishing of polyalfaolefin based lubricants under mild reaction condition using Pd on ligands decorated halloysite. *J. Colloid Interface Sci.* **2021**, *581*, 939–953. [CrossRef]
19. Dehghani, S.; Sadjadi, S.; Bahri-Laleh, N.; Nekoomanesh-Haghighi, M.; Poater, A. Study of the effect of the ligand structure on the catalytic activity of Pd@ligand decorated halloysite: Combination of experimental and computational studies. *Appl. Organomet. Chem.* **2019**, *33*, e4891. [CrossRef]
20. Alleshagh, M.; Sadjadi, S.; Arabi, H.; Bahri-Laleh, N.; Monflier, E. Pd on ligand-decorated chitosan as an efficient catalyst for hydrofinishing polyalphaolefins: Experimental and computational studies. *J. Phys. Chem. Solids* **2022**, *164*, 110611. [CrossRef]
21. Sun, Q.; Guo, C.Z.; Wang, G.H.; Li, W.C.; Bongard, H.J.; Lu, A.H. Fabrication of Magnetic Yolk–Shell Nanocatalysts with Spatially Resolved Functionalities and High Activity for Nitrobenzene Hydrogenation. *Chem.–A Eur. J.* **2013**, *19*, 6217–6220. [CrossRef] [PubMed]
22. Easterday, R.; Sanchez-Felix, O.; Losovyj, Y.; Pink, M.; Stein, B.D.; Morgan, D.G.; Rakitin, M.; Doluda, V.Y.; Sulman, M.G.; Mahmoud, W.E.; et al. Design of ruthenium/iron oxide nanoparticle mixtures for hydrogenation of nitrobenzene. *Catal. Sci. Technol.* **2015**, *5*, 1902–1910. [CrossRef]
23. Hajdu, V.; Prekob, Á.; Muránszky, G.; Kocserha, I.; Kónya, Z.; Fiser, B.; Viskolcz, B.; Vanyorek, L. Catalytic activity of maghemite supported palladium catalyst in nitrobenzene hydrogenation. *React. Kinet. Mech. Catal.* **2020**, *129*, 107–116. [CrossRef]
24. Fronczak, M.; Kasprzak, A.; Bystrzejewski, M. Carbon-encapsulated iron nanoparticles with deposited Pd: A high-performance catalyst for hydrogenation of nitro compounds. *J. Environ. Chem. Eng.* **2021**, *9*, 104673. [CrossRef]
25. Shokouhimehr, M. Magnetically Separable and Sustainable Nanostructured Catalysts for Heterogeneous Reduction of Nitroaromatics. *Catalysts* **2015**, *5*, 534–560. [CrossRef]
26. Peng, Y.; Wang, Z.; Liu, W.; Zhang, H.; Zuo, W.; Tang, H.; Chen, F.; Wang, B. Size- and shape-dependent peroxidase-like catalytic activity of MnFe$_2$O$_4$ Nanoparticles and their applications in highly efficient colorimetric detection of target cancer cells. *Dalton Trans.* **2015**, *44*, 12871–12877. [CrossRef]
27. Kavkhani, R.; Hajalilou, A.; Abouzari-Lotf, E.; Ferreira, L.P.; Cruz, M.M.; Yusefi, M.; Parvini, E.; Ogholbeyg, A.B.; Ismail, U.N. CTAB assisted synthesis of MnFe$_2$O$_4$@SiO$_2$ nanoparticles for magnetic hyperthermia and MRI application. *Mater. Today Commun.* **2022**, *31*, 103412. [CrossRef]
28. Liew, K.H.; Lee, T.K.; Yarmo, M.A.; Loh, K.S.; Peixoto, A.F.; Freire, C.; Yusop, R.M. Ruthenium supported on ionically cross-linked chitosan-carrageenan hybrid MnFe$_2$O$_4$ catalysts for 4-nitrophenol reduction. *Catalysts* **2019**, *9*, 254. [CrossRef]
29. Hazarika, M.; Chinnamuthu, P.; Borah, J.P. MWCNT decorated MnFe$_2$O$_4$ nanoparticles as an efficient photo-catalyst for phenol degradation. *J. Mater. Sci. Mater. Electron.* **2018**, *29*, 12231–12240. [CrossRef]
30. Facile hydrothermal synthesis of cubic spinel AB2O4 type MnFe2O4 nanocrystallites and their electrochemical performance. *Appl. Surf. Sci.* **2017**, *413*, 83–91. [CrossRef]
31. Chen, G.; Zhang, X.; Gao, Y.; Zhu, G.; Cheng, Q.; Cheng, X. Novel magnetic MnO$_2$/MnFe$_2$O$_4$ nanocomposite as a heterogeneous catalyst for activation of peroxymonosulfate (PMS) toward oxidation of organic pollutants. *Sep. Purif. Technol.* **2019**, *213*, 456–464. [CrossRef]
32. Liu, Z.; Chen, G.; Li, X.; Lu, X. Removal of rare earth elements by MnFe$_2$O$_4$ based mesoporous adsorbents: Synthesis, isotherms, kinetics, thermodynamics. *J. Alloys Compd.* **2021**, *856*, 158185. [CrossRef]
33. Li, Z.; Gao, K.; Han, G.; Wang, R.; Li, H.; Zhao, X.S.; Guo, P. Solvothermal synthesis of MnFe$_2$O$_4$ colloidal nanocrystal assemblies and their magnetic and electrocatalytic properties. *New J. Chem.* **2014**, *39*, 361–368. [CrossRef]
34. Akhtar, M.J.; Younas, M. Structural and transport properties of nanocrystalline MnFe$_2$O$_4$ synthesized by co-precipitation method. *Solid State Sci.* **2012**, *14*, 1536–1542. [CrossRef]
35. Vignesh, V.; Subramani, K.; Sathish, M.; Navamathavan, R. Electrochemical investigation of manganese ferrites prepared via a facile synthesis route for supercapacitor applications. *Colloids Surfaces A Physicochem. Eng. Asp.* **2018**, *538*, 668–677. [CrossRef]

36. Li, J.; Yuan, H.; Li, G.; Liu, Y.; Leng, J. Cation distribution dependence of magnetic properties of sol–gel prepared MnFe$_2$O$_4$ spinel ferrite nanoparticles. *J. Magn. Magn. Mater.* **2010**, *322*, 3396–3400. [CrossRef]
37. Baig, M.M.; Yousuf, M.A.; Agboola, P.O.; Khan, M.A.; Shakir, I.; Warsi, M.F. Optimization of different wet chemical routes and phase evolution studies of MnFe$_2$O$_4$ nanoparticles. *Ceram. Int.* **2019**, *45*, 12682–12690. [CrossRef]
38. Silambarasu, A.; Manikandan, A.; Balakrishnan, K.; Jaganathan, S.K.; Manikandan, E.; Aanand, J.S. Comparative Study of Structural, Morphological, Magneto-Optical and Photo-Catalytic Properties of Magnetically Reusable Spinel MnFe$_2$O$_4$ Nano-Catalysts. *J. Nanosci. Nanotechnol.* **2017**, *18*, 3523–3531. [CrossRef]
39. Singh Yadav, R.; Kuřitka, I.; Vilcakova, J.; Jamatia, T.; Machovsky, M.; Skoda, D.; Urbánek, P.; Masař, M.; Urbánek, M.; Kalina, L.; et al. Impact of sonochemical synthesis condition on the structural and physical properties of MnFe$_2$O$_4$ spinel ferrite nanoparticles. *Ultrason. Sonochem.* **2020**, *61*, 104839. [CrossRef]
40. Zipare, K.; Dhumal, J.; Bandgar, S.; Mathe, V.; Shahane, G. Superparamagnetic Manganese Ferrite Nanoparticles: Synthesis and Magnetic Properties. *J. Nanosci. Nanoeng.* **2015**, *1*, 178–182.

MDPI
St. Alban-Anlage 66
4052 Basel
Switzerland
www.mdpi.com

International Journal of Molecular Sciences Editorial Office
E-mail: ijms@mdpi.com
www.mdpi.com/journal/ijms

Disclaimer/Publisher's Note: The statements, opinions and data contained in all publications are solely those of the individual author(s) and contributor(s) and not of MDPI and/or the editor(s). MDPI and/or the editor(s) disclaim responsibility for any injury to people or property resulting from any ideas, methods, instructions or products referred to in the content.

www.ingramcontent.com/pod-product-compliance
Lightning Source LLC
LaVergne TN
LVHW070714100526
838202LV00013B/1095